An Introduction to Groups and Their Matrices for Science Students

Group theory, originating from algebraic structures in mathematics, has long been a powerful tool in many areas of physics, chemistry, and other applied sciences, but it has seldom been covered in a manner accessible to undergraduates. This book from renowned educator Robert Kolenkow introduces group theory and its applications starting with simple ideas of symmetry, through quantum numbers, and working up to particle physics. It features clear explanations, accompanying problems and exercises, and numerous worked examples from experimental research in the physical sciences. Beginning with key concepts and necessary theorems, topics are introduced systematically, including molecular vibrations and lattice symmetries; matrix mechanics; wave mechanics; rotation and quantum angular momentum; atomic structure; and finally particle physics. This comprehensive primer on group theory is ideal for advanced undergraduate topics courses, reading groups, or self-study, and it will help prepare graduate students for higher-level courses.

Robert Kolenkow was formerly Associate Professor of Physics at Massachusetts Institute of Technology and was awarded the Everett Moore Baker Award for Outstanding Teaching. He was the lead author of *Physical Geography Today* and coauthor, with Daniel Kleppner, of *An Introduction to Mechanics* (also published by Cambridge University Press).

AN INTRODUCTION TO GROUPS AND THEIR MATRICES FOR SCIENCE STUDENTS

ROBERT KOLENKOW

Formerly Massachusetts Institute of Technology

CAMBRIDGE
UNIVERSITY PRESS

University Printing House, Cambridge CB2 8BS, United Kingdom

One Liberty Plaza, 20th Floor, New York, NY 10006, USA

477 Williamstown Road, Port Melbourne, VIC 3207, Australia

314–321, 3rd Floor, Plot 3, Splendor Forum, Jasola District Centre,
New Delhi – 110025, India

103 Penang Road, #05–06/07, Visioncrest Commercial, Singapore 238467

Cambridge University Press is part of the University of Cambridge.

It furthers the University's mission by disseminating knowledge in the pursuit of
education, learning, and research at the highest international levels of excellence.

www.cambridge.org
Information on this title: www.cambridge.org/9781108831086
DOI: 10.1017/9781108923217

First published 2022

A catalogue record for this publication is available from the British Library.

Library of Congress Cataloging-in-Publication Data
Names: Kolenkow, Robert J., author.
Title: An introduction to groups and their matrices for science students /
Robert Kolenkow.
Description: Cambridge ; New York, NY : Cambridge University Press, 2022. |
Includes bibliographical references and index.
Identifiers: LCCN 2021051448 (print) | LCCN 2021051449 (ebook) | ISBN
9781108831086 (hardback) | ISBN 9781108923217 (ebook)
Subjects: LCSH: Group theory. | Matrices. | Science–Mathematics. | BISAC:
SCIENCE / Physics / Mathematical & Computational
Classification: LCC QA174.2 .K64 2022 (print) | LCC QA174.2 (ebook) | DDC
512/.2–dc23/eng/20220128
LC record available at https://lccn.loc.gov/2021051448
LC ebook record available at https://lccn.loc.gov/2021051449

ISBN 978-1-108-83108-6 Hardback

To Marcia
my help and support

Contents

PREFACE

This is not a math book, although a quick flip of the pages might give that impression. It is an introduction to group theory and matrix representations, a subject usually treated with mathematical rigor, theorems, and proofs. This book is intended to introduce the subject and to clear a path to more advanced treatments.

INTRODUCTION

Like typical undergraduates in science I took math courses every semester for three years, beginning with the wonders of calculus, differential equations, and advanced calculus. An elective course in linear algebra was my first contact with matrices. In graduate school I heard that group theory is a powerful tool for treating physical problems, so I took a course taught by a renowned theorist, but it was difficult for me. The term "transforms like" puzzled me – what was "transforming" and what was it "like"?

While on the physics faculty of the Massachusetts Institute of Technology I agreed to develop an elective course in group theory for juniors and seniors, having learned by then that a good way to understand a subject is to teach it. The course was popular, with 30–40 majors in physics, chemistry, and math attending.

This is the book I wish I had as a student and the book I would have wanted to teach from. A few proofs are omitted as unsuited to a text at this level, but often made plausible by examples. There are only a few uses of the crutch "it can be shown." Experimental results taken from original research papers are included to show how group theory helps us understand physical phenomena.

After studying this text, the student will be prepared to tackle more advanced texts and to understand, at least in part, research papers that employ group theory.

OVERVIEW

Most chapters end with a "Brief Bios" section to recognize the lives of experimentalists and theorists.

Chapter 1 Fundamental Concepts introduces the idea of symmetry by illustrations and elementary algebra of operations. The **32** group and the isomorphic permutation group illustrate the group axioms, and a matrix representation is derived using algebraic geometry. Matrix types are defined.

Chapter 2 Matrix Representations of Discrete Groups is entirely mathematical by necessity. The major concepts of basis functions, similarity transformations, character, and reducibility are defined and illustrated.

Chapter 3 Molecular Vibrations defines normal modes using Newtonian mechanics and group theory. The water molecule is taken as an example and its vibration modes are calculated and visualized in physical terms. IR active and Raman active modes are predicted from character tables.

Chapter 4 Crystalline Solids deals with ideal crystalline solids and their translation and rotation symmetries. Vibration of a 1-dimensional diatomic chain is calculated from mechanics to show the origin of branches.

Chapter 5 Bohr's Quantum Theory and Matrix Mechanics begins with a summary of Bohr's quantum theory followed by Born's recognition that Heisenberg's difference equations represent matrix algebra. The single quantization condition of matrix mechanics is based on a commutator and is applied to deriving physical principles such as conservation of energy and angular momentum. Heisenberg's uncertainty principle is made plausible by his thought experiments.

Chapter 6 Wave Mechanics, Measurement, and Entanglement Schrödinger's wave equation expresses the dispersion of matter waves. Quantization is illustrated by rotational spectra. A model 2×2 Hermitian matrix is diagonalized. Probability is challenged by the EPR thought experiment that local measurements should not produce distant results. Hidden variable theory and quantum mechanics are compared.

Chapter 7 Rotation uses algebraic geometry to develop matrices for rotation of a vector. Groups $U(1)$ and $SU(2)$ are defined, and $SO(3)$ is derived from $SU(2)$ by a similarity transformation. Euler angles are defined by sketches and matrices.

Chapter 8 Quantum Angular Momentum is a key chapter for applications of group theory. The Stern–Gerlach experiment introduces spatial quantization and angular momentum quantum numbers. Exponential operators are defined and commutators are calculated from matrices. Angular momentum labels for irreducible representations are developed. Spherical harmonics are generated using group theory. Combining quantum mechanical angular momentum is illustrated by positronium and Wigner *3-j* coefficients are generated. Selection rules for electric dipole transitions are derived from spherical harmonics, from the Wigner–Eckart theorem, and from parity.

Chapter 9 The Structure of Atoms summarizes Zeeman's experiments and uses the Zeeman effect to motivate numerous applications including diagonalization with spin-orbit coupling in any field. Group theory applied to He derives the singlet and triplet states. The Pauli principle applied to electron configurations determines allowed states. The building-up principle and Hund's rules give a qualitative account of multi-electron atoms and the periodic table.

Chapter 10 Particle Physics begins by listing the fundamental forces. SU(2) supports isospin of nucleons for strong interactions. Properties of Lagrangians are discussed. U(1) gauge invariance of Schrödinger's equation is demonstrated. The quark model is applied to hadrons. Conservation laws are applied to reactions. SU(3) is introduced and applied to the three-quark model and to color charge.

Appendix A Character Tables from Class Sums
Appendix B Born–Jordan Proof of the Quantization Condition
Appendix C Weyl Derivation of the Heisenberg Uncertainty Principle
Appendix D EPR Thought Experiment
Appendix E Photon Correlation Experiment
Appendix F Tables of Some *3-j* Coefficients
Appendix G Proof of the Wigner–Eckart Theorem

TO THE INSTRUCTOR

There is more than enough material for a one-semester junior-senior course. A solid introduction to group theory and applications would include Chapters 1 Fundamental Concepts, 2 Matrix Representations of Discrete Groups, 3 Molecular Vibrations, 6 Wave Mechanics, Measurement, and Entanglement, 7 Rotation, 8 Quantum Angular Momentum, and 9 The Structure of Atoms. Chapter 10 Particle Physics would be popular with students.

Chapter 5 Bohr's Quantum Theory and Matrix Mechanics and the Sections 6.6 and 6.7 on measurement and entanglement in Chapter 6 deal with topics not commonly treated in texts at this level. The inclusion of these special topics is intended to stimulate student interest, but they could be treated as material for a short course, for a reading course, or for self-study.

1

FUNDAMENTAL CONCEPTS

1.1 Introduction

The object of this chapter is to lay out the principal ideas and nomenclature of group theory in preparation for the physical applications discussed in later chapters. We shall look at what group theory deals with, we shall define the mathematical meaning of a group, we shall show examples of several groups, and we shall discuss the key subject of matrix representations of groups (with an example). A review of matrix algebra and definitions of some special matrices concludes the chapter.

As a student of science, you spent several years studying calculus, differential equations, and the properties of important mathematical functions (trigonometric functions, exponentials, Bessel functions, etc.). You used these tools to solve problems in Newtonian mechanics, electromagnetism, and maybe even problems in quantum mechanics.

At its heart, group theory is very different from calculus. It is more abstract and more fundamental, with little reliance on explicit mathematical functions. As we shall see in this text, group theory, though abstract, nevertheless has great power in dealing with a wide range of physical phenomena. One example is the angular momentum ("spin") of an electron that is experimentally a dimensionless "point" particle with no analogue in Newtonian mechanics.

In physical applications, group theory calculates numerical results by using mathematical functions in the group's representation matrices.

More profoundly, group theory can give deeper insight into subjects you may have already studied; for instance, the conservation of energy and the structure of hydrogen-atom wave functions in quantum mechanics. Newton invented calculus to explain how forces acting on an object determine its motion. In modern high-energy particle physics the forces are not well known, yet group theory in its abstract generality provides predictive schemes for classifying "strange" particles.

P. W. Anderson (1923–2020, Nobel laureate in physics 1977) wrote "It is only slightly overstating the case to say that physics is the study of symmetry."

1

1.2   Operations

Group theory deals with *operations*, also called *transformations*. In this book the symbols for operations are written in bold. We use the convention that an operation operates on the object (the *operand*) to its right.

Consider the simple example of a transformation $\mathbf{T}$ that operates on a variable x (the operand) to change its sign to $-x$. This is written symbolically as

$$\mathbf{T}x = -x.$$

If $\mathbf{T}$ operates on the function $ax + b$, where a and b are constants, $\mathbf{T}$ operates only on x and has no effect on constants. Hence

$$\mathbf{T}(ax + b) = a\mathbf{T}x + b$$
$$= -ax + b.$$

If $\mathbf{T}$ operates twice in succession, the sequence $\mathbf{TT}$ can be written as $\mathbf{T}^2$:

$$\mathbf{T}^2 x = \mathbf{TT}x$$
$$= \mathbf{T}(\mathbf{T}x)$$
$$= \mathbf{T}(-x)$$
$$= x.$$

An even simpler operator is the *identity* operator, which produces no change in the operand. The identity operator in group theory is conventionally given the symbol $\mathbf{E}$, from the German *Einheit*, unity or, literally, oneness:

$$\mathbf{E}x = x.$$

These simple examples illustrate the abstract nature of group theory. The operators are not expressed in terms of explicit mathematical functions; instead, operators are defined in terms of their effect on the operand.

1.2.1   Symmetry Operations

The figure shows an equilateral triangle in the *x-y* plane. The dot marks the location of the triangle's geometric center, the point equally distant from all three apexes.

Consider now three operations, $\mathbf{E}$, $\mathbf{A}$, and $\mathbf{B}$, that rotate the triangle about its geometric center through the specified angles.

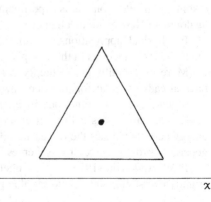

E: *rotate by* 0° (*equivalently, rotate by* 360°)
A: *rotate by* 120°
B: *rotate by* 240°

As the notation implies, operation **E** (rotation by 0°) clearly plays the role of the identity operation.

In this digital age, clocks with hands are no longer common but the terms *counterclockwise* and *clockwise* for the sense of a rotation are firmly entrenched. If a rotation when seen looking down the rotation axis toward the origin turns in the same sense as the hands of a clock, it is termed a clockwise rotation (cw), and if the rotation is in the opposite sense, it is counterclockwise (ccw) as illustrated by the sketches. This text follows the usual convention that counterclockwise rotations are positive.

The sketch shows the effect of the operations **E**, **A**, and **B** on the triangle. The operations have left the appearance of the triangle unchanged, which is the essence of the concept of symmetry. Frank Wilczek (Nobel laureate in physics 2004) coined a pithy phrase to describe the connection between operations and symmetry: "change without change." With reference to the triangle example, we have made a change – an operation was performed on the triangle by rotating it – but the triangle still looks the same.

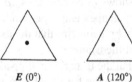

E (0°) A (120°) B (240°)

More generally, if an operation on an object leaves it unchanged, or *invariant*, the object must have a symmetry property. In the triangle example a 3-fold symmetry is revealed by rotation through the particular angles 0°, 120°, and 240°.

The symmetry of an equilateral triangle under certain rotations is an example of a *rotation* symmetry. There are many other examples of geometric symmetry. Consider the repeated pattern in the sketch, which could be a decorative frieze along the

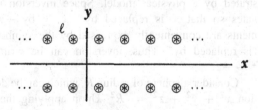

edge of a building. The two rows are parallel to the x-axis and are equidistant from the x-axis. The columns are all equally spaced along x by a distance ℓ, and the dots signify that the pattern extends far to the left and far to the right.

If the pattern is translated parallel to the x-axis by an integer multiple of ℓ, its appearance remains the same. This is an example of *translation* symmetry, important for the discussion of crystal lattices in Chapter 4.

If the pattern is folded along the x-axis, the two rows coincide. Each row is a mirror image of the other, an example of *reflection* symmetry, also called *mirror* symmetry.

Symmetry is appealing and has long played a role in art and architecture, from ancient rock carvings to mosaics in ancient Rome to ephemeral foam patterns on coffee drinks.

For the black-and-white geometric pattern in the figure, the z-axis is normal to the page and passes through the origin. Rotations about z by 0° and 180° and reflections about the diagonals are symmetry operations. Rotations about z by 90° and 240° and reflections about the x- and y-axes are not symmetry operations.

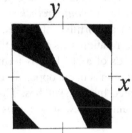

The equilateral triangle has additional symmetries revealed by no longer requiring the triangle to lie in the x-y plane. The figure shows three new axes aa, bb, and cc. Each axis passes through an apex and is perpendicular to the opposite edge. It follows by geometry that the axes intersect at the geometric center of the triangle.

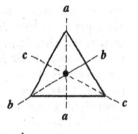

Suppose now that the triangle is "flipped" 180° about axis aa. The front becomes the back and vice versa; the appearance of the triangle is unchanged, so this is a symmetry operation on the triangle, and similarly for flips about axes bb and cc.

The three rotation operations in the plane and the three flip operations identify six symmetry operations for the equilateral triangle. These operations are easily demonstrated with a cardboard triangle.

Inversion symmetry is abstract and cannot be shown pictorially or demonstrated by a physical model. Space inversion reverses the signs of the coordinates so that x is replaced by $-x$, y by $-y$, and z by $-z$. These replacements are conveniently expressed by the symbol $\mapsto$, which means "maps to" or "is replaced by." Thus, inversion can be written $x \mapsto -x$, $y \mapsto -y$, and $z \mapsto -z$.

Consider a sphere of radius R, which can be described algebraically by the equation $x^2 + y^2 + z^2 = R^2$. Upon applying the space inversion operation to the coordinates, the equation is unchanged; the sphere is invariant under space inversion. We shall see important examples of inversion when symmetry and the quantum theory of atoms are discussed in Chapter 5.

But the use of symmetry in decorative arts and the description of geometric figures barely scratches the surface of its deep importance. Steven Weinberg (1933–2021, Nobel laureate in physics 1979) has written that symmetry is the "key to nature's secrets," which is why the application of symmetry principles to physical problems is the subject of this text.

1.2.2 Products of Operations

Consider again the set of three operations $\{\mathbf{E}, \mathbf{A}, \mathbf{B}\}$ from the triangle example discussed in Section 1.2.1. The *product* of two operations is the result of applying first one operation followed by a second. If, for example, $\mathbf{A}$ is applied first, followed by $\mathbf{B}$, the product is written symbolically as $\mathbf{BA}$. The operation on the right, here $\mathbf{A}$, is considered to be applied first. Note that although the product $\mathbf{BA}$ has the appearance of "multiplication" of $\mathbf{B}$ times $\mathbf{A}$, abstract group theory puts no restrictions on the method by which operations are actually combined. Some books on group theory use the term "multiplication" where we use "product." Such terms are symbolic only, with no reference to ordinary arithmetic.

In the example, $\mathbf{B}$ is applied to $\mathbf{A}$ "from the left." Alternatively, an operation can be applied "from the right" to give, in this case, $\mathbf{AB}$. These same ideas are also used with equations relating operations. Equations involving operations conform to the usual rule from algebra that both sides are to be treated equally. Consider, for example, the product of two operations $\mathbf{T_1}$ and $\mathbf{T_2}$ to give a third operation $\mathbf{T_3}$:

$$\mathbf{T_2 T_1} = \mathbf{T_3}.$$

Now apply an operation $\mathbf{C}$ from the left; $\mathbf{C}$ must act on both sides of the relation to maintain the equality.

$$\mathbf{C T_2 T_1} = \mathbf{C T_3}$$

Applying $\mathbf{C}$ from the right gives

$$\mathbf{T_2 T_1 C} = \mathbf{T_3} C.$$

The distinction between operations from the left and from the right is important. The reason is that for many group operations the order of combination makes a difference, unlike the multiplication of numbers or algebraic functions. If two operations $\mathbf{T_1}$ and $\mathbf{T_2}$ are combined, the two possible products $\mathbf{T_1 T_2}$ and $\mathbf{T_2 T_1}$ may not necessarily be equal. However, if $\mathbf{T_1 T_2} = \mathbf{T_2 T_1}$, then $\mathbf{T_1}$ and $\mathbf{T_2}$ are said to *commute*.

1.2.3 Product Tables

Rotation symmetry operations on the equilateral triangle are rotations through defined angles about defined axes, so it is easy to determine the product of any two operations. Consider, for example, the product $\mathbf{BA}$. First applying $\mathbf{A}$ produces an initial rotation of $120°$. The second operation $\mathbf{B}$ causes an additional rotation through $240°$, for a net result of $360°$ (equivalently $0°$). This is the same result as using operation $\mathbf{E}$ alone, so the product is written

$$\mathbf{BA} = \mathbf{E}.$$

Table 1.1 Products of **E**, **A**, and **B**

	E	A	B
E	EE = E	EA = A	EB = B
A	AE = A	AA = B	AB = E
B	BE = B	BA = E	BB = A

Table 1.2 Products of **E**, **A**, and **B**

	E	A	B
E	E	A	B
A	A	B	E
B	B	E	A

The same reasoning can be used to evaluate all of the nine possible products of **E**, **A**, and **B**, being sure as a general rule to maintain the order of the operations. The products are conveniently displayed in the form of a *product table*, where an operation in the top horizontal row is applied first followed by an operation from the left-hand vertical column. For clarity in this first illustration, both the product and the net result are given in Table 1.1, but after this a table will show only net results, as in Table 1.2.

The tables show that $\mathbf{AA} = \mathbf{A}^2 = \mathbf{B}$; geometrically, two successive counterclockwise rotations by 120° give the same result as a single counterclockwise rotation by 240°. All the members of this set are powers of a single member **A**. The triangle rotation operations **E**, **A**, and **B** are *cyclic* because they can all be written as powers of **A**: $\mathbf{E} = \mathbf{A}^0$, $\mathbf{A} = \mathbf{A}^1$, and $\mathbf{B} = \mathbf{A}^2$.

Table 1.2 shows that in this particular example the operations $\{\mathbf{E}, \mathbf{A}, \mathbf{B}\}$ all commute with one another, for instance, $\mathbf{AB} = \mathbf{BA}$. The identity operation **E** always commutes with any operation **T** because $\mathbf{ET} = \mathbf{TE} = \mathbf{T}$.

1.2.4   The Inverse of an Operation

For any operation **T** there may be an *inverse* operation, symbolized $\mathbf{T}^{-1}$, that undoes the effect of **T** on the operand. Because the identity operation **E** always signifies no change, it follows that $\mathbf{TT}^{-1} = \mathbf{T}^{-1}\mathbf{T} = \mathbf{E}$. An operation always commutes with its inverse.

In the triangle example, **A** is a counterclockwise rotation through 120°, so one way to undo the effect of **A** is by a further counterclockwise rotation through an additional 240°, to give a net rotation of 360° = 0°. In the set $\{\mathbf{E}, \mathbf{A}, \mathbf{B}\}$ the inverse of

A is identified as $\mathbf{A}^{-1} = \mathbf{B}$. By similar reasoning, $\mathbf{B}^{-1} = \mathbf{A}$. These results can also be read from Table 1.2. The entries $\mathbf{AB} = \mathbf{BA} = \mathbf{E}$ show, for example, that $\mathbf{B} = \mathbf{A}^{-1}$.

Another operation that undoes the effect of **A** is to rotate clockwise through 120° to bring the triangle back to the starting point. This clockwise rotation is equivalent to a counterclockwise rotation through −120°. This is a new operation and not a member of the operations **E**, **A**, and **B**, which are defined here only for counterclockwise rotations.

Here is an example involving inverses and a product table. Consider the set {**E, A**} with the following partial product tables:

	E	A
E	E	A
A	A	$\mathbf{A}^2$

What is the unidentified member $\mathbf{A}^2$? Try $\mathbf{A}^2 = \mathbf{A}$. Multiply both sides from the left by $\mathbf{A}^{-1}$.

$$\mathbf{A}^2 = \mathbf{A}$$
$$\mathbf{A}^{-1}\mathbf{A}^2 = \mathbf{A}^{-1}\mathbf{A}$$
$$(\mathbf{A}^{-1}\mathbf{A})\mathbf{A} = \mathbf{E}$$
$$\mathbf{EA} = \mathbf{E}$$
$$\mathbf{A} = \mathbf{E}$$

The result $\mathbf{A} = \mathbf{E}$ gives the dull and useless Table 1.3.

The alternative possibility $\mathbf{A}^2 = \mathbf{E}$ gives the more useful product Table 1.4 that has two distinctly different members.

In the product table for a set of operations, a given symmetry operation appears only once in each column as seen in the example. As a proof consider a set of distinctly different symmetry operations **A, B, C**, and **D**. Suppose that **A** occurs twice in the column headed by **B**, so that $\mathbf{BC} = \mathbf{A}$ and $\mathbf{BD} = \mathbf{A}$. Then $\mathbf{C} = \mathbf{B}^{-1}\mathbf{A}$ and $\mathbf{D} = \mathbf{B}^{-1}\mathbf{A}$ so that $\mathbf{C} = \mathbf{D}$, a contradiction because the operations are assumed to be different. Similarly, each operation occurs only once in a given row.

Table 1.3 $\mathbf{A} = \mathbf{E}$

	E	E
E	E	E
E	E	E

Table 1.4 $\mathbf{A}^2 = \mathbf{E}$

	E	A
E	E	A
A	A	E

1.3   What Is a Group?

With a solid foundation on the nature of operations, their products, and their inverses, it is time to take up the heart of the matter: the definition of a group. The definition is summarized in the following five axioms (i) to (v). They may seem a little dry, but they are needed because if a set of operations can be shown to form a group, a raft of useful theorems are then immediately applicable.

To illustrate the axioms, we shall show that the set of triangle rotation operations $\{\mathbf{E}, \mathbf{A}, \mathbf{B}\}$ form a group.

(i) A group consists of a set of operations called *members* of the group. We shall show that the set $\{\mathbf{E}, \mathbf{A}, \mathbf{B}\}$ are members of a group.

(ii) The product of any two members of a group is also a member of the group; products do not take us to new operations outside the set of group members. Table 1.2 shows that the products of $\mathbf{E}$, $\mathbf{A}$, and $\mathbf{B}$ are all members of the same set. Contrariwise, clockwise rotations of the triangle do not appear in Table 1.2 and are therefore not members of this group.

(iii) The group contains an identity member $\mathbf{E}$ that produces no change when combined with any group member. Table 1.2 for the triangle rotations show that $\mathbf{EE} = \mathbf{E}$, $\mathbf{EA} = \mathbf{AE} = \mathbf{A}$, and $\mathbf{EB} = \mathbf{BE} = \mathbf{B}$, showing that the notation is justified; $\mathbf{E}$ is truly the identity member in the set.

(iv) For every member $\mathbf{T}$ of a group, there is also a member $\mathbf{T}^{-1}$ in the group that is the inverse of $\mathbf{T}$, such that $\mathbf{T}\mathbf{T}^{-1} = \mathbf{T}^{-1}\mathbf{T} = \mathbf{E}$. As shown in Section 1.2.4 and also in Table 1.2, $\mathbf{E}^{-1} = \mathbf{E}$, $\mathbf{A}^{-1} = \mathbf{B}$, and $\mathbf{B}^{-1} = \mathbf{A}$.

(v) An additional axiom is that the products of operations are *associative* so that $\mathbf{T_1}(\mathbf{T_2}\mathbf{T_3}) = (\mathbf{T_1}\mathbf{T_2})\mathbf{T_3}$, where the products in parentheses are evaluated first, then combined with the remaining operation. This axiom will be satisfied automatically by the operations in applications.

Let $\boldsymbol{\Gamma}$ be the symbol for the group $\{\mathbf{E}, \mathbf{A}, \mathbf{B}\}$. The number of members in a group is called the *order* of the group: $\boldsymbol{\Gamma}$ is of order 3.

Note that $\mathbf{E}$ is always a member of any group and satisfies the group definition axioms. Therefore $\mathbf{E}$ is itself a group (of order 1). If a subset of group members are themselves a group, the subset is called a *subgroup*. $\mathbf{E}$ is a trivial subgroup of every group. The whole group itself is also a trivial subgroup of the group.

In the group $\boldsymbol{\Gamma}$, the set $\{\mathbf{E}, \mathbf{A}\}$ is not a subgroup, because the product $\mathbf{AA} = \mathbf{B}$, an operation not included in the set. In a subgroup, just as in a group, the product of two operations in the subgroup must also be a member of the subgroup.

The product table for a set of operations can be checked to see whether the group axioms are satisfied. The product table tells all.

1.3.1 Discrete and Continuous Groups

Groups with a finite (countable) number of members are called *discrete* or *finite* groups. The triangle rotation group $\boldsymbol{\Gamma} = \{\mathbf{E}, \mathbf{A}, \mathbf{B}\}$ has a finite number of members and is an example of a discrete group.

Consider now a flat circular disk with a perpendicular axis through its center, as suggested by the sketch. Rotation of the disk by an arbitrary counterclockwise angle θ leaves the disk invariant, so this leads us to suspect that there is a group involving rotations. The rotations form a group: rotation by $0°$ is the identity, two successive rotations by θ_1 and θ_2 give the same result as a single rotation by $\theta_1 + \theta_2$, and to every rotation θ there corresponds an inverse rotation $360° - \theta$.

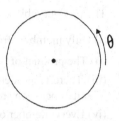

Because θ can be any angle, this group has an "infinite" (uncountable) number of members; it is an example of a *continuous group*. A continuous group depends on a continuous parameter, in this example the angle θ. Continuous groups are important in physics, for example, in the quantum-mechanical wave function of a hydrogen atom, which depends on two continuous parameters: the polar angle θ and the azimuthal angle ϕ.

1.4 Examples of Groups

1.4.1 Abelian Groups

A group in which all of the members commute is called an *Abelian* group, after the Norwegian mathematician Niels Henrik Abel (1802–29). The triangle rotation group composed of the set $\{\mathbf{E}, \mathbf{A}, \mathbf{B}\}$ is an Abelian group. This group is also a cyclic group and can be written as $\{\mathbf{E}, \mathbf{A}, \mathbf{A}^2\}$ as shown in Section 1.2.3.

All groups of order less than 6 are Abelian.

1.4.2 The 32 Group

Table 1.5 is the product table for a group of order 6, a popular example in textbooks on group theory. It is termed the **32** ("three-two") group.

Table 1.5 The **32** group (order 6)

	E	A	B	C	D	F
E	E	A	B	C	D	F
A	A	E	F	D	C	B
B	B	D	E	F	A	C
C	C	F	D	E	B	A
D	D	B	C	A	F	E
F	F	C	A	B	E	D

The product table shows that the group axioms are satisfied:

(i) Only members from the set appear in the product table.

(ii) The product of two members is a member of the set.

(iii) There is an identity member identified as **E** that obeys the properties of the identity operation such as $\mathbf{EA} = \mathbf{AE} = \mathbf{A}$.

(iv) Every member of a group has an inverse in the set as shown by products such as $\mathbf{DF} = \mathbf{E}$ so that $\mathbf{F} = \mathbf{D}^{-1}$.

Members of a given group may or may not commute. For example, $\mathbf{AB} = \mathbf{F}$ and $\mathbf{BA} = \mathbf{D}$. **A** and **B** do not commute, so **32** is not an Abelian group. It is the smallest group that is nonAbelian, accounting for its popularity as a teaching tool.

Table 1.5 shows that the **32** group has three nontrivial subgroups of order 2, namely {**E, A**}, {**E, B**}, and {**E, C**} and also a subgroup of order 3 {**E, D, F**}, but no others. A theorem in group theory states that for a group of order n each of its subgroups has an order that is a factor of n. The example of the **32** group demonstrates this theorem because $6 = 2 \cdot 3$ for the nontrivial subgroups of orders 2 and 3. $6 = 6 \cdot 1$ is satisfied by the trivial subgroup {**E**} of order 1 and by the group itself of order 6.

It follows from this theorem that if the order of a group is a prime number, the group has no nontrivial subgroups and must therefore be a cyclic group. The group Γ of order 3 is an example.

1.4.3 The Permutation (Symmetric) Group

This section discusses the apparently different example of the *permutation group*. Here is the permutation group of order 6.

$$\begin{pmatrix} 1 & 2 & 3 \\ 1 & 2 & 3 \end{pmatrix} \quad \begin{pmatrix} 1 & 2 & 3 \\ 2 & 1 & 3 \end{pmatrix} \quad \begin{pmatrix} 1 & 2 & 3 \\ 1 & 3 & 2 \end{pmatrix} \quad \begin{pmatrix} 1 & 2 & 3 \\ 3 & 2 & 1 \end{pmatrix} \quad \begin{pmatrix} 1 & 2 & 3 \\ 3 & 1 & 2 \end{pmatrix} \quad \begin{pmatrix} 1 & 2 & 3 \\ 2 & 3 & 1 \end{pmatrix}$$

$$\mathbf{P}_1 \qquad\qquad \mathbf{P}_2 \qquad\qquad \mathbf{P}_3 \qquad\qquad \mathbf{P}_4 \qquad\qquad \mathbf{P}_5 \qquad\qquad \mathbf{P}_6$$

A permutation rearranges a set of numbers. A permutation group of degree n has n different numbers, so there are n choices for the first number in the permutation, $n - 1$ for the second ... hence $n! = n \cdot (n - 1) \ldots 1$ different permutations. This example is a permutation group of degree 3: there are 3 numbers and $3! = 3 \cdot 2 \cdot 1 = 6$ permutations as shown.

A permutation can be viewed as a one-to-one mapping of n different numbers into a possibly different order. The permutation $\mathbf{P}_1$ does not cause a reordering, so it is evidently the identity operation $\mathbf{E}$. Using the $\mapsto$ "maps to" symbol, $j \mapsto k$ means that the number j is replaced by k. For example, the permutation $\mathbf{P}_5$ is the reordering $1 \mapsto 3, 2 \mapsto 1$, and $3 \mapsto 2$. As another example, consider a function of three variables $f(x_1, x_2, x_3) = x_3{}^2 + x_1 x_2$. If permutation $\mathbf{P}_5$ is applied to f, it becomes $x_2{}^2 + x_3 x_1$.

A permutation is classed as *even* if the bottom row of its matrix requires an even number of interchanges (*transpositions*) to bring it to standard ascending numerical order, and as *odd* if it involves an odd number. For example, $\mathbf{P}_2$ is odd, because in the bottom row only one interchange $1 \leftrightarrow 2$ brings the bottom row to the numerical order 1 2 3. $\mathbf{P}_5$ is even, because two interchanges $1 \leftrightarrow 3$ followed by $3 \leftrightarrow 2$ are required.

The permutation group of degree n is also called the *symmetric* group of degree n. The reason for the name "symmetric" is that permuting the labels of identical objects (such as electrons) does not alter the symmetry – another example of "change without change." The symmetric group of degree n has been given a variety of symbols, and this text uses $\mathbf{S}_n$. The factorial $n!$ increases more rapidly with n than either powers x^n or exponentials e^{nx}, so $\mathbf{S}_n$ can be a very large group even for relatively small n; for example, $\mathbf{S}_{10}$ has $3,628,800$ group elements.

Products of Permutations

To show how to find the product of two permutations, walk through the evaluation of $\mathbf{P}_4 \mathbf{P}_5$ as an example. Consider an operand x_{123} with three subscripts *123*. First apply the permutation $\mathbf{P}_5$, written as $\mathbf{P}_5 \, x_{123} = \left(\begin{smallmatrix} 1 & 2 & 3 \\ 3 & 1 & 2 \end{smallmatrix}\right) x_{123}$. $\mathbf{P}_5$ says $1 \mapsto 3, 2 \mapsto 1$, and $3 \mapsto 2$. The result is therefore $\mathbf{P}_5 \, x_{123} = x_{312}$. Now apply $\mathbf{P}_4$ to this result, $\mathbf{P}_4 \, x_{312} = \left(\begin{smallmatrix} 1 & 2 & 3 \\ 3 & 2 & 1 \end{smallmatrix}\right) x_{312}$. $\mathbf{P}_4$ says $1 \mapsto 3, 2 \mapsto 2$, and $3 \mapsto 1$ to give the final product $\mathbf{P}_4 \mathbf{P}_5 \, x_{123} = x_{132}$. This is the same permutation as $\mathbf{P}_3 \, x_{123} = x_{132}$, so $\mathbf{P}_4 \mathbf{P}_5 = \left(\begin{smallmatrix} 1 & 2 & 3 \\ 3 & 2 & 1 \end{smallmatrix}\right)\left(\begin{smallmatrix} 1 & 2 & 3 \\ 3 & 1 & 2 \end{smallmatrix}\right) = \left(\begin{smallmatrix} 1 & 2 & 3 \\ 1 & 3 & 2 \end{smallmatrix}\right) = \mathbf{P}_3$.

Here is a simple way to find the product of permutations. Consider again the product $\mathbf{P}_4 \mathbf{P}_5 = \left(\begin{smallmatrix} 1 & 2 & 3 \\ 3 & 2 & 1 \end{smallmatrix}\right)\left(\begin{smallmatrix} 1 & 2 & 3 \\ 3 & 1 & 2 \end{smallmatrix}\right)$. Rearrange the entries in $\mathbf{P}_4$ so that the top row of the $\mathbf{P}_4$ matrix duplicates the bottom row of $\mathbf{P}_5$: $\left(\begin{smallmatrix} 3 & 1 & 2 \\ 1 & 3 & 2 \end{smallmatrix}\right)\left(\begin{smallmatrix} 1 & 2 & 3 \\ 3 & 1 & 2 \end{smallmatrix}\right) = \left(\begin{smallmatrix} 1 & 2 & 3 \\ 1 & 3 & 2 \end{smallmatrix}\right)$. The result is $\mathbf{P}_3$. It is as if the top row of the second permutation "cancels" the bottom row of the first.

Proceeding in this fashion develops the product Table 1.6 for the permutation group of degree 3. Study of the product table verifies that the six permutations form a group because there is an identity $\mathbf{P}_1$, each permutation occurs only once in each row and column, and each permutation has an inverse, as shown in the table, wherever a product is $\mathbf{E}$.

Cycle Notation

Another way of representing a permutation is by *cycles*. A permutation cycle is developed by tracing the permutation from one number to another until the cycle begins to repeat. Consider permutation P_2: starting with 1 in the top row, $1 \mapsto 2$ and $2 \mapsto 1$, ending a cycle of length 2. Continuing to the next unused number 3, $3 \mapsto 3$ is a cycle of length 1. Combining, P_2 is written in cycle notation as $(1\,2)(3)$. Similarly,

$$P_1\ (1)(2)(3) \qquad P_2\ (1\,2)(3) \qquad P_3\ (1)(2\,3)$$
$$P_4\ (1\,3)(2) \qquad P_5\ (1\,3\,2) \qquad P_6\ (1\,2\,3).$$

A cycle of length 1 means that the permutation does not change that number. Cycles of length 1 are sometimes omitted in cycle notation.

To read a cycle, note that any number in the cycle maps to the number at its right, and the number at the right-hand end of the cycle is considered to return to the start of the cycle at the left-hand end. Take as an example the cycle $(3\,5\,2\,6\,1\,4)$. Starting from number 3, the permutation maps 3 to the number 5 at its right: $3 \mapsto 5$, etc. The last element 4 cycles around to the beginning so that $4 \mapsto 3$.

1.5 Matrix Representations of Groups

Mathematicians have proved a great number of theorems on the properties of abstract groups, but for work in physics it is a help to be able to relate abstract groups to calculable mathematics. For applications of group theory, the mathematical objects of choice are *matrices*, and the rule of combination (product) is matrix multiplication. For applications in physics, group theory is much less the theory of abstract groups and much more the theory of matrix representations, as this text amply demonstrates.

1.5.1 Isomorphism

Comparing the product Table 1.5 for **32** and Table 1.6 for S_3 shows they have exactly the same form except for the names of the members. When every member of one

Table 1.6 Permutation group of degree 3

	E	P_2	P_3	P_4	P_5	P_6
E	E	P_2	P_3	P_4	P_5	P_6
P_2	P_2	E	P_6	P_5	P_4	P_3
P_3	P_3	P_5	E	P_6	P_2	P_4
P_4	P_4	P_6	P_5	E	P_3	P_3
P_5	P_5	P_3	P_4	P_2	P_6	E
P_6	P_6	P_4	P_2	P_3	E	P_5

group corresponds to one and only one member of another and vice versa, the groups are said to be *isomorphic* (from Greek "identical form"). In the example of **32** and $\mathbf{S_3}$ the correspondence between the two sets of members is $\mathbf{E} \leftrightarrow \mathbf{P_1}$, $\mathbf{A} \leftrightarrow \mathbf{P_2}$, $\mathbf{B} \leftrightarrow \mathbf{P_3}$, $\mathbf{C} \leftrightarrow \mathbf{P_4}$, $\mathbf{D} \leftrightarrow \mathbf{P_5}$, and $\mathbf{F} \leftrightarrow \mathbf{P_6}$.

$$\begin{pmatrix} 1 & 2 & 3 \\ 1 & 2 & 3 \end{pmatrix} \quad \begin{pmatrix} 1 & 2 & 3 \\ 2 & 1 & 3 \end{pmatrix} \quad \begin{pmatrix} 1 & 2 & 3 \\ 1 & 3 & 2 \end{pmatrix} \quad \begin{pmatrix} 1 & 2 & 3 \\ 3 & 2 & 1 \end{pmatrix} \quad \begin{pmatrix} 1 & 2 & 3 \\ 3 & 1 & 2 \end{pmatrix} \quad \begin{pmatrix} 1 & 2 & 3 \\ 2 & 3 & 1 \end{pmatrix}$$

$$\quad\quad \mathbf{E} \quad\quad\quad\quad \mathbf{A} \quad\quad\quad\quad \mathbf{B} \quad\quad\quad\quad \mathbf{C} \quad\quad\quad\quad \mathbf{D} \quad\quad\quad\quad \mathbf{F}$$

Incidentally, these groups are also isomorphic to the group of six symmetry operations (rotations and flips) of the equilateral triangle.

If a set of matrices is isomorphic to the members of an abstract group, the matrices are called a *matrix representation* of the group.

1.5.2 Homomorphism

Sometimes the requirement is dropped that each representation has a unique 1-to-1 correspondence between members of two groups. Instead of an isomorphism, this is a *homomorphism*. The only requirement for a homomorphism is that the matrices have the same product table as the abstract group.

Here is a simple homomorphism of $\mathbf{\Gamma}$ (Table 1.2) in terms of 1-dimensional matrices.

$$(1) \quad\quad (1) \quad\quad (1)$$

$$\mathbf{E} \quad\quad \mathbf{A} \quad\quad \mathbf{B}$$

This set of permutations obviously obeys the product table for $\mathbf{\Gamma}$. The homomorphism where every group member is represented by the matrix (1) is called the *identity homomorphism* **1**. Every group has an identity homomorphism.

The **32** group of order 6 in Table 1.5 has, like every group, an identity homomorphism. However, it also has another distinctly different 1-dimensional homomorphism.

$$(1) \quad (-1) \quad (-1) \quad (-1) \quad (1) \quad (1)$$

$$\mathbf{E} \quad\quad \mathbf{A} \quad\quad \mathbf{B} \quad\quad \mathbf{C} \quad\quad \mathbf{D} \quad\quad \mathbf{F}$$

The permutation group $\mathbf{S_3}$ in Table 1.6 is isomorphic to the group of order 6, so this homomorphism applies to it also with a change of labels.

These homomorphisms are said to be *unfaithful*, because it is possible to form a matrix product such as $\mathbf{AB} = \mathbf{D}$ that does not agree with the group's product table. Put another way, it is impossible to reconstruct the group's product table from a homomorphism. In contrast, an isomorphism is *faithful*, because there is a 1-to-1 correspondence between group members and their matrices.

1.5.3 An Example of a Matrix Representation

This example sets up an isomorphism between the equilateral triangle group $\Gamma = \{\mathbf{E}, \mathbf{A}, \mathbf{B}\}$ and a set of matrices. Because the group operations of Γ describe rotations in the plane, we start by seeing how to use matrices to express a planar rotation.

To lay the groundwork, consider two Cartesian coordinate systems in the plane. The X'-Y' axes have the same origin as the X-Y axes, but they are rotated by angle θ as the figure shows.

A point in the plane has coordinates (x, y) with respect to the X-Y axes, as shown by the dashed lines. The same point has coordinates (x', y') with respect to the X'-Y' axes, as shown by the dotted lines.

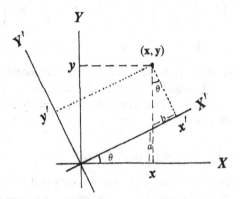

Express (x', y') in terms of (x, y) using trigonometry.

$$a = x \tan \theta \qquad b = (y - a) \sin \theta$$

$$x' = \frac{x}{\cos \theta} + b = \frac{x}{\cos \theta} + (y - a) \sin \theta$$

$$= \frac{x}{\cos \theta} + y \sin \theta - \frac{x \sin^2 \theta}{\cos \theta} = \frac{x}{\cos \theta} - \frac{x \left(1 - \cos^2 \theta\right)}{\cos \theta} + y \sin \theta$$

$$= x \cos \theta + y \sin \theta \tag{1.1}$$

$$y' = (y - a) \cos \theta = (y - x \tan \theta) \cos \theta$$

$$= -x \sin \theta + y \cos \theta \tag{1.2}$$

Writing Eqs. (1.1) and (1.2) in matrix form:

$$\begin{pmatrix} x' \\ y' \end{pmatrix} = \begin{pmatrix} \cos \theta & \sin \theta \\ -\sin \theta & \cos \theta \end{pmatrix} \begin{pmatrix} x \\ y \end{pmatrix}.$$

One-column matrices such as $\begin{pmatrix} x \\ y \end{pmatrix}$ are called *column vectors*. The matrix itself is square 2×2 with two rows and two columns.

The three members $\{\mathbf{E}, \mathbf{A}, \mathbf{B}\}$ of group Γ correspond geometrically to specific angles of rotation $0°, 120°, 240°$; evaluate the rotation matrix for each angle. Trigonometric identities such as $\cos(\theta_1 \pm \theta_2) = \cos(\theta_1) \cos(\theta_2) \pm \sin(\theta_1) \sin(\theta_2)$ and $\sin(\theta_1 \pm \theta_2) = \sin(\theta_1) \cos(\theta_2) \pm \cos(\theta_1) \sin(\theta_2)$ can be useful for referring sines and cosines to the first quadrant. To keep the signs straight, note that $\sin \theta \geq 0$ only in the first and second quadrants, and $\cos \theta \geq 0$ only in the first and fourth quadrants.

A common notation for the matrix representation of a group member $\mathbf{T}$ is $D(\mathbf{T})$, from the German *Darstellung*, representation. Here is a matrix representation of the members of Γ:

$$\begin{pmatrix} 1 & 0 \\ 0 & 1 \end{pmatrix} \quad \begin{pmatrix} -\frac{1}{2} & \frac{\sqrt{3}}{2} \\ -\frac{\sqrt{3}}{2} & -\frac{1}{2} \end{pmatrix} \quad \begin{pmatrix} -\frac{1}{2} & -\frac{\sqrt{3}}{2} \\ \frac{\sqrt{3}}{2} & -\frac{1}{2} \end{pmatrix}$$

$$D(\mathbf{E}) \qquad\qquad D(\mathbf{A}) \qquad\qquad D(\mathbf{B}).$$

The matrix $D(\mathbf{E})$ clearly acts as the group identity because its product with any matrix in the set leaves that matrix unchanged.

This representation is faithful. If $\mathbf{C}$ and $\mathbf{F}$ are members of a group, a faithful representation $D(\mathbf{C})D(\mathbf{F}) = D(\mathbf{CF})$ agrees with the product table entry $\mathbf{CF}$ because of the isomorphism between the representation matrices and the group members. As an example, consider the product $D(\mathbf{A})D(\mathbf{B})$ for the group Γ.

$$\begin{pmatrix} -\frac{1}{2} & \frac{\sqrt{3}}{2} \\ -\frac{\sqrt{3}}{2} & -\frac{1}{2} \end{pmatrix} \begin{pmatrix} -\frac{1}{2} & -\frac{\sqrt{3}}{2} \\ \frac{\sqrt{3}}{2} & -\frac{1}{2} \end{pmatrix} = \begin{pmatrix} \frac{1}{4}+\frac{3}{4} & \frac{\sqrt{3}}{4}-\frac{\sqrt{3}}{4} \\ \frac{\sqrt{3}}{4}-\frac{\sqrt{3}}{4} & \frac{3}{4}+\frac{1}{4} \end{pmatrix} = \begin{pmatrix} 1 & 0 \\ 0 & 1 \end{pmatrix}$$

The product is $D(\mathbf{E})$ in agreement with the Γ product table $\mathbf{AB} = \mathbf{E}$. The result also shows that the matrix $D(\mathbf{A})$ has an inverse equal to $D(\mathbf{B})$, and vice versa.

A mathematical theorem states that a matrix has a unique inverse only if the determinant of the matrix $\neq 0$. The determinants of all three representation matrices of Γ are $\neq 0$ assuring that each of the matrices has an inverse.

In this text the matrix representation $D(\mathbf{A})$ of a group member $\mathbf{A}$ is often written simply A.

In summary, we found a faithful matrix representation of group Γ by establishing an isomorphism with the abstract group members, in this example with the help of the algebraic geometry of rotations. We showed that there is a correspondence between the matrices and the abstract group members and also showed that the matrices agree with the group's product table, hence they obey the criterion for a faithful representation.

1.6 Matrix Algebra

Matrix notation, matrix multiplication, and determinants of matrices are often studied in math courses. This section is intended as a refresher.

A general notation is needed to describe matrices of arbitrary size. Every element of a matrix is given two subscripts as, for example, a_{ij}. The first subscript (here i) signifies the row, and the second subscript (here j) signifies the column. If matrix $D(\mathbf{A})$ has n rows and n columns, it is termed a square $n \times n$ matrix and in element notation is written

$$D(\mathbf{A}) = \begin{pmatrix} a_{11} & a_{12} & \cdots & a_{1n} \\ a_{21} & a_{22} & \cdots & a_{2n} \\ \vdots & \vdots & \ddots & \vdots \\ a_{n1} & a_{n2} & \cdots & a_{nn} \end{pmatrix}.$$

If a matrix has m rows and n columns, it is described as $m \times n$. Here is a 2×3 matrix:

$$\begin{pmatrix} 4 & 7 & 0 \\ 1 & 3 & -1 \end{pmatrix}.$$

1.6.1   A Constant Times a Matrix

If a constant multiplies a matrix, every element in the matrix is multiplied by that constant. Let $B=cA$ where c is a constant and A, B are matrices. In matrix element notation:

$$b_{ij} = ca_{ij}.$$

Every element of A is multiplied by c as shown:

$$c\begin{pmatrix} 0 & 2 & 0 \\ 1 & 0 & -1 \\ 0 & 0 & 1 \end{pmatrix} = \begin{pmatrix} 0 & 2c & 0 \\ c & 0 & -c \\ 0 & 0 & c \end{pmatrix}.$$

1.6.2   Addition of Matrices

Let $C = A + B$ be the sum of two matrices, all with the same number of rows and columns. In matrix element notation:

$$c_{ij} = a_{ij} + b_{ij}.$$

Corresponding elements are added. Here is an example:

$$\begin{pmatrix} 1 & 4 & 0 \\ 3 & 0 & -1 \\ 3 & 0 & 1 \end{pmatrix} + \begin{pmatrix} 4 & 2 & -3 \\ -2 & 0 & 2 \\ 0 & 0 & 4 \end{pmatrix} = \begin{pmatrix} 5 & 6 & -3 \\ 1 & 0 & 1 \\ 3 & 0 & 5 \end{pmatrix}.$$

1.6.3   Products of Matrices

Consider the product $C = AB$ of $n \times n$ matrices. The element c_{ik} of the product matrix is written

$$c_{ik} = \sum_{j=1}^{n} a_{ij} b_{jk}.$$

The sum runs from $j = 1$ to $j = n$ as indicated on the summation sign. Often the range of the summation is clear from the context, so that only the subscript summed over (a dummy variable) needs to be shown on the summation sign:

$$c_{ik} = \sum_{j=1}^{n} a_{ij} b_{jk} \quad \longrightarrow \quad \sum_{j} a_{ij} b_{jk}.$$

As an example, consider the product of two 2×2 matrices $AB = C$. Written with explicit subscripts:

$$\begin{pmatrix} a_{11} & a_{12} \\ a_{21} & a_{22} \end{pmatrix} \begin{pmatrix} b_{11} & b_{12} \\ b_{21} & b_{22} \end{pmatrix} = \begin{pmatrix} a_{11} b_{11} + a_{12} b_{21} & a_{11} b_{12} + a_{12} b_{22} \\ a_{21} b_{11} + a_{22} b_{21} & a_{21} b_{12} + a_{22} b_{22} \end{pmatrix}$$

$$= \begin{pmatrix} c_{11} & c_{12} \\ c_{21} & c_{22} \end{pmatrix}.$$

When multiplying ordinary numbers, the result is independent of the order: $2 \times 3 = 3 \times 2$. This may not be the case with matrices, where the order can be important as the following example demonstrates:

$$AB = \begin{pmatrix} 0 & 1 \\ 1 & 0 \end{pmatrix} \begin{pmatrix} 1 & 0 \\ 0 & -1 \end{pmatrix}$$

$$= \begin{pmatrix} 0 & -1 \\ 1 & 0 \end{pmatrix}$$

$$BA = \begin{pmatrix} 1 & 0 \\ 0 & -1 \end{pmatrix} \begin{pmatrix} 0 & 1 \\ 1 & 0 \end{pmatrix}$$

$$= \begin{pmatrix} 0 & 1 \\ -1 & 0 \end{pmatrix}$$

$$\neq AB.$$

Matrices can be multiplied even if they are not square, if the number of columns in the first matrix is equal to the number of rows in the second, so that the first matrix is $m \times s$ and the second is $s \times n$. The product matrix is then $m \times n$. Here is the product of a 2×3 matrix and a 3×1 matrix to give a 2×1 result:

$$\begin{pmatrix} a_{11} & a_{12} & a_{13} \\ a_{21} & a_{22} & a_{23} \end{pmatrix} \begin{pmatrix} b_{11} \\ b_{21} \\ b_{31} \end{pmatrix} = \begin{pmatrix} a_{11} b_{11} + a_{12} b_{21} + a_{13} b_{31} \\ a_{21} b_{11} + a_{22} b_{21} + a_{23} b_{31} \end{pmatrix}.$$

1.6.4 Determinant of a Matrix

The determinant of a $n \times n$ matrix has $n!$ terms, where each term is the product of n different elements of the matrix. The determinant of a square matrix A with coefficients a_{ij} is symbolized $|A|$. Equations (1.3) show the determinants of a 2×2 matrix ($2! = 2$ terms) and a 3×3 matrix ($3! = 6$ terms).

$$|A| = \begin{vmatrix} a_{11} & a_{12} \\ a_{21} & a_{22} \end{vmatrix} = a_{11} a_{22} - a_{12} a_{21}$$

$$|A| = \begin{vmatrix} a_{11} & a_{12} & a_{13} \\ a_{21} & a_{22} & a_{23} \\ a_{31} & a_{32} & a_{33} \end{vmatrix}$$

$$= a_{11} (a_{22} a_{33} - a_{23} a_{32}) - a_{21} (a_{12} a_{33} - a_{13} a_{32})$$

$$+ a_{31} \left(a_{12} a_{23} - a_{13} a_{22} \right)$$
$$= a_{11} a_{22} a_{33} - a_{11} a_{23} a_{32} + a_{13} a_{21} a_{32} - a_{12} a_{21} a_{33}$$
$$+ a_{12} a_{23} a_{31} - a_{13} a_{22} a_{31} \tag{1.3}$$

To illustrate one method of determining the algebraic signs of the terms, consider the case $n = 3$ as an example. In the last line of Eq. (1.3), the terms of the expansion are rearranged so that they all have the form $a_{1i} a_{2j} a_{3k}$. If the permutation $\left(\begin{smallmatrix} 1 & 2 & 3 \\ i & j & k \end{smallmatrix} \right)$ is even, the term has a plus sign; if the permutation is odd, the sign is minus. For the second term in Eq. (1.3), the permutation $\left(\begin{smallmatrix} 1 & 2 & 3 \\ 1 & 3 & 2 \end{smallmatrix} \right)$ is odd (only one interchange $2 \leftrightarrow 3$ is required), so the term has a minus sign as shown. Using this method to evaluate the determinant of a matrix doesn't apply if the elements are strictly numeric, but it is fine if the elements are symbolic with explicit subscripts. The abstract form is important in the quantum theory of multi-electron atoms where abstract permutations are directly related to the interchange of identical electrons.

1.6.5   The Kronecker Delta

Matrix notation presents an opportunity to introduce a useful new symbol. Consider $\mathbf{TT}^{-1} = \mathbf{E}$. The matrix for the identity has 1 everywhere along the main diagonal and 0 everywhere else. In matrix form the product is therefore written as

$$\sum_j t_{ij} t_{jk}^{-1} = \begin{pmatrix} 1 & 0 & \cdots & 0 \\ 0 & 1 & \cdots & 0 \\ \vdots & \vdots & \ddots & \vdots \\ 0 & 0 & \cdots & 1 \end{pmatrix}.$$

The *Kronecker delta symbol* δ_{ij} is useful to denote the matrix elements of the identity matrix. It is defined as

$$\delta_{ij} = \delta_{ji} = \begin{cases} 1 & i = j \\ 0 & i \neq j. \end{cases}$$

Using the Kronecker delta the matrix elements are

$$\sum_j t_{ij} t_{jk}^{-1} = \delta_{ik}.$$

1.7   Special Matrices

This section defines some special matrices that arise in applications.

1.7.1   The Complex Conjugate of a Matrix

The range of matrix elements can be enlarged to include complex numbers of the form $\alpha + \beta i$ where α and β are real numbers that have a real part α and an imaginary part βi and where $i = \sqrt{-1}$. The *complex conjugate* is $(\alpha + \beta i)^* = \alpha - \beta i$. If u is a complex number, its complex conjugate is denoted by u^*. The complex conjugate of a matrix A is A^*, and its elements are a_{ij}^* as in this example:

$$A = \begin{pmatrix} 1 + 2i & 4 - i \\ 3 + 4i & 6 \end{pmatrix} \qquad A^* = \begin{pmatrix} 1 - 2i & 4 + i \\ 3 - 4i & 6 \end{pmatrix}.$$

1.7.2 The Transpose of a Matrix

Consider a matrix A with elements a_{ij}. The *transpose* $\tilde{A}$ of A is the matrix formed by interchanging the rows and columns of A so that the elements of $\tilde{A}$ are $\tilde{a}_{ij} = a_{ji}$. It follows that elements on the main diagonal are unchanged. Here is an example of a matrix and its transpose:

$$A = \begin{pmatrix} 1 & 4 & 7 \\ 3 & 6 & 0 \\ 2 & 1 & 5 \end{pmatrix} \qquad \tilde{A} = \begin{pmatrix} 1 & 3 & 2 \\ 4 & 6 & 1 \\ 7 & 0 & 5 \end{pmatrix}.$$

The transpose of a matrix product AB is $\tilde{B}\tilde{A}$. Proof is left to the problems.

1.7.3 The Adjoint of a Matrix

The *adjoint* of a matrix A is its transpose with complex conjugates of its elements. The adjoint is symbolized $A^\dagger$. The result is the same whether the transpose or the complex conjugate is done first:

$$A = \begin{pmatrix} 1 + 2i & 4 - i \\ 3 + 4i & 6 \end{pmatrix} \qquad A^\dagger = \begin{pmatrix} 1 - 2i & 3 - 4i \\ 4 + i & 6 \end{pmatrix}.$$

1.7.4 Hermitian Matrices

If a matrix A is equal to its adjoint, so that $A = A^\dagger$, the matrix is said to be *Hermitian*. Here are two examples of Hermitian matrices, one real, the other complex. All Hermitian matrices are square:

$$\begin{pmatrix} 1 & 3 & 2 \\ 3 & 6 & 0 \\ 2 & 0 & 5 \end{pmatrix} \qquad \begin{pmatrix} 2 & 3 - i & 2 \\ 3 + i & 0 & -4i \\ 2 & 4i & 3 \end{pmatrix}.$$

The left-hand matrix is a *real symmetric* matrix; all real symmetric matrices are Hermitian. Note also that for every Hermitian matrix, the diagonal elements must be real even if the matrix is complex.

Hermitian matrices are fundamental in quantum mechanics. They guarantee that physical quantities calculated according to quantum mechanics are always real numbers as they must be.

1.7.5　Unitary and Orthogonal Matrices

If the adjoint of a matrix A is equal to its inverse so that $A^\dagger = A^{-1}$, the matrix is said to be *unitary*. It follows that the product of a unitary matrix with its adjoint is the identity. If A is real $A = A^*$, then its transpose equals its inverse $\tilde{A} = A^{-1}$, and A is said to be *real orthogonal* or simply *orthogonal*. All unitary and orthogonal matrices are square.

The identity matrix E is an orthogonal matrix because $E^\dagger = E$ and the product $EE = E$ shows that E is its own inverse, $E^{-1} = E$.

Here is the product of a matrix with its adjoint. It is equal to the identity, showing that the matrix is orthogonal.

$$\begin{pmatrix} \frac{1}{2} & -\frac{\sqrt{3}}{2} & 0 \\ \frac{\sqrt{3}}{2} & \frac{1}{2} & 0 \\ 0 & 0 & 1 \end{pmatrix}\begin{pmatrix} \frac{1}{2} & \frac{\sqrt{3}}{2} & 0 \\ -\frac{\sqrt{3}}{2} & \frac{1}{2} & 0 \\ 0 & 0 & 1 \end{pmatrix} = \begin{pmatrix} 1 & 0 & 0 \\ 0 & 1 & 0 \\ 0 & 0 & 1 \end{pmatrix}$$

Orthogonal matrices have the property that if their rows or columns are considered to be the components of n-dimensional vectors with respect to a set of unit basis vectors $\hat{e}_1, \hat{e}_2, \ldots, \hat{e}_n$, the row and column vectors are both unit orthogonal vectors (called an *orthonormal set*). Take, for instance, the first row of the left-hand matrix in the example. Its scalar product with itself is $(\frac{1}{2}\hat{e}_1 - \frac{\sqrt{3}}{2}\hat{e}_2 + 0) \cdot (\frac{1}{2}\hat{e}_1 - \frac{\sqrt{3}}{2}\hat{e}_2 + 0) = (\frac{1}{4} + \frac{3}{4}) = 1$. The scalar product of its first and second column vectors is $(\frac{1}{2}\hat{e}_1 + \frac{\sqrt{3}}{2}\hat{e}_2 + 0) \cdot (-\frac{\sqrt{3}}{2}\hat{e}_1 + \frac{1}{2}\hat{e}_2 + 0) = (-\frac{\sqrt{3}}{4} + \frac{\sqrt{3}}{4}) = 0$.

1.8　A Brief History of Group Theory

The major part of group theory was developed by mathematicians in the nineteenth and early twentieth centuries. Space does not permit citing all the famous names, but a few are especially worthy of note. The French mathematician and political activist Évariste Galois (1811–32) applied the name "group" to this branch of mathematics and is considered to be the founder of group theory. Sadly, he died at the young age of 20 after losing a duel. Ferdinand Georg Frobenius (1849–1917) in Germany introduced the fundamental concepts of characters and representations.

Amalie ("Emmy") Noether (1882–1935), "the greatest woman mathematician who ever lived," should also be mentioned. German-born, she taught in Germany until forced from her post in 1933. She spent the brief remaining years of her life at Bryn Mawr College in Pennsylvania and at the Institute for Advanced Study near Princeton University. Her greatest achievement was her "wonderful theorem" proving that the symmetries of a physical system lead to conservation laws obeyed by the system, another indication of the importance of group theory in physics.

The advent of quantum mechanics in the 1920s transformed group theory from an abstract mathematical discipline to a powerful tool for understanding nature, particularly phenomena in atomic physics, solid state physics, nuclear physics, and strange particle physics. In the early days of quantum mechanics some physicists did not appreciate the power of group theory in physics and called it a "pest," perhaps because their universities had not offered physics courses in group theory.

Mathematical physicist Eugene Wigner (1902–95, Nobel laureate in physics 1963) was one of the first to realize the value of group theory, which he used to explain the energy levels of electrons in atoms. Born in Budapest, he later became an American citizen. He was at Princeton University for much of his career and he shared the 1963 Nobel Prize for applying group theory to nuclear structure.

1.9 Brief Bios

Arthur Cayley (1821–95), a British mathematician, was the first to establish criteria for defining a group. In his earlier days as a lawyer he wrote many papers in mathematics as a hobby but later left his law practice, eventually becoming a professor of mathematics at Cambridge University.

The German mathematician Leopold Kronecker (1823–91) obtained his doctorate in mathematics at the age of 22, but then spent the next eight years managing valuable properties inherited from a wealthy uncle. By age 30, Kronecker had enough money for a comfortable life and returned to mathematics.

Hermitian matrices are named after the French mathematician Charles Hermite (1822–1901). Hermite spent his time studying great mathematicians like Lagrange and Gauss. He neglected humdrum course material and as a consequence did poorly on the standard examinations. Luckily, professors recognized his abilities, and Hermite went on to become one of the greatest mathematicians.

Summary of Chapter 1

Chapter 1 introduces fundamental concepts of group theory.

a) Group theory deals with abstract operations. Combinations of operations ("products") can be displayed in a product table.

b) The meaning of symmetry has been termed "change without change" because applying a symmetry operation leaves an object the same as at the start.

c) To be classed as a group, a set of operations must satisfy certain axioms. The product of any two members of the set must also be a member of the set. There must be an identity member in the set. Each member of the set must have an inverse in the set.

d) Examples of groups are presented, especially the triangle rotation group, the **32** group, and the permutation group S_3 of order 6.

e) Groups are described as discrete if they have a finite number of group members, or as continuous with an uncountable number of members that depend on a continuous parameter, for example, a rotation angle.

f) Isomorphism and homomorphism are possible relations between different groups. Two groups are isomorphic if there is a unique one-to-one correspondence between the members of one group and the members of the other, so that each obeys the same product table. A homomorphism between two groups also obeys the same product table, but a member of one group corresponds to more than one member of the other.

g) A group has a faithful representation by matrices where there is an isomorphism between the group elements and the matrices so that the matrices obey the group's product table.

h) Matrix algebra is the principal mathematical tool in applications of group theory. Examples are presented of matrix notation, matrix addition $A + B$, matrix multiplication AB, and determinant of a matrix $|A|$. Several special matrices are defined: complex conjugate A^*, transpose $\tilde{A}$, adjoint $A^\dagger = \tilde{A}^*$, Hermitian $A = A^\dagger$, unitary $A^{-1} = A^\dagger$, orthogonal $A^{-1} = \tilde{A}$.

i) The Kronecker delta $\delta_{ij} = \delta_{ji} = 1$ for $i = j$ or 0 for $i \neq j$.

Problems and Exercises

1.1   Evaluate $\mathbf{T}e^x$ for the transformation $\mathbf{T}x = -x$.

1.2   Evaluate $\mathbf{T}\cos x$ for the transformation $\mathbf{T}x = -x$.

1.3   Express the inverse $(\mathbf{AB})^{-1}$ of the product $\mathbf{AB}$ in terms of $\mathbf{A}$ and $\mathbf{B}$.

1.4   Show that the positive and negative real integers (including 0) form a group under the operation of addition.

1.5   Show that the real integers $1, 2, \ldots$ do not form a group under the operation of multiplication.

1.6   Show that the group members $\{\mathbf{E}, \mathbf{A}, \mathbf{B}\}$ for the 3-fold rotation of an equilateral triangle described in Section 1.2.1 can be written $\{\mathbf{E}, \mathbf{B}, \mathbf{B}^2\}$.

1.7   Write the product table for a group of order 2.

1.8   Write a product table for the 4-fold rotations of a square about its geometric center. Is this a cyclic group?

   E: *rotate by* $0°$
   A: *rotate by* $90°$
   B: *rotate by* $180°$
   C: *rotate by* $270°$

1.9   Prove that a given group member occurs only once in a given column of the product table.

1.10 Cayley proved that every discrete group of order n can be found in the product table for S_n. Illustrate this result for S_3 (Table 1.5 or Table 1.6). What does your result say about the number of distinct groups of order 3?

1.11 There are only two distinct groups of order 4. One of them is the group for rotations of a square, Problem 8. Here is the product table of the other.

	E	K	L	M
E	E	K	L	M
K	K	E	M	L
L	L	M	E	K
M	M	L	K	E

Show that $\{\mathbf{E, K, L, M}\}$ indeed form a group. Is it an Abelian group? Is it a cyclic group? Explain.

1.12 Find three subgroups of order 2 in the product table for the permutation group of order 6, Table 1.6.

1.13 Class each of the following permutations as even or odd.

$$\begin{pmatrix} 1 & 2 & 3 \\ 2 & 3 & 1 \end{pmatrix} \qquad \begin{pmatrix} 1 & 2 & 3 & 4 \\ 2 & 3 & 1 & 4 \end{pmatrix} \qquad \begin{pmatrix} 1 & 2 & 3 & 4 \\ 4 & 3 & 2 & 1 \end{pmatrix}$$

$$(a) \qquad\qquad\qquad (b) \qquad\qquad\qquad (c)$$

1.14 In the permutation $(5\,3\,6\,1\,4\,2)$ what number does 6 map to?

1.15 In the permutation $(3\,2\,6\,4\,1\,5)$ what number does 5 map to?

1.16 Find a matrix representation for the "flip" of an equilateral triangle about its aa axis.

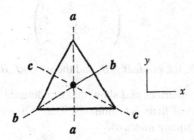

1.17 Consider the group $\boldsymbol{\Gamma} = \{\mathbf{E, A, B}\}$ for the rotations of an equilateral triangle, Table 1.2. Are the following matrices a homomorphic representation of $\boldsymbol{\Gamma}$?

$$(1) \quad (-1) \quad (-1)$$
$$\mathbf{E} \qquad \mathbf{A} \qquad \mathbf{B}$$

1.18 A *diagonal* matrix is a matrix that has zero elements except on its diagonal. Show that the product of two diagonal matrices is a diagonal matrix.

1.19 Find matrix representations for the 4-fold rotations of a square as described in Problem 8.

1.20 Which of the following matrices has an inverse?

$$\begin{pmatrix} 1 & 0 \\ 0 & -1 \end{pmatrix} \qquad \begin{pmatrix} 1 & -1 \\ 1 & -1 \end{pmatrix} \qquad \begin{pmatrix} 1 & 3 & 2 \\ 3 & 0 & -1 \\ 2 & -2 & 1 \end{pmatrix}$$

$$(a) \qquad\qquad\qquad (b) \qquad\qquad\qquad (c)$$

1.21 Consider the matrices A and B.

$$A = \begin{pmatrix} 3 & 2 & -3 \\ -2 & 3 & 2 \\ 1 & 0 & 4 \end{pmatrix}$$

$$B = \begin{pmatrix} 5 & 2 & -3 \\ 1 & 3 & 1 \\ -2 & 0 & 5 \end{pmatrix}$$

Find $A + B$, $A - B$, the product AB, and the product BA.

1.22 Consider the matrices A and B.

$$A = \begin{pmatrix} -1 & 5 & 4 \\ -3 & 3 & 4 \\ 2 & 0 & -3 \end{pmatrix}$$

$$B = \begin{pmatrix} 2 & -2 & -1 \\ 5 & 1 & 4 \\ -3 & 0 & 2 \end{pmatrix}$$

Find $A + B$, $A - B$, the product AB, and the product BA.

1.23 Consider these two matrices and show that the determinant of their product is equal to the product of their determinants. This is a general result true for the product of any two square matrices.

$$\begin{pmatrix} 2 & -1 & 2 \\ 2 & 0 & 3 \\ 3 & -1 & -1 \end{pmatrix} \qquad \begin{pmatrix} 1 & 2 & -2 \\ 0 & 3 & -1 \\ 2 & 1 & -2 \end{pmatrix}$$

1.24 For $n \times n$ matrices A and B, show that $\widetilde{AB} = \tilde{B}\tilde{A}$.

1.25 For the matrix A, find A^*, $\tilde{A}$, and $A^\dagger$.

$$A = \begin{pmatrix} 1 & 5 & 2 \\ 3 & 0 & -1 \\ 4 & -2 & 1 \end{pmatrix}$$

1.26 For the matrix A, find A^*, $\tilde{A}$, and $A^\dagger$.

$$A = \begin{pmatrix} 1 & 3+i & 2i \\ 3 & 0 & -1+2i \\ 4-3i & -2 & 1-i \end{pmatrix}$$

2

MATRIX REPRESENTATIONS OF DISCRETE GROUPS

2.1 Introduction

We begin with a word of apology because this chapter is purely mathematical without any physical applications. This book is intended for students of science, but the material in this chapter is essential when applying group theory to physical problems. Most physical applications rely more on matrix representation theory than on abstract group theory; this will become apparent when we turn to applications in later chapters.

As we build up the concepts of representation theory in this chapter, important theorems are stated, their meaning is discussed, and they are illustrated by examples. Some proofs are given to show the techniques involved. Refer to more advanced texts for complete detailed proofs.

2.2 Basis Functions and Representations

A matrix representation of an abstract group is a set of matrices that is isomorphic to the abstract group. Because of the isomorphism, the matrices obey the group's product table with matrix multiplication as the rule of combination. In this section we shall discuss the relation between the group operations and the group's matrix representations by looking at how group operations affect functions.

To start with a familiar example, consider the basis vectors $\{\hat{\mathbf{i}}, \hat{\mathbf{j}}, \hat{\mathbf{k}}\}$ of a 3-dimensional Cartesian frame, where the *hat* (caret) is the symbol for unit vectors.

The basis vectors are said to *span* the space, because the location of a particle anywhere in 3-dimensional space can be specified by coordinates with respect to the basis vectors. If we transform to a different set of basis vectors $\{\widehat{\mathbf{i'}}, \widehat{\mathbf{j'}}, \widehat{\mathbf{k'}}\}$, the coordinate values change, but the particle has not moved – only our description of its location is different.

By analogy, a member of an n-dimensional group can be thought of as operating on an abstract function space spanned by a set of *basis functions* $\{\phi_1, \phi_2, \ldots, \phi_n\}$. If $\mathbf{T}$ is a member of the group, the relation between the basis functions and the α matrix representation $D_{ij}^{(\alpha)}(\mathbf{T})$ of $\mathbf{T}$ is expressed by

$$\mathbf{T}\phi_j = \sum_{i=1}^{h_\alpha} D_{ij}^{(\alpha)}(\mathbf{T})\phi_i, \tag{2.1}$$

where h_α is the dimension of the α representation. Equation (2.1) does not look like matrix multiplication where the ij subscripts are normally seen reversed. The reason for the notation is that matrices written according to Eq. (2.1) obey the group's product table, hence are a valid faithful representation. As a proof, let $\mathbf{R}, \mathbf{S}, \mathbf{T}$ be members of a group such that $\mathbf{R} = \mathbf{ST}$ according to the group's product table. Then

$$\mathbf{R}\phi_j = \mathbf{ST}\phi_j$$
$$= \mathbf{S} \sum_i D_{ij}(\mathbf{T})\phi_i$$
$$= \sum_i D_{ij}(\mathbf{T})\mathbf{S}\phi_i$$
$$= \sum_k \sum_i D_{ij}(\mathbf{T}) D_{ki}(\mathbf{S})\phi_k.$$

Matrix elements are just numbers, so they can be multiplied with other matrix elements in any order, unlike the matrices themselves:

$$\mathbf{R}\phi_j = \sum_k \sum_i D_{ki}(\mathbf{S}) D_{ij}(\mathbf{T})\phi_k$$
$$= \sum_k D_{kj}(\mathbf{ST})\phi_k$$
$$= \sum_k D_{kj}(\mathbf{R})\phi_k,$$

as required. Without transposing the subscripts, the result would have been $D(\mathbf{TS})$, in disagreement with the group's product table. The matrices would not be isomorphic to the group members, hence would not be a faithful representation of the group.

Some texts omit summation symbols $\sum$ and instead use the *Einstein summation convention* that a repeated index in a factor is to be summed over. This text explicitly shows the summation symbol with a subscript for the index to be summed over, for example, $\sum_i$.

Consider a group consisting of two members, the identity **E** and an operation **T**. The group's product table is

	E	**T**
E	**E**	**T**.
T	**T**	**E**

Let $\mathbf{T}x = -x$. The function x spans the entire x-axis from $-\infty$ to $+\infty$, and $\mathbf{T}x$ is also in the same space. Applying Eq. (2.1) to the single basis function x, $D_{ij}(\mathbf{T})$ has only a single entry, $D_{11}(\mathbf{T})$,

$$\mathbf{T}x = -x = D_{11}(\mathbf{T})x$$
$$D_{11}(\mathbf{T}) = -1,$$

so that the matrix representation of **T** is the 1-dimensional matrix (-1). Together with the identity matrix $D(\mathbf{E}) = (1)$, these matrices obey the product table and are indeed a faithful matrix representation of the group.

Now perform the same calculation using the basis function x^2:

$$\mathbf{T}x^2 = (-x)^2 = x^2 = D_{11}(\mathbf{T})x^2$$
$$D_{11}(\mathbf{T}) = 1,$$

so that the matrix representation of **T** is the 1-dimensional matrix (1). Together with the identity matrix $D(\mathbf{E}) = (1)$, these matrices obey the product table and are a homomorphic representation of the group but not a faithful representation.

A group can have more than one representation. Our examples showed that the basis function x and the basis function x^2 lead to different matrix representations of the group $\{\mathbf{E}, \mathbf{T}\}$. Different sets of basis functions may give rise to different matrix representations with possibly different dimensions.

Consider the representation $D(\mathbf{E}) = (1)$ and $D(\mathbf{T}) = (1)$, in which every group member is represented by the matrix (1). Every group has this homomorphic *identity* representation, usually given the symbol **1** in modern texts.

As a counterexample, take e^x to be a tentative basis function, and apply the same calculation.

$$\mathbf{T}e^x = e^{-x} = D_{11}(\mathbf{T})e^x$$
$$D_{11}(\mathbf{T}) = e^{-2x}$$

However, the matrix $D(\mathbf{T}) = (e^{-2x})$ is not a representation of the group member $\mathbf{T}$. The product table is not obeyed; for instance, $\mathbf{T}^2 = e^{-4x} \neq \mathbf{E}$. e^x is not a basis function for the group $\{\mathbf{E}, \mathbf{T}\}$.

If a group's representation α has dimension h_α, there will be h_α^2 matrix elements and several basis functions; so if $h_\alpha > 1$, the sum in Eq. (2.1) will have several terms. For $h_\alpha = 2$ the matrix elements for an arbitrary group member $\mathbf{T}$ are

$$\mathbf{T}\phi_1 = D_{11}^{(\alpha)}(\mathbf{T})\phi_1 + D_{21}^{(\alpha)}(\mathbf{T})\phi_2$$
$$\mathbf{T}\phi_2 = D_{12}^{(\alpha)}(\mathbf{T})\phi_1 + D_{22}^{(\alpha)}(\mathbf{T})\phi_2,$$

so its representation matrix is the 2×2 matrix

$$D^\alpha(\mathbf{T}) = \begin{pmatrix} D_{11} & D_{12} \\ D_{21} & D_{22} \end{pmatrix}.$$

For a concrete example, suppose that operator $\mathbf{T}$ changes ϕ_1 to ϕ_2 and changes ϕ_2 to ϕ_1.

$$\mathbf{T}\phi_1 = \phi_2$$
$$\mathbf{T}\phi_2 = \phi_1$$

Hence $D_{11} = 0$, $D_{12} = 1$, $D_{21} = 1$, $D_{22} = 0$ so that the representation matrix for $\mathbf{T}$ is

$$\begin{pmatrix} 0 & 1 \\ 1 & 0 \end{pmatrix}.$$

The product table says that $\mathbf{T}^2 = \mathbf{E}$. The representation matrix obeys this product, so the representation is indeed faithful.

$$\begin{pmatrix} 0 & 1 \\ 1 & 0 \end{pmatrix} \begin{pmatrix} 0 & 1 \\ 1 & 0 \end{pmatrix} = \begin{pmatrix} 1 & 0 \\ 0 & 1 \end{pmatrix} \quad \checkmark$$

2.2.1 "Transforms Like ..."

Lecturers on group theory often use the phrase "transforms like" when discussing the relation between basis functions and matrix representations. This phrase is a verbal description of the relation stated above in Eq. (2.1) among group operations, basis functions, and matrix representations: $\mathbf{T}\phi_j = \sum_i D_{ij}(\mathbf{T})\phi_i$, where $\mathbf{T}$ is any member of the group, ϕ_i and ϕ_j are basis functions, and $D(\mathbf{T})$ is the matrix representation for $\mathbf{T}$ generated by the basis functions. The ϕ_i *transform like* the group members or, in other words, according to the representation whose matrix elements are D_{ij}.

2.3   Similarity Transformations

To set the stage for this concept, turn again to an example from algebraic geometry. Consider a 2-dimensional Cartesian frame (x, y) and a second Cartesian frame
(x', y') that has the same origin as the first so that

$$x' = a_{11}x + a_{12}y \qquad y' = a_{21}x + a_{22}y,$$

or in matrix form

$$\begin{pmatrix} x' \\ y' \end{pmatrix} = \begin{pmatrix} a_{11} & a_{12} \\ a_{21} & a_{22} \end{pmatrix} \begin{pmatrix} x \\ y \end{pmatrix} = A \begin{pmatrix} x \\ y \end{pmatrix}.$$

Now introduce a vector function that has components f_x and f_y, respectively,
in the first frame and components f_x' and f_y' in the second. Let the components be
linear functions of the respective coordinates. As an explicit example, the s_{ij} would
be trigonometric functions for rotation of the frames.

$$f_x = s_{11}x + s_{12}y \qquad f_y = s_{21}x + s_{22}y$$

$$f_x' = s_{11}x' + s_{12}y' \qquad f_y' = s_{21}x' + s_{22}y'$$

$$\begin{pmatrix} f_x \\ f_y \end{pmatrix} = \begin{pmatrix} s_{11} & s_{12} \\ s_{21} & s_{22} \end{pmatrix} \begin{pmatrix} x \\ y \end{pmatrix} = S \begin{pmatrix} x \\ y \end{pmatrix}$$

$$\begin{pmatrix} f_x' \\ f_y' \end{pmatrix} = \begin{pmatrix} s_{11} & s_{12} \\ s_{21} & s_{22} \end{pmatrix} \begin{pmatrix} x' \\ y' \end{pmatrix} = S \begin{pmatrix} x' \\ y' \end{pmatrix}$$

To express the components f_x' and f_y' in terms of f_x and f_y, note that

$$\begin{pmatrix} x \\ y \end{pmatrix} = S^{-1} \begin{pmatrix} f_x \\ f_y \end{pmatrix},$$

so that

$$\begin{pmatrix} f_x' \\ f_y' \end{pmatrix} = S \begin{pmatrix} x' \\ y' \end{pmatrix} = SA \begin{pmatrix} x \\ y \end{pmatrix} = SAS^{-1} \begin{pmatrix} f_x \\ f_y \end{pmatrix}.$$

The structure SAS^{-1} is called a *similarity transformation* of A with respect to
S. With the substitution $P \mapsto S^{-1}$ this similarity transformation can also be written
$P^{-1}AP$.

The fundamental meaning of a similarity transformation is a change to a different
set of basis functions, as shown in this vector function example where the similarity
transformation is from basis functions f_x, f_y to f_x', f_y'.

When all the group's operations undergo a similarity transformation with respect
to the same **S**, the group continues to be a valid group satisfying the group axioms
and with the same product table. Consider the group $\{\mathbf{E}, \mathbf{T}\}$ for which $\mathbf{T}^2 = \mathbf{E}$. In
terms of representation matrices,

$$STS^{-1}STS^{-1} = STTS^{-1}$$
$$= SES^{-1}$$
$$= E.$$

The group's product table is satisfied by the transformed operations STS^{-1} and SES^{-1}.

Here is an example of a similarity transformation in quantum mechanics. The fundamental equation of quantum mechanics is $\mathcal{H} \psi_n = E_n \psi_n$, where $\mathcal{H}$ is the Hamiltonian operator for the system of interest, ψ_n is the *wave eigenfunction* for the nth quantum state, and E_n is the *energy eigenvalue* of the nth state. Note that here E_n is a number, not an operator.

Suppose that ψ_n is replaced by $\psi'_n = S\psi_n$, where S is an operator. Because $S^{-1}S$ is the identity, inserting it between $\mathcal{H}$ and ψ_n in the original equation causes no change: $\mathcal{H}S^{-1}S\psi_n = E_n\psi_n$. Multiply both sides of the equation from the left by S to give $S\mathcal{H}S^{-1}S\psi_n = E_nS\psi_n$, again the fundamental equation with the same energy eigenvalue E_n, but with the original Hamiltonian replaced by its similarity transformation $S\mathcal{H}S^{-1}$ and with a new wave function $\psi'_n = S\psi_n$.

2.4 Equivalent Representations

A group can have many different matrix representations, each with a different set of basis functions. Let $\mathbf{T}$ be a member of a group and let the ϕ be basis functions for the group with matrix representation $D(\mathbf{T})$.

$$\mathbf{T}\phi_j = \sum_k D_{kj}(\mathbf{T})\phi_k \tag{2.2}$$

Now let the ϕ' be a linear combination of the ϕ according to matrix C.

$$\phi'_i = \sum_j C_{ji}\phi_j \tag{2.3}$$

Transposed subscripts are used in Eq. (2.3) because basis functions must "transform like" the representation matrices.

Any linear combination of the original basis functions is in the same function space and is therefore a basis function for the group. Applying $\mathbf{T}$ to Eq. (2.3) and using Eq. (2.2) gives

$$\mathbf{T}\phi'_i = \sum_j C_{ji}\mathbf{T}\phi_j$$
$$= \sum_j \sum_k C_{ji} D_{kj}(\mathbf{T})\phi_k. \tag{2.4}$$

Formally solving Eq. (2.3) for the ϕ basis functions in terms of the ϕ' gives

$$\phi_k = \sum_m A_{mk} \phi'_m, \tag{2.5}$$

where the A_{mk} are the elements of a coefficient matrix $\mathbf{A}$ to be determined. Multiply Eq. (2.5) by C_{kj} and sum over k.

$$\sum_k C_{kj}\phi_k = \sum_k \sum_m C_{kj} A_{mk}\, \phi'_m = \sum_k \sum_m A_{mk} C_{kj}\, \phi'_m$$

By Eq. (2.3) the sum on the left is ϕ'_j.

$$\phi'_j = \sum_k \sum_m C_{kj} A_{mk}\, \phi'_m$$

Summing over k, and noting that both sides must equal ϕ'_j:

$$\phi'_j = \sum_m (AC)_{mj}\, \phi'_m$$
$$(AC)_{mj} = \delta_{mj}$$
$$AC = E$$
$$A = C^{-1},$$

so from Eq. (2.5):

$$\phi_k = \sum_m C^{-1}_{mk}\, \phi'_m.$$

Inserting Eq. (2.4) gives

$$\mathbf{T}\phi'_i = \sum_j \sum_k \sum_m C_{ji}\, D_{kj}(\mathbf{T}) C^{-1}_{mk}\, \phi'_m$$

$$\mathbf{T}\phi'_i = \sum_j \sum_k \sum_m C^{-1}_{mk} D_{kj}(\mathbf{T}) C_{ji}\, \phi'_m. \tag{2.6}$$

Summing over k and then over j,

$$\mathbf{T}\phi'_i = \sum_m D'_{mi}(\mathbf{T})\phi'_m, \tag{2.7}$$

where

$$D' = C^{-1}DC. \tag{2.8}$$

Two results from this calculation should be noted. First, Eq. (2.7) proves that the ϕ' are basis functions for a representation with new matrices $D'(\mathbf{T})$. Second, Eq. (2.8) expresses the transformed representation matrices $D'(\mathbf{T}) = C^{-1}D(\mathbf{T})C$ in terms of the original representation matrices. It is not surprising that Eq. (2.8) is a similarity transformation. The similarity transformation is executed by the coefficient matrix C that expresses the ϕ' basis functions in terms of the ϕ basis functions in Eq. (2.3). The matrices D and D' may be different, but they are both valid representations of the group. They are *equivalent matrix representations* of the group. As we shall see later, equivalent representations describe the same physics.

2.5   Similarity Transformations and Unitary Matrices

When converting from one representation to another, similarity transformations are always executed by unitary matrices, as we now show. Recall that a matrix U is unitary if the complex conjugate of its transpose is equal to its inverse $U^\dagger = \tilde{U}^* = U^{-1}$. In terms of matrix elements, $u_{ij}^\dagger = u_{ji}^* = u_{ij}^{-1}$. If the matrix elements are real $u_{ij}^* = u_{ij}$, the matrix is real orthogonal (called orthogonal for short), as discussed in Section 1.7.5.

Consider a set of n basis functions $\{\phi_1, \phi_2, \ldots, \phi_n\}$. We suppose that this is an orthonormal set with some sort of scalar product denoted by angle brackets $\langle \phi_i^*, \phi_j \rangle = \delta_{ij}$. The explicit nature of the scalar product is immaterial for our purposes here – it could be a dot product of vectors $\mathbf{u}^* \cdot \mathbf{v}$ or it might be an integral $\int \phi_i^* \phi_j \, d\tau$ over some space. A familiar example of an orthonormal set is the Cartesian unit basis vectors $\{\hat{\mathbf{i}}, \hat{\mathbf{j}}, \hat{\mathbf{k}}\}$ in physical space.

The set of n basis functions is considered to be *linearly independent*. By definition a set of functions is linearly independent if the equation

$$\sum_{i=1}^{n} c_i \phi_i = 0$$

has only the solution $c_i = 0$ for all i, which implies that none of the functions ϕ_i can be expressed in terms of the other functions of the set.

Let an operation $\mathbf{T}$ with matrix representation $D(\mathbf{T})$ transform a basis set ϕ to a new set ψ assumed to be orthonormal with a defined scalar product. Then, using Eq. (2.3),

$$\psi_i = \sum_k D_{ki}(\mathbf{T})\phi_k,$$

$$
\begin{aligned}
\delta_{ij} &= \langle \psi_i^*, \psi_j \rangle \\
&= \left\langle \sum_k D_{ki}^*(\mathbf{T})\phi_k^*, \sum_m D_{mj}(\mathbf{T})\phi_m \right\rangle \\
&= \sum_k \sum_m D_{ki}^*(\mathbf{T})D_{mj}(\mathbf{T})\langle \phi_k^*, \phi_m \rangle = \sum_k \sum_m D_{ki}^*(\mathbf{T})D_{mj}(\mathbf{T})\delta_{km} \\
&= \sum_k D_{ki}^*(\mathbf{T})D_{kj}(\mathbf{T}),
\end{aligned}
$$

so that

$$\sum_k D_{ik}^{\dagger}(\mathbf{T}) D_{kj}(\mathbf{T}) = \delta_{ij}.$$

This result implies $D^{\dagger}(\mathbf{T}) = D^{-1}(\mathbf{T})$, showing that $D(\mathbf{T})$ is unitary.

Theorem 1 *A similarity transformation from one orthonormal set to another is always executed by a unitary matrix.*

This proof has neglected to show that the ψ set is indeed orthonormal, but it can in any case be made orthonormal using *Gram–Schmidt orthogonalization*, a systematic procedure that converts any set of linearly independent functions to an orthonormal set. Refer to other texts for details.

2.6   Character and Its Invariance under Similarity Transformations

The *character*, or *trace*, of a matrix is the sum of its diagonal elements. The character of a matrix A is symbolized by $\chi(A)$ and in terms of matrix elements as $\chi(A) = \sum_i a_{ii}$. This would be written $\chi(A) = a_{ii}$, using the Einstein summation convention.

Under a similarity transformation of A by some matrix S, the character of $B = SAS^{-1}$ is equal to the character of A so that $\chi(B) = \chi(A)$. Here is a proof.

$$\chi(A) = \sum_i a_{ii}$$

$$\chi(B) = \sum_i b_{ii}$$

$$= \sum_i \sum_j \sum_k s_{ij} a_{jk} s_{ki}^{-1} = \sum_i \sum_j \sum_k s_{ki}^{-1} s_{ij} a_{jk}$$

The sum over i is

$$\sum_i s_{ki}^{-1} s_{ij} = \delta_{kj} \quad \text{so that}$$

$$\chi(B) = \sum_j \sum_k \delta_{kj} a_{jk} = \sum_j a_{jj} = \chi(A).$$

The last step follows because $\delta_{kj} = 1$ for $k = j$ and 0 for $k \neq j$, so in the sum over k only the term $k = j$ makes a contribution.

As an application of character invariance, consider the matrix for physical rotation by θ about the z-axis.

$$\begin{pmatrix} \cos\theta & -\sin\theta & 0 \\ \sin\theta & \cos\theta & 0 \\ 0 & 0 & 1 \end{pmatrix}$$

Its character is $2\cos\theta + 1$. Rotation by θ about an arbitrary axis has the same character, because the change of axis is executed by a similarity transformation.

Theorem 2 *The matrix for physical rotation by θ about any axis has character $2\cos\theta + 1$.*

Any two matrices related by a similarity transformation therefore have the same character. The German word for character is *Spur*, like the English word *spoor* for the track of an animal. A matrix cannot hide its character by undergoing a similarity transformation. If the matrices for two representations have the same characters, they are related by a similarity transformation and are equivalent.

If two sets of representation matrices for the same group do not have the same character set, they are not related by a similarity transformation and do not span the same space. Take as an example the group $\{E, T\}$ introduced earlier that has two 1-dimensional representations. In both of them $D(E) = (1)$ and the representations for **T** were $D(T) = (1)$ and (-1). The matrices are 1-dimensional, so the characters are just the matrix entries. It is convenient to display the characters in a *character table*: The column headings are the two group members. The row headings are the symbols for the representations, where 1 always labels the identity representation and A is the label for the other 1-dimensional representation. A and B are the standard labels for 1-dimensional representations that are not the identity representation. The characters are not the same, showing that these two representations are not equivalent; there is no similarity transformation that can change one of the representations into the other.

	E	**T**
1	1	1
A	1	-1

Put another way, the basis functions for the different representations span different spaces.

Theorem 3 *If two outwardly different matrix representations of the same group have the same character tables, the representations are related by a similarity transformation. They are equivalent and not fundamentally different, because their basis functions both span the same function subspace.*

If the rows of a character table are considered to be vectors, the rows, and also the columns are orthogonal. Taking the two rows in this character table, their scalar product is $1 \times 1 + 1 \times (-1) = 0$. This behavior is generalized in Section 2.7.4.

Appendix A shows how to find characters directly from group properties and various theorems without needing to know the matrices explicitly.

The determinant of a matrix is also invariant under a similarity transformation. It is shown in problem 1.23 that for square matrices the determinant of a matrix product equals the product of the matrix determinants. Consider the similarity transformation of matrix V with respect to matrix U to give matrix V'.

$$V' = UVU^{-1}$$
$$|V'| = |UVU^{-1}| = |U||V||U^{-1}|$$
$$= |V||UU^{-1}| = |V||E|$$
$$= |V|$$

By expanding in cofactors of the first row it is easy to show that $|E| = 1$ for E of any dimension.

2.7   Irreducible Representations

Much of what we have discussed so far has been in preparation for this section. This is the most important section of the chapter, because irreducible representations crop up time and time again in physical applications such as molecular vibrations and quantum mechanics. This section explains the meaning of irreducible representations and why they are important. Section 2.7.3 develops the related concept of class that extends the meaning of character, and a number of theorems are stated, some with formal proofs and the others illustrated by examples.

Consider an n-dimensional group with basis functions $\{\phi_1, \phi_2, \ldots, \phi_n\}$ so that its representations are $n \times n$ matrices. Suppose there is a similarity transformation with some S that changes the original matrix representations to matrices that have smaller nonoverlapping square "block" matrices arrayed along the diagonal and 0 elsewhere, as in this illustrative example.

$$\begin{pmatrix} a_{11} & 0 & 0 & 0 & 0 & 0 \\ 0 & b_{11} & 0 & 0 & 0 & 0 \\ 0 & 0 & c_{11} & c_{12} & 0 & 0 \\ 0 & 0 & c_{21} & c_{22} & 0 & 0 \\ 0 & 0 & 0 & 0 & d_{11} & d_{12} \\ 0 & 0 & 0 & 0 & d_{21} & d_{22} \end{pmatrix}$$

This matrix is said to be in *block diagonal form*.

Because a similarity transformation is a transformation of the group's basis functions to a new set, a similarity transformation by the same $\mathbf{S}$ applies to every member of the group and puts all of the original representation matrices into block diagonal forms of the same structure. The original matrix representation is said to be *reducible*. If no further similarity transformation exists to break some or all of the block matrices into even smaller blocks, the representation is *irreducible*. Some authors use the neologism *irrep* to denote an irreducible representation.

The block matrices of an irreducible representation are themselves each a representation of the group member. These representations might include homomorphisms as well as faithful isomorphic representations. Also, some representations might be repeated among the blocks.

Suppose that the members of a group have representations by matrices in block diagonal form. Because all the matrices for the group have the same block diagonal structure, if two of the matrices are multiplied, the result is in the same block diagonal form, with each block just the matrix product of the corresponding blocks, as in this example.

$$\begin{pmatrix} 2 & 0 & 0 & 0 & 0 \\ 0 & 1 & 3 & 0 & 0 \\ 0 & 2 & 2 & 0 & 0 \\ 0 & 0 & 0 & 2 & 1 \\ 0 & 0 & 0 & 1 & 1 \end{pmatrix} \begin{pmatrix} 4 & 0 & 0 & 0 & 0 \\ 0 & 2 & 1 & 0 & 0 \\ 0 & 1 & 2 & 0 & 0 \\ 0 & 0 & 0 & 1 & 1 \\ 0 & 0 & 0 & 3 & 2 \end{pmatrix} = \begin{pmatrix} 8 & 0 & 0 & 0 & 0 \\ 0 & 5 & 7 & 0 & 0 \\ 0 & 6 & 6 & 0 & 0 \\ 0 & 0 & 0 & 5 & 4 \\ 0 & 0 & 0 & 4 & 3 \end{pmatrix}$$

The blocks retain their individuality, guaranteeing that the representation continues to satisfy the group's product table.

The meaning of an irreducible representation can be understood in terms of basis functions. A similarity transformation with S transforms the original basis set to a new set that spans the same function space. But now the block diagonal form of the representation matrices means that the new basis set has been subdivided into subsets that each span only a portion of the function space. Put another way, any function in one of these subspaces can be expressed as a linear combination of the basis functions in the new basis subset. If a subset has dimension μ, let the new basis subset be $\{\phi_1, \phi_2, \ldots, \phi_\mu\}$. Then any function ϕ' in that subspace can be expressed as a linear combination: $\phi' = c_1\phi_1 + c_2\phi_2 + \cdots + c_\mu\phi_\mu$.

In terms of group operations, a group member operating on an irreducible subspace can only result in a combination of the existing basis functions that span the subspace. The subspace is said to be *invariant* under the action of the associated group members, because no group operation can lead to a function outside the invariant subspace. We shall see in later chapters that physical meaning is attached to invariant subspaces.

2.7.1 The Regular Representation

One way to construct an irreducible representation is to start from a reducible representation. Suppose a group does not have a geometrical interpretation so that a reducible representation is not easy to construct. However, a representation of a group, called the *regular representation*, can always be constructed from the group's product table.

The product table below is for the **32** group as laid out in Section 1.4.2 but with columns interchanged to put all identity operations **E** on the diagonal.

To form the matrices of the regular representation, use the modified product table as a template with 1 entered wherever the symbol for the operation appears and 0

	E	A	B	C	F	D
E	E	A	B	C	F	D
A	A	E	F	D	B	C
B	B	D	E	F	C	A
C	C	F	D	E	A	B
D	D	B	C	A	E	F
F	F	C	A	B	D	E

otherwise. Here are the regular representations for group members **A** and **C**:

$$D^{(reg)}(\mathbf{A}) = \begin{pmatrix} 0 & 1 & 0 & 0 & 0 & 0 \\ 1 & 0 & 0 & 0 & 0 & 0 \\ 0 & 0 & 0 & 0 & 0 & 1 \\ 0 & 0 & 0 & 0 & 1 & 0 \\ 0 & 0 & 0 & 1 & 0 & 0 \\ 0 & 0 & 1 & 0 & 0 & 0 \end{pmatrix} \qquad D^{(reg)}(\mathbf{C}) = \begin{pmatrix} 0 & 0 & 0 & 1 & 0 & 0 \\ 0 & 0 & 0 & 0 & 0 & 1 \\ 0 & 0 & 0 & 0 & 1 & 0 \\ 1 & 0 & 0 & 0 & 0 & 0 \\ 0 & 0 & 1 & 0 & 0 & 0 \\ 0 & 1 & 0 & 0 & 0 & 0 \end{pmatrix}.$$

In any representation the matrix for **E** has entry 1 everywhere on the main diagonal and 0 elsewhere because this is the only way **E** can commute with all the other matrices.

The matrices of the regular representation are distinct because the product table is distinct. In the regular representation, there is only a single 1 in each row or column, because in the product table a given group member occurs only once in each row and column. No matrix in the regular representation except **E** has 1 on the main diagonal. The matrices in the regular representation are easy to multiply together to show that the product table is satisfied; it is, therefore, a faithful isomorphic representation of the group.

2.7.2   Example: Reducing the Regular Representation

The regular representation of the **32** group can be transformed to irreducible representations by a similarity transformation with the following matrix S.

$$S = \frac{1}{\sqrt{6}} \begin{pmatrix} 1 & 1 & 1 & 1 & 1 & 1 \\ 1 & -1 & -1 & -1 & 1 & 1 \\ \sqrt{\frac{3}{2}} & -\sqrt{\frac{3}{2}} & \sqrt{\frac{3}{2}} & 0 & 0 & -\sqrt{\frac{3}{2}} \\ \frac{1}{\sqrt{2}} & \frac{1}{\sqrt{2}} & \frac{1}{\sqrt{2}} & -\sqrt{2} & -\sqrt{2} & \frac{1}{\sqrt{2}} \\ \frac{1}{\sqrt{2}} & -\frac{1}{\sqrt{2}} & -\frac{1}{\sqrt{2}} & \sqrt{2} & -\sqrt{2} & \frac{1}{\sqrt{2}} \\ -\sqrt{\frac{3}{2}} & -\sqrt{\frac{3}{2}} & \sqrt{\frac{3}{2}} & 0 & 0 & \sqrt{\frac{3}{2}} \end{pmatrix}$$

S is a unitary matrix, in this case a real orthogonal matrix, so its inverse is the transpose: $S^{-1} = \tilde{S}$.

$$S^{-1} = \frac{1}{\sqrt{6}}\begin{pmatrix} 1 & 1 & \sqrt{\frac{3}{2}} & \frac{1}{\sqrt{2}} & \frac{1}{\sqrt{2}} & -\sqrt{\frac{3}{2}} \\ 1 & -1 & -\sqrt{\frac{3}{2}} & \frac{1}{\sqrt{2}} & -\frac{1}{\sqrt{2}} & -\sqrt{\frac{3}{2}} \\ 1 & -1 & \sqrt{\frac{3}{2}} & \frac{1}{\sqrt{2}} & -\frac{1}{\sqrt{2}} & \sqrt{\frac{3}{2}} \\ 1 & -1 & 0 & -\sqrt{2} & \sqrt{2} & 0 \\ 1 & 1 & 0 & -\sqrt{2} & -\sqrt{2} & 0 \\ 1 & 1 & -\sqrt{\frac{3}{2}} & \frac{1}{\sqrt{2}} & \frac{1}{\sqrt{2}} & \sqrt{\frac{3}{2}} \end{pmatrix}$$

As an example, applying a similarity transformation with S to $D^{(reg)}(\mathbf{C})$ given above reduces $D^{(reg)}(\mathbf{C})$ to irreducible representations. After lengthy matrix multiplications, the result for $S D^{(reg)}(\mathbf{C}) S^{-1}$ is

$$SCS^{-1} = \begin{pmatrix} 1 & 0 & 0 & 0 & 0 & 0 \\ 0 & -1 & 0 & 0 & 0 & 0 \\ 0 & 0 & \frac{1}{2} & -\frac{\sqrt{3}}{2} & 0 & 0 \\ 0 & 0 & -\frac{\sqrt{3}}{2} & -\frac{1}{2} & 0 & 0 \\ 0 & 0 & 0 & 0 & \frac{1}{2} & -\frac{\sqrt{3}}{2} \\ 0 & 0 & 0 & 0 & -\frac{\sqrt{3}}{2} & -\frac{1}{2} \end{pmatrix}.$$

This result has several features. There are three distinct matrix representations: two different 1-dimensional matrix representations and a repeated 2-dimensional matrix representation. The 1-dimensional matrix at the extreme upper left is the representation of $\mathbf{C}$ in the homomorphic identity representation 1, where all group members have the matrix representation (1).

$$\begin{array}{cccccc} (1) & (1) & (1) & (1) & (1) & (1) \\ \mathbf{E} & \mathbf{A} & \mathbf{B} & \mathbf{C} & \mathbf{D} & \mathbf{F} \end{array}$$

It is evident that these matrices will occur in the irreducible representation of every group.

The second block (-1) belongs to the second 1-dimensional homomorphic representation of the **32** group introduced in Section 1.4.

$$\begin{array}{cccccc} (1) & (-1) & (-1) & (-1) & (1) & (1) \\ \mathbf{E} & \mathbf{A} & \mathbf{B} & \mathbf{C} & \mathbf{D} & \mathbf{F} \end{array}$$

This representation is given the symbol A for a 1-dimensional representation.

The two 2×2 repeated blocks belong to an irreducible isomorphic representation of the **32** group. They can be obtained by applying the similarity transformation with S to the regular representation of each group member.

$$\begin{pmatrix} 1 & 0 \\ 0 & 1 \end{pmatrix} \quad \begin{pmatrix} -1 & 0 \\ 0 & 1 \end{pmatrix} \quad \begin{pmatrix} \frac{1}{2} & \frac{\sqrt{3}}{2} \\ \frac{\sqrt{3}}{2} & -\frac{1}{2} \end{pmatrix} \quad \begin{pmatrix} \frac{1}{2} & -\frac{\sqrt{3}}{2} \\ -\frac{\sqrt{3}}{2} & -\frac{1}{2} \end{pmatrix} \quad \begin{pmatrix} -\frac{1}{2} & \frac{\sqrt{3}}{2} \\ -\frac{\sqrt{3}}{2} & -\frac{1}{2} \end{pmatrix} \quad \begin{pmatrix} -\frac{1}{2} & -\frac{\sqrt{3}}{2} \\ \frac{\sqrt{3}}{2} & -\frac{1}{2} \end{pmatrix}$$

$$\quad\quad \text{E} \quad\quad\quad \text{A} \quad\quad\quad\quad \text{B} \quad\quad\quad\quad\quad \text{C} \quad\quad\quad\quad\quad \text{D} \quad\quad\quad\quad\quad \text{F}$$

The symbol for this representation is Γ.

The matrices of the irreducible representation Γ are real orthogonal. The row vectors and column vectors of each matrix form a set of orthonormal vectors. To illustrate, consider the scalar product of the rows of matrix B:

$$\left(\tfrac{1}{2}\right)\left(\tfrac{\sqrt{3}}{2}\right) + \left(\tfrac{\sqrt{3}}{2}\right)\left(-\tfrac{1}{2}\right) = 0.$$

The scalar product of the first row of B with itself is

$$\left(\tfrac{1}{2}\right)\left(\tfrac{1}{2}\right) + \left(\tfrac{\sqrt{3}}{2}\right)\left(\tfrac{\sqrt{3}}{2}\right) = 1.$$

Section 2.7.4 proves that *all* of a group's irreducible representations are contained in the regular representation. Another proof will show that irreducible representations occur in the reduced regular representation a number of times equal to their dimension, as illustrated by representation Γ of the **32** group. Γ has matrices of dimension 2 and occurs twice in the reduction.

Any matrix representation of a group can be expressed in terms of one or more of the irreducible representations by using a suitable similarity transformation. As an example from the **32** group, let $D^{(reg)}(\mathbf{C})$ be the regular representation of $\mathbf{C}$ and let $D^{(red)}(\mathbf{C})$ be its reduced block diagonal form. $D^{(reg)}(\mathbf{C})$ and $D^{(red)}(\mathbf{C})$ have the same dimensions. As our example showed,

$$S D^{(reg)}(\mathbf{C}) S^{-1} = D^{(red)}(\mathbf{C}).$$

Multiply both sides by S^{-1} from the left and by S from the right to express $D^{(reg)}(\mathbf{C})$ in terms of irreducible representations.

$$D^{(reg)}(\mathbf{C}) = S^{-1} D^{(red)}(\mathbf{C}) S$$

Any representation $D^{(\alpha)}$ that is in a different reduced block form is related to the original reduced block form by a similarity transformation with some U.

$$D^{(\alpha)}(\mathbf{C}) = U^{-1} D^{(red)}(\mathbf{C}) U$$

If all the irreducible representations of a group are known, that is all there is to know about any of its other irreducible matrix representations. They are all related by similarity transformations.

The example using the **32** group illustrates how similarity transformations can generate irreducible representations of a group. It is difficult to calculate the needed transformation matrix such as S in the example. Luckily, mathematicians have worked out the irreducible representations for groups of interest.

2.7.3 Conjugates and Classes

The example showed that the **32** group has three distinct irreducible representations. The number of irreducible representations for any group can be predicted from its product table using the concept of *classes*.

Suppose that **A** and **R** are both members of the same group. The product **RAR**$^{-1}$ is called the *conjugate* to **A**. According to the group axioms, a conjugate must be a member of the same group. The use of conjugates "unpacks" the structure of a group, dividing it into classes.

Again, take the example of the **32** group, which has the following product table.

	E	A	B	C	D	F
E	E	A	B	C	D	F
A	A	E	F	D	C	B
B	B	D	E	F	A	C
C	C	F	D	E	B	A
D	D	B	C	A	F	E
F	F	C	A	B	E	D

The inverse of each group member is readily found from the product table.
$$E^{-1} = E \quad A^{-1} = A \quad B^{-1} = B \quad C^{-1} = C \quad D^{-1} = F \quad F^{-1} = D$$

To determine the classes of this group, take each member of the group and evaluate its conjugate with every member of the group.

$EEE^{-1} = $ **E**	$EDE^{-1} = $ **D**	$EFE^{-1} = $ **F**
$AEA^{-1} = $ **E**	$ADA^{-1} = $ **F**	$AFA^{-1} = $ **D**
$BEB^{-1} = $ **E**	$BDB^{-1} = $ **F**	$BFB^{-1} = $ **D**
$CEC^{-1} = $ **E**	$CDC^{-1} = $ **F**	$CFC^{-1} = $ **D**
$DED^{-1} = $ **E**	$DDD^{-1} = $ **D**	$DFD^{-1} = $ **F**
$FEF^{-1} = $ **E**	$FDF^{-1} = $ **D**	$FFF^{-1} = $ **F**

$$
\begin{array}{lll}
\mathbf{EAE}^{-1} = \mathbf{A} & \mathbf{EBE}^{-1} = \mathbf{B} & \mathbf{ECE}^{-1} = \mathbf{C} \\
\mathbf{AAA}^{-1} = \mathbf{A} & \mathbf{ABA}^{-1} = \mathbf{C} & \mathbf{ACA}^{-1} = \mathbf{B} \\
\mathbf{BAB}^{-1} = \mathbf{C} & \mathbf{BBB}^{-1} = \mathbf{B} & \mathbf{BCB}^{-1} = \mathbf{A} \\
\mathbf{CAC}^{-1} = \mathbf{B} & \mathbf{CBC}^{-1} = \mathbf{A} & \mathbf{CCC}^{-1} = \mathbf{C} \\
\mathbf{DAD}^{-1} = \mathbf{C} & \mathbf{DBD}^{-1} = \mathbf{A} & \mathbf{DCD}^{-1} = \mathbf{B} \\
\mathbf{FAF}^{-1} = \mathbf{B} & \mathbf{FBF}^{-1} = \mathbf{C} & \mathbf{FCF}^{-1} = \mathbf{A}
\end{array}
$$

Conjugation has resolved the group members into three distinct *classes*: $\{\mathbf{E}\}$, $\{\mathbf{D}, \mathbf{F}\}$, and $\{\mathbf{A}, \mathbf{B}, \mathbf{C}\}$. Note that every group must have a class $\{\mathbf{E}\}$ with only one member, because for any member $\mathbf{R}$ of the group, $\mathbf{RER}^{-1} = \mathbf{RR}^{-1} = \mathbf{E}$, as the example illustrates.

All members of the same class have the same character. As a proof, let $\mathbf{A}$, $\mathbf{B}$, and $\mathbf{R}$ be members of a group. If $\mathbf{B} = \mathbf{RAR}^{-1}$, then $\mathbf{A}$ and $\mathbf{B}$ are members of the same class and have the same character because the similarity transformation leaves character unchanged (Section 2.6).

Theorem 4 *The number of nonequivalent irreducible representations of a group is equal to the number of classes.*

Without a formal proof, Theorem 4 is plausible from the example of the **32** group, which has three classes and three irreducible representations.

For any n-dimensional representation, $\chi(\mathbf{E}) = n$ because the identity has 1 everywhere along the main diagonal so that $\chi(\mathbf{E})$ equals the dimension of the representation. Every class in irreducible representation 1 has character 1. For group **32** matrices in irreducible representation A, the classes $\{\mathbf{E}\}$ and $\{\mathbf{D}, \mathbf{F}\}$ have character 1, and matrices in the class $\{\mathbf{A}, \mathbf{B}, \mathbf{C}\}$ have character -1. In irreducible representation Γ, the matrix in the class $\{\mathbf{E}\}$ has character 2, the matrices in the class $\{\mathbf{D}, \mathbf{F}\}$ have character -1, and in the class $\{\mathbf{A}, \mathbf{B}, \mathbf{C}\}$ the matrices have character 0.

Here is the character table for the **32** group.

	$\{\mathbf{E}\}$	$\{\mathbf{A}, \mathbf{B}, \mathbf{C}\}$	$\{\mathbf{D}, \mathbf{F}\}$
1	1	1	1
A	1	-1	1
Γ	2	0	-1

The column headings are the classes, and the row headings are the designations of the irreducible representations.

For another application of the invariance of character under a similarity transformation, suppose as an illustrative example that an 8×8 reducible matrix U is transformed to irreducible block diagonal form by S. The block diagonal matrix is also 8×8, and it has irreducible representations of U strung along its main diagonal.

The process might look like this:

$$S \begin{pmatrix} u_{11} & u_{12} & \cdots & u_{18} \\ u_{21} & u_{22} & \cdots & u_{28} \\ \vdots & \vdots & \ddots & \vdots \\ u_{81} & u_{82} & \cdots & u_{88} \end{pmatrix} S^{-1} = \begin{pmatrix} a_1 & & & & & & \\ & a_2 & & & & & \\ & & b_1 & & & & \\ & & & b_2 & & & \\ & & & & c_1 & c_2 & \\ & & & & c_3 & c_4 & \\ & & & & & & c_1 & c_2 \\ & & & & & & c_3 & c_4 \end{pmatrix}.$$

The 0 entries in the block diagonal matrix have been omitted to focus attention on the reduced block matrices along the main diagonal.

The similarity transformation has reduced $D(\mathbf{U})$ to four 1-dimensional irreducible representations and a 2-dimensional irreducible representation that occurs twice. Because a similarity transformation does not change the character of $D(\mathbf{U})$,

$$\sum_i u_{ii} = a_1 + a_2 + b_1 + b_2 + 2(c_1 + c_4).$$

In symbolic form,

$$\chi(\mathbf{U}) = \sum_\alpha c_\alpha \chi^{(\alpha)}(\mathbf{U}), \tag{2.9}$$

where, on the left, $\chi(\mathbf{U})$ is the character of the matrix before the similarity transformation. On the right, $\chi^{(\alpha)}(\mathbf{U})$ is the character of the α irreducible representation in the reduced block diagonal matrix and c_α is the number of times the α irreducible representation appears in the reduction.

The example in Section 2.7.2 showed the reduction of the regular representation of group member $\mathbf{C}$ of the **32** group. The character of $D^{(reg)}(\mathbf{C})$ in the regular representation is 0 because in the regular representation only $D^{(reg)}(\mathbf{E})$ has nonzero elements on the main diagonal. Check that the character of the reduced block diagonal matrix $\mathbf{C}$ is also 0.

$$\sum_\alpha c_\alpha \chi^{(\alpha)}(\mathbf{C}) = \chi^{(1)}(\mathbf{C}) + \chi^{(A)}(\mathbf{C}) + 2\chi^{(\Gamma)}(\mathbf{C})$$

$$= 1 + (-1) + 2\,(0)$$

$$= 0 \quad \checkmark$$

2.7.4 Orthogonality Theorems

The following theorem leads to important applications, and many authors call it the "wonderful orthogonality theorem." Its proof is a long road and is omitted here, but the proof given in more advanced texts can be followed using the material discussed so far.

Theorem 5

$$\sum_{\mathbf{R}} D_{ij}^{(\alpha)*}(\mathbf{R}) D_{i'j'}^{(\beta)}(\mathbf{R}) = \frac{n}{h}\delta_{\alpha\beta}\delta_{ii'}\delta_{jj'},$$

About the notation: Theorem 5 applies to a group with members $\mathbf{R}$, with the sum taken over all the group members. $D^{(\alpha)}$ and $D^{(\beta)}$ are any irreducible matrix representations of the group, either the same or different. n is the order of the group, and h is the dimension of the irreducible representation. The right-hand side is 0 unless the representations are the same $\alpha = \beta$, so the matrices are $h \times h$ and h is uniquely defined.

Theorem 5 can be viewed as a generalized orthogonality relation analogous to the scalar product of Cartesian vectors, where the scalar product of two orthogonal vectors is 0.

To demonstrate the theorem, take both representations to be the irreducible representation Γ of the **32** group, and take the 1,1 element of each matrix. The **32** group is of order $n = 6$, and Γ has dimension $h = 2$.

$$(1)^2 + (-1)^2 + \left(\tfrac{1}{2}\right)^2 + \left(\left(\tfrac{1}{2}\right)\right)^2 + \left(\left(-\tfrac{1}{2}\right)\right)^2 + \left(\left(-\tfrac{1}{2}\right)\right)^2 \overset{?}{=} \frac{6}{2}$$

$$1 + 1 + \frac{1}{4} + \frac{1}{4} + \frac{1}{4} + \frac{1}{4} \overset{?}{=} 3$$

$$3 = 3 \quad \checkmark$$

The theorem is satisfied.

Now use Theorem 5 to find a relation involving the characters of irreducible representations. The character of a matrix is the sum of its diagonal elements, so in Theorem 5 set $j = i$ and $j' = i'$ and sum over i and i'.

$$\sum_i \sum_{i'} \sum_R D_{ii}^{(\alpha)*}(\mathbf{R}) D_{i'i'}^{(\beta)}(\mathbf{R}) = \frac{n}{h} \sum_i \sum_{i'} \delta_{\alpha\beta} \delta_{ii'} \delta_{ii'}$$

$$\sum_R \chi^{(\alpha)*}(\mathbf{R}) \chi^{(\beta)}(\mathbf{R}) = \frac{n}{h} \delta_{\alpha\beta} \sum_i \delta_{ii}$$

$\delta_{ii} = 1$, so the sum on the right is $\sum_i 1 = h$. The result is worth listing as a theorem.

Theorem 6

$$\sum_R \chi^{(\alpha)*}(\mathbf{R}) \chi^{(\beta)}(\mathbf{R}) = n \delta_{\alpha\beta}$$

Theorem 6 says that the character sets of two different irreducible representations are mutually orthogonal, a result illustrated earlier for the $\{\mathbf{E}, \mathbf{T}\}$ group in Section 2.2.

Now use Theorem 6 to calculate the number of times an irreducible representation occurs in the block diagonal reduction of a matrix $D(\mathbf{R})$. From Eq. (2.9),

$$\chi(\mathbf{R}) = \sum_\alpha c_\alpha \chi^{(\alpha)}(\mathbf{R}).$$

Multiply by $\chi^{(\beta)*}(\mathbf{R})$ from the left, where $\chi^{(\beta)}(\mathbf{R})$ is the character of the β irreducible representation of $\mathbf{R}$. Sum over the group members.

$$\sum_R \chi^{(\beta)*}(\mathbf{R})\chi(\mathbf{R}) = \sum_R \chi^{(\beta)*}(\mathbf{R}) \sum_\alpha c_\alpha \chi^{(\alpha)}(\mathbf{R})$$

$$= \sum_\alpha c_\alpha \sum_R \chi^{(\beta)*}(\mathbf{R})\chi^{(\alpha)}(\mathbf{R})$$

Use Theorem 6 to evaluate the sum over $\mathbf{R}$ on the right.

$$\sum_R \chi^{(\beta)*}(\mathbf{R})\chi^{(\alpha)}(\mathbf{R})) = n \sum_\alpha c_\alpha \delta_{\alpha\beta} = n c_\beta$$

$$c_\beta = \frac{1}{n} \sum_R \chi^{(\beta)*}(\mathbf{R})\chi^{(\alpha)}(\mathbf{R}) \tag{2.10}$$

The result implies that the number of each irreducible representation in the reduced matrices $D(\mathbf{R})$ is unique – it can only happen one way.

Now apply Theorem 6 to the regular representation $\alpha = reg$. In the regular representation, the characters of all the group members except $\mathbf{E}$ are 0, so the sum reduces to a single term: $\chi^{(\beta)*}(\mathbf{E})\chi^{(reg)}(\mathbf{E})$. Here $\chi^{(reg)}(\mathbf{E}) = n$ is the order of the group, and $\chi^{(\beta)}(\mathbf{E}) = h_\beta$ is the dimension of the β irreducible representation. Substituting,

$$c_\beta = \frac{1}{n}\chi^{(\beta)*}(\mathbf{E})\chi^{(reg)}(\mathbf{E})$$

$$= \frac{1}{n}(nh_\beta)$$

$$= h_\beta.$$

The regular representation contains every irreducible representation of the group a number of times equal to the dimension of the irreducible representation. For example, the irreducible representation Γ of the **32** group has dimension 2 and occurs twice in the reduction of the regular representation.

Applied to the same irreducible representation $\alpha = \beta$, the left-hand side of Theorem 6 becomes the squared magnitude of $\chi^\alpha(\mathbf{R})$ summed over the group members:

Theorem 7

$$\sum_R \left|\chi^{(\alpha)}(\mathbf{R})\right|^2 = n$$

If instead we sum over classes keeping in mind that all members of a class have the same character, Theorem 7 can be written

$$\sum_\beta h_\beta \left|\chi^{(\beta)}\right|^2 = n,$$

where h_β is the number of group members in the class β.

Now apply Theorem 7 to the reduction of the regular representation into its constituent irreducible representations.

$$\chi^{(reg)}(\mathbf{R}) = \sum_{\alpha} c_{\alpha} \chi^{(\alpha)}(\mathbf{R})$$

Take $\mathbf{R} = \mathbf{E}$ and use $c_{\alpha} = h_{\alpha}$, $\chi^{(reg)}(\mathbf{E}) = n$, and $\chi^{(\alpha)}(\mathbf{E}) = h_{\alpha}$ to obtain the result.

Theorem 8

$$n = \sum_{\alpha} h_{\alpha} \chi^{(\alpha)}(\mathbf{E})$$
$$= \sum_{\alpha} (h_{\alpha})^2$$

The sum of the squares of the dimensions of all the different irreducible representations is equal to the order of the group. For the **32** group, $n = 6$ and there are only two possible ways to satisfy Theorem 8:

$$6 = (1)^2 + (1)^2 + (1)^2 + (1)^2 + (1)^2 + (1)^2$$
$$6 = (1)^2 + (1)^2 + (2)^2.$$

But the **32** group has three classes, hence only three irreducible representations according to Theorem 4. Only the second choice fits: there are two distinct 1-dimensional irreducible representations, and one 2-dimensional irreducible representation.

Theorem 9 *As a corollary to Theorem 8, all the irreducible representations of an Abelian group are 1-dimensional.*

To prove Theorem 9, let $\mathbf{A}$ be a member of an Abelian group. The class of $\mathbf{A}$ is $\mathbf{R}\mathbf{A}\mathbf{R}^{-1}$, where $\mathbf{R}$ is every member of the group. Because the group is Abelian, its members commute: $\mathbf{R}\mathbf{A} = \mathbf{A}\mathbf{R}$. Hence the class of $\mathbf{A}$ is

$$\mathbf{R}\mathbf{A}\mathbf{R}^{-1} = \mathbf{A}\mathbf{R}\mathbf{R}^{-1} = \mathbf{A}\mathbf{E} = \mathbf{A}.$$

Every member of an Abelian group is therefore its own class. If an Abelian group is of order n, there are n classes, hence n irreducible representations by Theorem 3. Theorem 8 is satisfied only if all the irreducible representations are 1-dimensional.

$$\underbrace{(1)^2 + (1)^2 + \cdots + (1)^2}_{n \ terms} = n$$

Tests for Reducibility

The first test for reducibility uses Theorem 6:

$$c_\beta = \frac{1}{n} \sum_R \chi^{(\beta)*}(\mathbf{R}) \chi^{(\alpha)}(\mathbf{R}).$$

Here c_β is the number of times $D(\mathbf{R})$ contains the β irreducible representation. If $c_\beta \neq 0$ for two or more different values of β, it follows that the original representation must contain at least two different irreducible representations and is therefore reducible.

For a second test of reducibility, note that the character of a group member $\mathbf{R}$ is the sum of the characters of its irreducible representations, some possibly repeated.

$$\chi(\mathbf{R}) = \sum_\alpha c_\alpha \chi^{(\alpha)}(\mathbf{R})$$

Take the squared magnitude of both sides and sum over $\mathbf{R}$:

$$\sum_R \chi^*(\mathbf{R}) \chi(\mathbf{R}) = \sum_R \sum_\alpha \sum_\beta c_\alpha c_\beta \chi^{(\alpha)}(\mathbf{R}) \chi^{(\beta)*}(\mathbf{R})$$

$$= \sum_\alpha \sum_\beta c_\alpha c_\beta \sum_R \chi^{(\alpha)}(\mathbf{R}) \chi^{(\beta)*}(\mathbf{R})$$

$$= n \sum_\alpha \sum_\beta c_\alpha c_\beta \delta_{\alpha\beta},$$

where the last step follows from Theorem 6 for the orthogonality of the characters of irreducible representations. Hence

$$\sum_R \chi^*(\mathbf{R}) \chi(\mathbf{R}) = \sum_R |\chi(\mathbf{R})|^2 = n \sum_\alpha c_\alpha^2.$$

If $D(\mathbf{R})$ is irreducible, $c_\alpha = 1$ for a particular value of α and $c_\alpha = 0$ otherwise. Hence

$$\sum_R \left| \chi^{(\beta)}(\mathbf{R}) \right|^2 = n \quad \rightarrow \quad \text{representation } \beta \text{ is irreducible.}$$

However, if $D(\mathbf{R})$ is reducible, then two or more of the $c_\alpha \neq 0$ and

$$\sum_R \left| \chi^{(\beta)}(\mathbf{R}) \right|^2 > n \quad \rightarrow \quad \text{representation } \beta \text{ is reducible.}$$

As an illustration of this inequality, consider the regular representation of a group of order n. All characters of the regular representation are 0 except for $\chi(\mathbf{E}) = n$.

The sum over **R** in the inequality reduces to one term to give $\left|\chi^{(\beta)}(\mathbf{E})\right|^2 = n^2 > n$. The inequality is satisfied in this example because the regular representation is reducible.

An Easy Way to Find Character Tables

Appendix A shows how to calculate the characters of irreducible representations using algebra and the product table without knowing explicit matrices. The method is not needed for later developments in this text. The lengthy calculations needed to construct character tables can be avoided by noting that the work has already been done by others and tabulated for groups of interest. Character tables are displayed in references under the name of the group.

2.8   Kronecker (Direct) Product

It is sometimes useful to combine two groups into a larger group; for example, combining physical rotation with inversion. Such a combination is performed by the *Kronecker (direct) product*. Consider a group $\mathcal{Q} = \{\mathbf{A}_1, \mathbf{A}_2, \dots\}$ of order n_a and a second group $\mathcal{B} = \{\mathbf{B}_1, \mathbf{B}_2, \dots\}$ of order n_b. The direct product is symbolized $\mathcal{Q} \otimes \mathcal{B}$. The members of the direct product are all possible combinations $A_i B_j$ and therefore number $n_a n_b$.

Assume that the members from the two groups commute so that $\mathbf{A}_i \mathbf{B}_j = \mathbf{B}_j \mathbf{A}_i$. This is not a serious restriction because in applications the two groups normally span different spaces and therefore do not interact – they don't "talk" to each other. With this assumption the members of the direct product set form a group. Consider the product of two members of the direct product set.

$$(\mathbf{A}_i \mathbf{B}_j)(\mathbf{A}_{i'} \mathbf{B}_{j'}) = (\mathbf{A}_i \mathbf{A}_{i'})(\mathbf{B}_j \mathbf{B}_{j'})$$

The product $(\mathbf{A}_i \mathbf{A}_{i'})$ is some member $\mathbf{A}_{i''}$ of the group $\mathcal{Q}$, and $(\mathbf{B}_j \mathbf{B}_{j'})$ is some member $\mathbf{B}_{j''}$ of the group $\mathcal{B}$, and it follows that $(\mathbf{A}_{i''} \mathbf{B}_{j''})$ is a member of the direct product set, satisfying the closure axiom for a group. The other axioms are also easily proved.

Let group $\mathcal{Q}$ have a representation matrix $D^{(\mu)}(\mathbf{A}_m)$ with elements a_{ij}, and let group $\mathcal{B}$ have a representation matrix $D^{(\nu)}(\mathbf{B}_n)$ with elements $b_{k\ell}$. The matrix elements c of the direct product representation are

$$D_{ik,j\ell}^{(\mu\nu)}(\mathbf{A}_m \mathbf{B}_n) = D_{ij}^{(\mu)}(\mathbf{A}_m) D_{k\ell}^{(\nu)}(\mathbf{B}_n) \tag{2.11}$$

$$c_{ik,j\ell} = a_{ij} b_{k\ell}.$$

Note the order in which the subscripts are written.

Let an element from each group have a 2×2 matrix representation so the direct product $\mathbb{C}$ is 4×4.

$$\mathbb{C} = \mathbb{Q} \otimes \mathbb{B}$$

$$= \begin{pmatrix} a_{11} & a_{12} \\ a_{21} & a_{22} \end{pmatrix} \otimes \begin{pmatrix} b_{11} & b_{12} \\ b_{21} & b_{22} \end{pmatrix}$$

$$= \begin{pmatrix} a_{11}b_{11} & a_{11}b_{12} & a_{12}b_{11} & a_{12}b_{12} \\ a_{11}b_{21} & a_{11}b_{22} & a_{12}b_{21} & a_{12}b_{22} \\ a_{21}b_{11} & a_{21}b_{12} & a_{22}b_{11} & a_{22}b_{12} \\ a_{21}b_{21} & a_{21}b_{22} & a_{22}b_{21} & a_{22}b_{22} \end{pmatrix}$$

$$= \begin{pmatrix} c_{11} & c_{12} & c_{13} & c_{14} \\ c_{21} & c_{22} & c_{23} & c_{24} \\ c_{31} & c_{32} & c_{33} & c_{34} \\ c_{41} & c_{42} & c_{43} & c_{44} \end{pmatrix}$$

Matrices need not be square to form a direct product. For example, if A is $m \times n$ and B is $m' \times n'$, the direct product matrix $A \otimes B$ is $mm' \times nn'$.

We can use Eq. (2.11) to find the character of a direct product matrix.

$$\chi(\mathbf{A}_m \mathbf{B}_n) = \sum_i \sum_j D_{ij,ij}(\mathbf{A}_m \mathbf{B}_n)$$

$$= \sum_i \sum_j D_{ii}(\mathbf{A}_m) D_{jj}(\mathbf{B}_n)$$

$$= \chi(\mathbf{A}_m) \chi(\mathbf{B}_n)$$

The characters of direct product matrices are the products of the corresponding characters.

A lengthy proof, not given here, shows that the irreducible representations of a direct product group are the products of the irreducible representations of the corresponding constituent groups. Assuming this to be proven, we can use Theorem 8 and the definition of the direct product to show that these are *all* the irreducible representations of the direct product.

Let group $\mathbb{Q}$ of order n_a have α irreducible representations of dimensions h_α, and let group $\mathbb{B}$ of order n_b have β irreducible representations of dimensions h_β. Let their direct product have irreducible representations of dimension $h_{\alpha,\beta}$:

$$\sum_\alpha \sum_\beta (h_{\alpha,\beta})^2 = \sum_\alpha \sum_\beta (h_\alpha)^2 (h_\beta)^2$$

$$= \sum_\alpha (h_\alpha)^2 \sum_\beta (h_\beta)^2$$

$$= n_a n_b$$

$$= \text{the order of the direct product.}$$

Hence, according to Theorem 8 the direct product has no other irreducible representations.

2.8.1 Example: The Klein Four-Group

There are only two groups of order 4, one of them cyclic and the other noncyclic.
The noncyclic group is called the *Klein Four-group* symbolized $\mathbf{K_4}$ or **mm2**. Table
2.1 shows its product table.

Table 2.1 Product table for $\mathbf{K_4}$

	E	A	B	C
E	E	A	B	C
A	A	E	C	B
B	B	C	E	A
C	C	B	A	E

The product table shows that the members of $\mathbf{K_4}$ commute. It is therefore an
Abelian group and each member is its own class. There are four classes, hence four
irreducible representations all of dimension 1 according to Theorem 9.

$$(1)^2 + (1)^2 + (1)^2 + (1)^2 = n = 4 \quad \checkmark$$

Table 2.2 shows the characters of $\mathbf{K_4}$. These are also the 1-dimensional irreducible
representation matrices.

Table 2.2 Character table for $\mathbf{K_4}$

	{E}	{A}	{B}	{C}
A_1	1	1	1	1
A_2	1	1	-1	-1
B_1	1	-1	1	-1
B_2	1	-1	-1	1

Consider now $\mathbf{C_2}$, a group of order 2. Its product table is shown in Table 2.3.

Table 2.3 Product table for $\mathbf{C_2}$

	{E}	{T}
A	1	1
B	1	-1

This is also the character table because the representation is 1-dimensional.

Form the direct product of $\mathbf{C_2}$ with itself.

$$\mathbf{C_2} \otimes \mathbf{C_2} = \begin{pmatrix} 1 & 1 \\ 1 & -1 \end{pmatrix} \otimes \begin{pmatrix} 1 & 1 \\ 1 & -1 \end{pmatrix}$$

$$= \begin{pmatrix} (1)(1) & (1)(1) & (1)(1) & (1)(1) \\ (1)(1) & (1)(-1) & (1)(1) & (1)(-1) \\ (1)(1) & (1)(1) & (-1)(1) & (-1)(1) \\ (1)(1) & (1)(-1) & (-1)(1) & (-1)(-1) \end{pmatrix}$$

The product table and also the character table is shown in Table 2.4.

Table 2.4 Character table for $\mathbf{C_2} \otimes \mathbf{C_2}$

1	1	1	1
1	-1	1	-1
1	1	-1	-1
1	-1	-1	1

Comparing the character tables for $\mathbf{K_4}$ and for $\mathbf{C_2} \otimes \mathbf{C_2}$, they are seen to be identical aside from row ordering. Two groups with the same character table are equivalent, so $\mathbf{K_4} = \mathbf{C_2} \otimes \mathbf{C_2}$.

2.9 Kronecker Sum

The *Kronecker sum* is symbolized $\oplus$. Unlike the Kronecker (direct) product, the matrices in a Kronecker sum must be square. Let matrix A be $m \times m$ and matrix B be $n \times n$. Let E_m be the $m \times m$ identity matrix, and let E_n be $n \times n$. The Kronecker sum is defined as

$$A \oplus B = A \otimes E_n + E_m \otimes B.$$

The Kronecker sum matrices $A \oplus B$ in this example are square $mn \times mn$.

Summary of Chapter 2

Chapter 2 lays out fundamental principles of matrix representations of group operations. The format is numerous theorems, many with proofs. Chapter 2 is mathematical to provide the groundwork for physical applications in later chapters.

a) Group representations can be understood as the action of group members on basis functions that span certain abstract spaces.

b) The relation between basis functions and representations is that basis functions "transform like" the group members and its matrix representations.

c) The similarity transformation of **T** by **S** is denoted $\mathbf{STS}^{-1}$.

d) A group may have several different matrix representations; but if they are related by similarity transformations, they are all equivalent.

e) Similarity transformations are carried out by unitary matrices.

f) The character of a matrix is the sum of its diagonal elements and is unchanged by a similarity transformation. If two matrix representations of a group have the same character set, they are equivalent and are related by a similarity transformation. Their basis functions span the same abstract space.

g) A representation is irreducible, if it cannot be broken down into simpler block diagonal form.

h) The regular representation can be constructed from the group's product table. Mirroring the product table, there is only one entry 1 in every row and in every column, with 0 elsewhere. Only the identity matrix $D(\mathbf{E})$ has entries on the main diagonal.

i) When matrices of the regular representation are put into block diagonal form by a suitable similarity transformation, the blocks are all the irreducible representations of the group, appearing a number of times equal to the representation's dimension.

j) The conjugate of a group member **A** is formed by applying a similarity transformation to **A** using members **X** of the group, $\mathbf{XAX}^{-1}$. Conjugates divide the group members into classes. The identity member **E** is always a class of its own. The number of irreducible representations equals the number of classes.

k) In a given representation, every group member in the same class has the same character.

l) In the regular representation, the character of **E** is $\chi^{(reg)}(\mathbf{E}) = n$, where n is the order of the group. In an irreducible representation, $\chi^{(\alpha)}(\mathbf{E}) = h_\alpha$, where h_α is the dimension of the α irreducible representation.

m) The matrices of irreducible representations and their characters obey orthogonality relations.

n) A matrix representation can be tested for reducibility using its characters if the characters of the irreducible representations are known.

o) The character table of a group's irreducible representations can be found using the group's product table to calculate class sums and their products, but it is much easier just to look in references.

Stated Theorems in Chapter 2

Theorem 1 The similarity transformation from one orthonormal set to another is carried out by a unitary matrix.

Theorem 2 The matrix for physical rotation by θ has character $2\cos\theta + 1$ for any axis of rotation.

Theorem 3 If two apparently different matrix representations of the same group have matrices with the same characters, the representations are related by a similarity transformation. They are equivalent: the basis functions for two equivalent representations both span the same space.

Theorem 4 The number of nonequivalent irreducible representations of a group is equal to the number of classes. For the **32** group, this is 3.

Theorem 5

$$\sum_{\mathbf{R}} D_{ij}^{(\alpha)*}(\mathbf{R}) D_{i'j'}^{(\beta)}(\mathbf{R}) = \frac{n}{h}\delta_{\alpha\beta}\delta_{ii'}\delta_{jj'}$$

Theorem 6

$$\sum_{\mathbf{R}} \chi^{(\alpha)*}(\mathbf{R})\chi^{(\beta)}(\mathbf{R}) = n\delta_{\alpha\beta}$$

Theorem 7

$$\sum_{\mathbf{R}} \left|\chi^{(\alpha)}(\mathbf{R})\right|^2 = n$$

If instead we sum over classes, keeping in mind that all members of a class have the same character, Theorem 7 is written as

$$\sum_{\beta} h_\beta \left|\chi^{(\beta)}\right|^2 = n,$$

where h_β is the number of group members in the class β.

Theorem 8

$$n = \sum_{\beta} h_\beta \chi^{(\beta)}(\mathbf{E})$$

$$= \sum_{\beta} (h_\beta)^2$$

Theorem 9 As a corollary to Theorem 8, all the irreducible representations of an Abelian group are 1-dimensional.

Problems and Exercises

2.1 Consider the triangle rotation group $\{\mathbf{E}, \mathbf{A}, \mathbf{B}\}$. The group member $\mathbf{A}$ is a rotation by $120°$. Show that x, y are basis functions for $\mathbf{A}$, and develop a 2×2 matrix representation for $\mathbf{A}$. Compare with the representation given in Section 1.4.

2.2 Consider the triangle rotation group $\{\mathbf{E}, \mathbf{A}, \mathbf{B}\}$. The group member $\mathbf{A}$ is a rotation by $120°$. Show that $x^2 - y^2, xy$ are basis functions for $\mathbf{A}$, and develop a 2×2 matrix representation for $\mathbf{A}$. Compare with the representation given in Section 1.4. Why is your result different?

2.3 For the group of an equilateral triangle, let operation $\mathbf{A}$ be a "flip" – a rotation by $180°$ about the y-axis. Show that $\phi_1 = x^2 + y^2$ and $\phi_2 = z$ are basis functions for $\mathbf{A}$ by finding the corresponding matrix representation for $\mathbf{A}$.

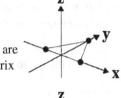

2.4 For the group of an equilateral triangle, let operation $\mathbf{A}$ be a "flip" – a rotation by $180°$ about the y-axis. Show that $\phi_1 = x^2 - y^2$ and $\phi_2 = yz$ are basis functions for $\mathbf{A}$ by finding the corresponding matrix representation for $\mathbf{A}$.

2.5 A 2-dimensional representation of a group member $\mathbf{T}$ is generated by the α set of basis functions to give $D^{(\alpha)}(\mathbf{T}) = \left(\begin{smallmatrix} 1 & 0 \\ 0 & -1 \end{smallmatrix}\right)$ and by the β set of basis functions to give $D^{(\beta)}(\mathbf{T}) = \left(\begin{smallmatrix} -1 & 0 \\ 0 & 1 \end{smallmatrix}\right)$. Show that these representations are equivalent, related by a similarity transformation with $\mathbf{S}$. Show that your $\mathbf{S}$ is unitary $\mathbf{S}\mathbf{S}^\dagger = \mathbf{E}$.

2.6 Under a similarity transformation with $\mathbf{S}$, group members $\mathbf{A}$ and $\mathbf{B}$ transform as $\mathbf{S}\mathbf{A}\mathbf{S}^{-1} = \mathbf{R}$ and $\mathbf{S}\mathbf{B}\mathbf{S}^{-1} = \mathbf{T}$. Show that $\mathbf{S}\mathbf{A}\mathbf{B}\mathbf{S}^{-1} = \mathbf{R}\mathbf{T}$.

2.7 For the group $\{\mathbf{E}, \mathbf{T}\}$ introduced in Section 2.2, the regular representation for $\mathbf{T}$ is $\left(\begin{smallmatrix} 0 & 1 \\ 1 & 0 \end{smallmatrix}\right)$. A similarity transformation with a matrix $\mathbf{S}$ reduces it to the block diagonal form $\left(\begin{smallmatrix} 1 & 0 \\ 0 & a \end{smallmatrix}\right)$. Show that $a = \pm 1$ and choose the correct value by calculating the matrix $\mathbf{S}$. Show that your matrix is unitary $\mathbf{S}\mathbf{S}^\dagger = \mathbf{E}$.

2.8 Consider the 2×2 matrix

$$\mathbf{S} = \begin{pmatrix} \frac{1}{\sqrt{2}} & -\frac{1}{\sqrt{2}} \\ \frac{1}{\sqrt{2}} & \frac{1}{\sqrt{2}} \end{pmatrix}$$

(a) Show that $\mathbf{S}$ is orthogonal.
(b) Perform a similarity transformation with $\mathbf{S}$ on each of the following matrices:

$$D(\mathbf{G1}) = \begin{pmatrix} 1 & 0 \\ 0 & 1 \end{pmatrix} \quad D(\mathbf{G2}) = \begin{pmatrix} \frac{1}{2} & \frac{\sqrt{3}}{2} \\ \frac{\sqrt{3}}{2} & -\frac{1}{2} \end{pmatrix} \quad D(\mathbf{G3}) = \begin{pmatrix} -\frac{1}{2} & \frac{\sqrt{3}}{2} \\ -\frac{\sqrt{3}}{2} & -\frac{1}{2} \end{pmatrix}.$$

(c) Use the results from (b) to calculate the characters of the three transformed matrices.

2.9 Consider the 2×2 matrix

$$\mathbf{T} = \begin{pmatrix} \frac{\sqrt{3}}{2} & -\frac{1}{2} \\ \frac{1}{2} & \frac{\sqrt{3}}{2} \end{pmatrix}$$

(a) Show that $\mathbf{T}$ is orthogonal.

(b) Perform a similarity transformation with $\mathbf{T}$ on each of the following matrices:

$$D(\mathbf{H1}) = \begin{pmatrix} -1 & 0 \\ 0 & 1 \end{pmatrix} \quad D(\mathbf{H2}) = \begin{pmatrix} \frac{1}{2} & -\frac{\sqrt{3}}{2} \\ -\frac{\sqrt{3}}{2} & -\frac{1}{2} \end{pmatrix} \quad D(\mathbf{H3}) = \begin{pmatrix} -\frac{1}{2} & -\frac{\sqrt{3}}{2} \\ \frac{\sqrt{3}}{2} & -\frac{1}{2} \end{pmatrix}.$$

(c) Use the results from (b) to calculate the characters of the three transformed matrices.

2.10 Find the character of each of the following matrices:

$$(-2) \quad \begin{pmatrix} 2 & 1 \\ -3 & -1 \end{pmatrix} \quad \begin{pmatrix} a & b \\ c & d \end{pmatrix} \quad \begin{pmatrix} -1+i & 0 & 3 \\ 2 & 2 & 4i \\ 3-i & 3+2i & 1 \end{pmatrix} \quad \begin{pmatrix} 1 & 0 & 0 & 0 \\ 0 & 1 & 0 & 0 \\ 0 & 0 & 1 & 0 \\ 0 & 0 & 0 & 1 \end{pmatrix}$$

(a) (b) (c) (d) (e)

2.11 Here is the product table for the triangle rotation group $\{\mathbf{E}, \mathbf{A}, \mathbf{B}\}$. Write the matrices for the regular representation of group members $\mathbf{E}$, $\mathbf{A}$, and $\mathbf{B}$.

	E	A	B
E	E	A	B
A	A	B	E
B	B	E	A

2.12 The noncyclic group of order 4 has members $\{\mathbf{E}, \mathbf{K}, \mathbf{L}, \mathbf{M}\}$, and its product table is

	E	K	L	M
E	E	K	L	M
K	K	E	M	L
L	L	M	E	K
M	M	L	K	E

Write the regular representation of group member $\mathbf{K}$.

2.13 The cyclic group of order 4 has members $\{E, A, B = A^2, C = A^3\}$, and its product table is.

	E	A	B	C
E	E	A	B	C
A	A	B	C	E
B	B	C	E	A
C	C	E	A	B

Write the regular representation of group member **A**.

2.14 The **422** group describes the symmetries of a square, including "flips." It has eight members and five classes.
(a) How many irreducible representations does the **422** group have?
(b) What are the dimensions of this group's irreducible representations?

2.15 The noncyclic group of order 4 has members $\{E, K, L, M\}$, with product table

	E	K	L	M
E	E	K	L	M
K	K	E	M	L
L	L	M	E	K
M	M	L	K	E

(a) What are the classes of this group?
(b) How many irreducible representations does this group have?
(c) What are the dimensions of its irreducible representations?

2.16 The cyclic group of order 4 has members $\{E, A, B = A^2, C = A^3\}$, and its product table is

	E	A	B	C
E	E	A	B	C
A	A	B	C	E
B	B	C	E	A
C	C	E	A	B

(a) What are the classes of this group?
(b) How many irreducible representations does this group have?
(c) What are the dimensions of its irreducible representations?

2.17 Consider a group with general members $\mathbf{T}$. Starting from the relation

$$T\phi_j = \sum_k D_{kj}^{(\alpha)}(\mathbf{T})\phi_k,$$

use Theorem 5 to prove

$$\phi_i = \frac{h_\alpha}{n} \sum_{\mathbf{T}} D_{ij}^{(\alpha)*}(\mathbf{T})\mathbf{T}\phi_j,$$

where h_α is the dimension of the α irreducible representation and n is the order of the group.

2.18 The three irreducible matrix representations of the **32** group are given in Section 2.7.2. Demonstrate Theorem 5.

2.19 Here is the character table of the **32** group.

	{E}	{A, B, C}	{D, F}
1	1	1	1
A	1	−1	1
Γ	2	0	−1

Demonstrate Theorem 6 using this character table.

2.20 Here is the character table of the **32** group.

	{E}	{A, B, C}	{D, F}
1	1	1	1
A	1	−1	1
Γ	2	0	−1

Demonstrate Theorem 7 using this character table.

2.21 Consider the permutation group of 4 numbers (1234). There are two irreducible representations of dimension 1, and the other irreducible representations all have dimensions greater than 1. What are the dimensions of the other irreducible representations?

2.22 The cyclic group $\{\mathbf{E}, \mathbf{A}, \mathbf{B}\}$ for the rotation of an equilateral triangle by $0°$, $120°$, and $240°$ has a matrix representation

$$\begin{pmatrix} 1 & 0 \\ 0 & 1 \end{pmatrix} \qquad \begin{pmatrix} -\frac{1}{2} & \frac{\sqrt{3}}{2} \\ -\frac{\sqrt{3}}{2} & -\frac{1}{2} \end{pmatrix} \qquad \begin{pmatrix} -\frac{1}{2} & -\frac{\sqrt{3}}{2} \\ \frac{\sqrt{3}}{2} & -\frac{1}{2} \end{pmatrix}.$$

$D(\mathbf{E}) \qquad\qquad D(\mathbf{A}) \qquad\qquad D(\mathbf{B})$

Show that this representation is reducible.

2.23 Show that $(A \otimes B)(A' \otimes B') = (AA') \otimes (BB')$. For simplicity let the matrices A, A', B, B' all be $n \times n$. The general requirement on matrix dimensions for this relationship is that the number of columns in A must equal the number of rows in A' to allow matrix multiplication AA', and similarly for BB'.

3

MOLECULAR VIBRATIONS

3.1 Introduction

Chapter 2 was a necessary introduction to the representations of discrete groups with emphasis on the meaning of basis functions and irreducible representations. Now we are prepared to see how these ideas are used in physical applications. This chapter treats vibrating systems, and Chapter 4 treats crystalline solids.

3.2 Oscillating Systems and Newton's Laws

The discussion of vibrating systems begins prosaically with masses and springs to establish the groundwork for the problem of molecular vibrations, where group theory comes into its own.

A run-through of the classical Newtonian treatment of vibrating systems helps establish concepts and nomenclature. A physics student's first encounter with a serious differential equation is probably when applying Newton's second law to a *harmonic oscillator*, modeled by a mass m acted on by a spring with spring constant k (the force exerted by the spring per unit length of stretch).

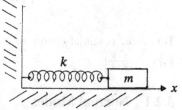

Assume that the table in the sketch is frictionless so that the only horizontal force on m is due to the spring. The equation of motion from Newton's second law is

$$m\ddot{x} + kx = 0$$

$$\ddot{x} + \frac{k}{m}x = 0$$

$$\ddot{x} + \omega^2 x = 0.$$

The general solution is

$$x = A \sin(\omega t) + B \cos(\omega t),$$

where $\omega = \sqrt{k/m}$ is the *angular frequency* of the oscillation in radians $\cdot \text{s}^{-1}$ and A and B are constants to be determined by initial conditions. Newton's dot notation is used here for the time derivative instead of Leibniz's: $\ddot{x} \equiv \frac{d^2 x}{dt^2}$.

The general solution can be written in the equivalent form

$$C \sin(\omega t + \phi) = C \cos\phi \sin(\omega t) + C \sin\phi \cos(\omega t)$$
$$= A \sin(\omega t) + B \cos(\omega t)$$

where $A = C \cos\phi$ and $B = C \sin\phi$ and where ϕ is called the *phase constant*. The *circular frequency* f is the number of cycles in 1 second. f is measured in *hertz*, where 1 hertz (Hz) = 1 cycle per second. The system completes one cycle when ωt increases by 2π radians, so $f = \omega/2\pi$. The *period* τ is the time for 1 cycle $\tau = 1/f$.

The harmonic oscillator model is widely applicable to bound systems. To see why, consider the 1-dimensional example of a particle on the x-axis bound in a potential well $V(x)$. Expanding $V(x)$ in Taylor's series about the equilibrium position x_0 gives

$$V(x) \approx V(x_0) + \left.\frac{dV}{dx}\right|_{x_0} (x - x_0) + \frac{1}{2}\left.\frac{d^2 V}{dx^2}\right|_{x_0} (x - x_0)^2 + \dots.$$

The term linear in $(x - x_0)$ vanishes because $\left.\frac{dV}{dx}\right|_{x_0} = 0$ at the equilibrium position x_0 where the force is 0. The constant term $V(x_0)$ does not lead to any forces and can be neglected. As far as the dynamics are concerned, a good approximation to $V(x)$ for small-amplitude vibrations x near x_0 is then

$$V(x) \approx \frac{1}{2}\left.\frac{d^2 V}{dx^2}\right|_{x_0} (x - x_0)^2.$$

This is the potential energy of a harmonic oscillator with effective spring constant $k_{eff} = \left.\frac{d^2 V}{dx^2}\right|_{x_0}$.

3.2.1 Normal Modes

We now show that systems of masses and springs can oscillate only at a finite number of discrete frequencies; these motions are called *normal modes*. All possible vibrational motions of a system are linear combinations of its normal modes. The simple system of one mass and one spring has only one possible oscillation frequency, hence only one normal mode, but a system of several masses and several springs has several normal modes.

Generalizing to a system of several masses and springs leads to a set of equations of the form

$$\ddot{x}_i + \sum_j b_{ij} x_j = 0. \tag{3.1}$$

The normal modes x'_n are certain combinations of the x_i:

$$x'_n = \sum_i c_{ni} x_i, \tag{3.2}$$

such that the normal modes x'_n satisfy uncoupled harmonic oscillator equations

$$\ddot{x}'_n + \omega_n^2 x'_n = 0, \tag{3.3}$$

where ω_n is the angular frequency of normal mode n.

Multiply Eq. (3.1) by c_{ni} and sum over i:

$$\sum_i c_{ni} \ddot{x}_i + \sum_i \sum_j c_{ni} b_{ij} x_j = 0. \tag{3.4}$$

By Eq. (3.2) the first term on the left is just $\ddot{x}'_n$. Comparing with Eq. (3.3) we require

$$\sum_i \sum_j c_{ni} b_{ij} x_j = \omega_n^2 x'_n$$

$$= \omega_n^2 \sum_j c_{nj} x_j$$

so that

$$\sum_i \sum_j \left(c_{ni} b_{ij} - \omega_n^2 c_{nj} \right) x_j = 0.$$

The x_j are assumed to be linearly independent, so the individual coefficients of x_j must equal 0. The condition becomes

$$\sum_i \left(c_{ni} b_{ij} - \omega_n^2 c_{nj} \right) = \sum_i \left(b_{ij} - \omega_n^2 \delta_{ij} \right) c_{ni} = 0. \tag{3.5}$$

Note the substitution $c_{nj} = \delta_{ij} c_{ni}$.

At first glance, Eq. (3.5) looks like a way of calculating the normal mode coefficients c_{ni}, but straightforward application of Cramer's rule for arbitrary values of ω_n yields the trivial solution $c_{ni} = 0$. There are nontrivial solutions only for particular values of the discrete frequencies ω_n of the normal modes. To calculate the discrete frequencies, the determinant of the coefficients must be 0 so that Cramer's rule does not apply. Let B be the matrix with elements b_{ij}. The determinant condition is called the *secular equation*,

$$\det (B - \Omega) = 0, \tag{3.6}$$

where Ω is a matrix with squared normal mode frequencies down the main diagonal and 0 elsewhere:

$$\Omega = \begin{pmatrix} \omega_1^2 & 0 & \cdots & 0 \\ 0 & \omega_2^2 & \cdots & 0 \\ \vdots & \vdots & \ddots & \vdots \end{pmatrix}.$$

More generally, Ω could be in block diagonal form with blocks of dimension greater than 1, a case to be discussed in Section 3.3.1.

If C is the matrix with elements c_{ij}, Eq. (3.5) can be written in matrix form:

$$CB - \Omega C = 0.$$

Multiplying from the right by C^{-1} gives

$$CBC^{-1} = \Omega. \tag{3.7}$$

Equation (3.7) should look familiar – it is a similarity transformation of B by C to give the diagonal matrix Ω. Put another way, the C matrix *diagonalizes* the B matrix.

3.2.2   Example: Two Masses and Three Springs

This example is a system with two normal modes.

Consider two identical masses m connected to three springs as shown. The center spring has spring constant k, and the two outer springs have spring constant k'. Neglect gravity.

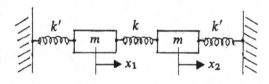

The equations of motion are

$$m\ddot{x}_1 = -k' x_1 + k(x_2 - x_1)$$
$$m\ddot{x}_2 = -k' x_2 - k(x_2 - x_1)$$

$$\ddot{x}_1 + \left(\frac{k'+k}{m}\right) x_1 - \left(\frac{k}{m}\right) x_2 = 0$$
$$\ddot{x}_2 - \left(\frac{k}{m}\right) x_1 + \left(\frac{k'+k}{m}\right) x_2 = 0.$$

Referring to Eq. (3.1) for the definition of the B matrix, the secular equation Eq. (3.6) is

$$\begin{vmatrix} \frac{k'+k}{m} - \omega^2 & -\frac{k}{m} \\ -\frac{k}{m} & \frac{k'+k}{m} - \omega^2 \end{vmatrix} = 0.$$

The equation is biquadratic in ω^2 and has solutions $\omega_1^2 = \frac{k'}{m}$ for normal mode 1 and $\omega_2^2 = \frac{k'+2k}{m}$ for normal mode 2. Any possible oscillation of the system can only be a linear combination of these normal modes.

$$x_1 = A_1 \sin(\omega_1 t) + A_2 \cos(\omega_1 t) + A_3 \sin(\omega_2 t) + A_4 \cos(\omega_2 t)$$

$$x_2 = B_1 \sin(\omega_1 t) + B_2 \cos(\omega_1 t) + B_3 \sin(\omega_2 t) + B_4 \cos(\omega_2 t)$$

Once the normal mode frequencies are known, Eq. (3.5) can be solved for the c_{ij}, which gives $c_{12} = c_{11}$ for normal mode 1 and $c_{21} = -c_{22}$ for normal mode 2. From Eq. (3.2):

$$x_1' = c_{11}x_1 + c_{12}x_2 = c_{11}(x_1 + x_2)$$

$$x_2' = c_{21}x_1 + c_{22}x_2 = c_{22}(-x_1 + x_2).$$

In the language of group theory, $x_1 + x_2$ and $x_1 - x_2$ are basis functions for the two modes. This will become clearer in Section 3.3 where normal modes are treated according to group theory.

Assume that the masses are at maximum amplitude in their motion. To visualize the motion in normal mode 1, note that both terms in $x_1 + x_2$ have the same sign – the two masses are moving together *in phase* as the sketch indicates. In mode 2, x_1 and x_2 have opposite signs so that the masses move in opposite directions 180° *out of phase*. Perhaps you can see qualitatively from the sketches why mode 1 has a lower oscillation frequency than mode 2. Such mode diagrams are often used in the discussion of molecular vibrations.

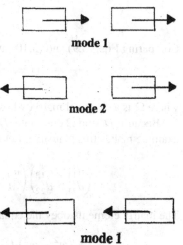

mode 1

mode 2

The masses are constantly in motion, and half a period later the amplitudes will again be at maximum but reversed as shown in the sketch.

mode 1

3.3 Normal Modes and Group Theory

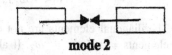

mode 2

An oscillating system has normal modes n that satisfy equations for simple harmonic motion.

$$\ddot{x}_n' + \omega_n^2 x_n' = 0 \tag{3.8}$$

The object of this section is to show that the normal modes of an oscillating system are in fact the basis functions of the irreducible representations of the system's symmetry group.

Suppose that the oscillating system being considered has symmetry properties described by a symmetry group with members $\mathbf{T}$. This idea will be key for the discussion of molecular vibrations in Section 3.4. Suppose further that the normal modes x' are basis functions for the group so that they generate a matrix representation with matrices $D'(\mathbf{T})$.

$$\mathbf{T}x'_n = \sum_j D'_{jn}(\mathbf{T})x'_j$$

$$\mathbf{T}\ddot{x}'_n = \sum_j D'_{jn}(\mathbf{T})\ddot{x}'_j$$

$$= -\sum_j D'_{jn}(\mathbf{T})\omega_j^2 x'_j \tag{3.9}$$

using the equation of motion Eq. (3.8). Now apply $\mathbf{T}$ to Eq. (3.8).

$$\mathbf{T}\ddot{x}'_n = -\omega_n^2 \mathbf{T}x'_n$$

$$= -\omega_n^2 \sum_j D'_{jn}(\mathbf{T})x'_j \tag{3.10}$$

Comparing Eqs. (3.9) and (3.10), it follows that

$$D'\Omega = \Omega D',$$

where Ω is a diagonal matrix whose diagonal elements are ω_n^2.

Because D' and Ω commute, D' must be in block diagonal form. A 2-dimensional example makes this plausible. Let d'_{ij} be the matrix elements of D'.

$$\begin{pmatrix} d'_{11} & d'_{12} \\ d'_{21} & d'_{22} \end{pmatrix} \begin{pmatrix} \omega_1^2 & 0 \\ 0 & \omega_2^2 \end{pmatrix} = \begin{pmatrix} \omega_1^2 & 0 \\ 0 & \omega_2^2 \end{pmatrix} \begin{pmatrix} d'_{11} & d'_{12} \\ d'_{21} & d'_{22} \end{pmatrix}$$

The matrix elements obey the equations

$$(\omega_1^2 - \omega_1^2)d'_{11} = 0 \qquad (\omega_1^2 - \omega_2^2)d'_{12} = 0$$
$$(\omega_2^2 - \omega_1^2)d'_{21} = 0 \qquad (\omega_2^2 - \omega_2^2)d'_{22} = 0.$$

The diagonal elements are not determined by these relations, but the off-diagonal elements $= 0$ if $\omega_1 \neq \omega_2$. If all the normal mode frequencies are different, as in this example, D' has only diagonal elements and cannot be reduced further. The normal modes x' are therefore basis functions for two 1-dimensional irreducible representations of the group.

Two questions remain. What happens if two or more of the normal mode frequencies are equal? Second, what are the basis functions for irreducible representations of dimension higher than 1? Both these questions have the same answer, as the following section shows.

3.3.1 Degeneracy

If an irreducible representation of a system's symmetry group has dimension $n > 1$, there will be n normal modes with different motions but the same frequency; these modes are said to be *degenerate*. Putting $\omega_2 = \omega_1$ in our example:

$$\begin{pmatrix} d'_{11} & d'_{12} \\ d'_{21} & d'_{22} \end{pmatrix} \begin{pmatrix} \omega_1^2 & 0 \\ 0 & \omega_1^2 \end{pmatrix} = \begin{pmatrix} \omega_1^2 & 0 \\ 0 & \omega_1^2 \end{pmatrix} \begin{pmatrix} d'_{11} & d'_{12} \\ d'_{21} & d'_{22} \end{pmatrix}.$$

Now all four matrix elements d'_{ij} are indeterminate. Put another way, the two degenerate frequencies give rise to a 2×2 block in the D' matrices of the representation. Similarly, three equal frequencies would give a 3×3 block.

Taking the example of two equal frequencies, the resulting 2×2 block is irreducible. The equations of motion for the corresponding basis functions x'_1 and x'_2 are

$$\ddot{x}'_1 + \omega_1^2 x'_1 = 0$$
$$\ddot{x}'_2 + \omega_1^2 x'_2 = 0.$$

Multiply the second equation by an arbitrary constant α and add.

$$(\ddot{x}'_1 + \alpha \ddot{x}'_2) + \omega_1^2(x'_1 + \alpha x'_2) = 0$$

There is no unique way of separating x'_1 and x'_2. These basis functions are tied together, and it is not possible to find a unique linear combination of x'_1 and x'_2 that reduces the 2×2 block. Put another way, the two degenerate frequencies generate a 2-dimensional irreducible representation.

3.4 Normal Modes of a Water Molecule

Consider the water molecule, H_2O. It consists of a central oxygen atom of mass M_O bound to two hydrogen atoms, each of mass m_H. The O-H separation is 9.6 nm, the H-O-H bond angle 104°, and the H-H distance 25 nm.

A water molecule at rest is planar (three points determine a plane). Take it to lie in the y-z plane as shown. The directions of the axes are in agreement with the right-hand rule, $\hat{\mathbf{i}} \times \hat{\mathbf{j}} = \hat{\mathbf{k}}$. The "hat" (caret) symbolizes a unit vector.

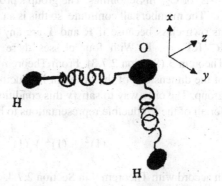

The H-O interatomic forces are modeled by springs, because to a first approximation binding forces are linear with distance (like springs) for small amplitudes of vibration. The force between the H-H atoms is neglected because the H atoms are relatively far

Table 3.1 Product table for **mm2**

	1	**2_z**	**m_y**	**m_x**
1	1	2_z	m_y	m_x
2_z	2_z	1	m_x	m_y
m_y	m_y	m_x	1	2_z
m_x	m_x	m_y	2_z	1

apart and their electrons needed for bonding are already mostly shared in the bonds with the oxygen atom.

The first task is to determine the symmetry group of the molecule. The symmetry operations are named according to the *Hermann–Mauguin* symbols (also called the *international notation*) widely used to describe crystal lattices as well as molecules. An alternative is the *Schönflies* notation, used particularly in spectroscopy.

With reference to the diagram, there are four symmetry operations:

1. The identity operation E – don't do anything. This is always a symmetry operation. Its standard symbol is **1**.

2. Rotate the molecule by $180°$ about the z-axis. Its symbol is 2_z: 2 because it is a rotation by $360°/2$, and z because it is a rotation about the z-axis.

3. Reflect the molecule in a mirror normal to the y-axis and passing through the x-z plane. Its symbol is m_y because it is a mirror plane m normal to the y-axis.

4. Reflect the molecule in a mirror normal to the x-axis. Its symbol is m_x, analogous to the symbol m_y.

The Schönflies notation names the operations $2_z \mapsto C_{2z}, m_y \mapsto \sigma_y, m_x \mapsto \sigma_x$.

These operations form a group of four members, labeled **mm2** (**2mm** in some texts) or C_{2v} in Schönflies. The group's product table is shown in Table 3.1.

The members all commute, so this is an Abelian group. Each member is therefore its own class because if **R** and **T** are any members of the group, then $\mathbf{R}^{-1}\mathbf{TR} = \mathbf{R}^{-1}\mathbf{RT} = \mathbf{T}$. With four classes there are four irreducible representations by Theorem 3 (Section 2.7.3). From Theorem 8 (Section 2.7.4) the sum of the squares of the dimensions of the different irreducible representations equals the order of the group. The only way to satisfy this condition with four irreducible representations is for all of the irreducible representations to be 1-dimensional:

$$(1)^2 + (1)^2 + (1)^2 + (1)^2 = 4 \quad \checkmark$$

in accord with Theorem 9 in Section 2.7.4.

The goal is to determine the basis functions that serve as normal modes for the vibration of the water molecule. Use the character table (shown in Table 3.2) for the irreducible representations of **mm2**.

Table 3.2 Character table for **mm2**

	1	**2_z**	**m_y**	**m_x**
A_1	1	1	1	1
A_2	1	1	−1	−1
B_1	1	−1	1	−1
B_2	1	−1	−1	1

The character table can be derived solely from the product table using class sums (Appendix A). An easier way is to consult references. Table 3.2 is in standard form.

One-dimensional irreducible representations are labeled A when rotation about the principal axis (here z) has character +1, and B when that character is −1. Two-dimensional representations are labeled E (not to be confused with the group's identity operation), and 3-dimensional representations are labeled T. Irreducible representations of higher dimensionality rarely occur when calculating vibration modes of molecules.

The basis functions of the irreducible representations will be combinations of displacements of the individual atoms in the molecule. It is convenient to decompose the general displacements into orthogonal components $\delta x_n, \delta y_n, \delta z_n$ as shown in the sketch. The task is to find the combinations of the nine displacements that act as normal modes for the vibrations of the molecule.

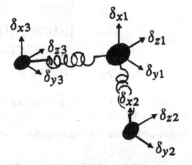

There is a stumbling block. No external forces act on the molecule, so it is free to translate and to rotate as a whole. These motions do not correspond to oscillations of the system. More precisely, they are trivial normal modes with $\omega = 0$. The basis functions corresponding to translation and rotation must be identified in order to discard them. The plan of attack is to reduce the 9×9 matrices using characters to determine the irreducible representations involved and then subtract out the translational and rotational modes to leave the normal mode basis functions for the molecule's vibrations.

Only atoms that "stay at home" and remain unmoved by an operation can give rise to nonzero values on the main diagonal, hence only unmoved atoms can contribute to the character. For the identity symmetry operation **1**, all atoms stay at home, so its representation matrix has 1 down the main diagonal 9 times $\chi(\mathbf{1}) = 9$.

Consider now the operation $\mathbf{2_z}$ (refer to the sketch to follow changes in the displacements). The 180° rotation about the z-axis leaves only the oxygen atom unmoved, so only its displacements contribute to the character.

$$\delta x_1 \mapsto -\delta x_1$$
$$\delta y_1 \mapsto -\delta y_1$$
$$\delta z_1 \mapsto \delta z_1$$

The sum of the diagonal elements gives $\chi(2_z) = -1 - 1 + 1 = -1$.

The mirror operation $\mathbf{m_y}$ leaves only the oxygen atom in place. The y-displacement δy_1 is reversed, and δx_1 and δz_1 are unchanged, so the character is $\chi(\mathbf{m_y}) = -1 + 1 + 1 = +1$.

The mirror operation $\mathbf{m_x}$ leaves all the atoms at home. For each atom, δx is reversed and δy and δz are unchanged, giving $\chi(\mathbf{m_x}) = 3(-1 + 1 + 1) = 3$.

Knowing the characters of the 9×9 reducible matrices and the characters of the irreducible representations, use Eq. (2.10) in Section 2.7.4 to calculate c_α, the number of times the irreducible representation α occurs in the reducible matrices:

$$c_\alpha = \frac{1}{n} \sum_{\mathbf{T}} \chi^{(\alpha)}(\mathbf{T}) \chi(\mathbf{T}),$$

where $n = 4$ is the order of the group, and the sum is over all group members.

$$c_{A_1} = \tfrac{1}{4}[(1)(9) + (1)(-1) + (1)(1) + (1)(3)] = 3$$
$$c_{A_2} = \tfrac{1}{4}[(1)(9) + (1)(-1) + (-1)(1) + (-1)(3)] = 1$$
$$c_{B_1} = \tfrac{1}{4}[(1)(9) + (-1)(-1) + (1)(1) + (-1)(3)] = 2$$
$$c_{B_2} = \tfrac{1}{4}[(1)(9) + (-1)(-1) + (-1)(1) + (1)(3)] = 3$$

As expected, there is a total of nine 1-dimensional irreducible representations $\Gamma^{(irr)} = 3A_1 + A_2 + 2B_1 + 3B_2$. There will be three modes for mass motion translation and three for mass motion rotation, leaving three nontrivial vibration modes.

To subtract the representations that correspond to mass motion translation and rotation, go more deeply into the concept of "transforms like" from Section 2.2.1. The tool is the relation among group operations $\mathbf{T}$, water molecule displacement basis functions δ_i, and representation matrices $D(\mathbf{T})$ according to the equation

$$\mathbf{T}\delta_i = \sum_j D_{ji}(\mathbf{T})\delta_j. \qquad (3.11)$$

For the water molecule, all of the irreducible representations are 1-dimensional, so for any irreducible representation α, set $j = i = 1$ and write

$$\frac{(\mathbf{T}\delta_1)}{\delta_1} = D_{11}^{(\alpha)}(\mathbf{T}) = \chi^{(\alpha)}(\mathbf{T}). \qquad (3.12)$$

For translations, the vector $\mathbf{r}$ specifies a position with Cartesian coordinates (x, y, z),

$$\mathbf{r} = x\hat{\mathbf{i}} + y\hat{\mathbf{j}} + z\hat{\mathbf{k}},$$

where $\hat{\mathbf{i}}, \hat{\mathbf{j}}, \hat{\mathbf{k}}$ are the usual Cartesian unit vectors. Appropriate basis functions for translations are therefore (x, y, z). Applying Eq. (3.12) to x gives

$$\frac{(1x)}{x} = \frac{x}{x} = 1 \qquad = D_{11}(1) \qquad \longrightarrow \chi(1) \quad = +1$$
$$\frac{(2_z x)}{x} = \frac{-x}{x} = -1 \qquad = D_{11}(2_z) \qquad \longrightarrow \chi(2_z) \quad = -1$$
$$\frac{(\mathbf{m}_y x)}{x} = \frac{x}{x} = 1 \qquad = D_{11}(\mathbf{m}_y) \qquad \longrightarrow \chi(\mathbf{m}_y) \quad = +1$$
$$\frac{(\mathbf{m}_x x)}{x} = \frac{-x}{x} = -1 \qquad = D_{11}(\mathbf{m}_y) \qquad \longrightarrow \chi(\mathbf{m}_y) \quad = -1.$$

Comparing with the character table for **mm2** shows that x transforms like irreducible representation B_1. Similarly y transforms like B_2 and z like A_1. Hence $\Gamma^{(trans)} = A_1 + B_1 + B_2$.

To deal with mass motion rotation of the molecule, use angular momentum $\mathbf{L}$ to see how coordinates are involved. Writing $\mathbf{L}$ in determinant form,

$$\mathbf{L} = \mathbf{r} \times \mathbf{p} = m \begin{vmatrix} \hat{\mathbf{i}} & \hat{\mathbf{j}} & \hat{\mathbf{k}} \\ x & y & z \\ \dot{x} & \dot{y} & \dot{z} \end{vmatrix},$$

where $\mathbf{L}_x = m(y\dot{z} - z\dot{y})$, which shows that the combination $\phi = yz$ is a basis function. Proceeding as before,

$$\frac{(1(yz))}{(yz)} = \frac{(yz)}{(yz)} = 1 \qquad = D_{11}(1) \qquad \longrightarrow \chi(1) \quad = +1$$
$$\frac{(2_z(yz))}{(yz)} = \frac{(-y)(z)}{(yz)} = -1 \qquad = D_{11}(2_z) \qquad \longrightarrow \chi(2_z) \quad = -1$$
$$\frac{(\mathbf{m}_y(yz))}{(yz)} = \frac{((-y)z)}{(yz)} = -1 \qquad = D_{11}(\mathbf{m}_y) \qquad \longrightarrow \chi(\mathbf{m}_y) \quad = -1$$
$$\frac{(\mathbf{m}_x(yz))}{(yz)} = \frac{(yz)}{(yz)} = 1 \qquad = D_{11}(\mathbf{m}_x) \qquad \longrightarrow \chi(\mathbf{m}_x) \quad = +1,$$

so yz transforms like B_2. Proceeding similarly, xz transforms like B_1, and xy transforms like A_2. Hence $\Gamma^{(rot)} = A_2 + B_1 + B_2$. Subtracting the mass motion translation and rotation modes leaves $\Gamma^{(vib)} = 2A_1 + B_2$. There are three normal modes, hence three vibration frequencies, which are nondegenerate in the case of the water molecule because all the normal modes correspond to 1-dimensional irreducible representations. The two A_1 modes are physically different with different vibration motions and different vibration frequencies – they just happen to have the same symmetry. A molecule with a 2-dimensional irreducible representation would in contrast be doubly degenerate with two different normal modes having the same vibration frequency.

Out-of-plane vibrations are neglected. Note that in modes A_1 and B_2, $\mathbf{m}_x \phi = (+1)\phi$, so no displacements are out of plane.

3.5 Visualizing Normal Modes

Before proceeding to visualize the water molecule's vibration modes, a *projection operator* tool is helpful to generate appropriate basis functions for the normal modes.

3.5.1 Projection Operator

An example from vector analysis demonstrates the fundamental principle of projection. The scalar product of vector $\mathbf{A}$ with Cartesian unit vector $\mathbf{i}$ yields A_x. The scalar product operation has *projected out* the portion of $\mathbf{A}$ that lies along the x-axis.

A projection operator in group theory can in analogous fashion select out functions that have desired transformation properties from a larger abstract subspace. A projection operator enables us to find basis functions that transform like $\Gamma^{(vib)} = 2A_1 + B_2$ to allow the water molecule's vibration modes to be visualized.

The projection operator is

$$\phi_i^{(\alpha)} = \sum_{\mathbf{T}} D_{ij}^{(\alpha)*}(\mathbf{T})\mathbf{T}\phi. \tag{3.13}$$

Some explanation may be helpful. $\mathbf{T}$ is a member of the symmetry group involved, and $D_{ij}^{(\alpha)*}(\mathbf{T})$ is a complex conjugate matrix element for the α irreducible representation of $\mathbf{T}$. The symbol ϕ (no subscript, no superscript) is some trial function from the abstract space. The manipulations on the right-hand side of Eq. (3.13) deliver $\phi_i^{(\alpha)}$, a basis function that transforms like the α irreducible representation. If the trial function ϕ does not contain such a basis function, the result is $\phi_i^{(\alpha)} = 0$, analogous to the zero scalar product of two orthogonal vectors.

The basis function $\phi_i^{(\alpha)}$ delivered by the projection operator are not necessarily normalized, but this is only a minor inconvenience.

The unitary matrix $D^{(\alpha)}(\mathbf{T})$ is assumed to have dimension h. The subscript j in Eq. (3.13) is arbitrary and can have any value from 1 to h. Use this capability to rephrase Eq. (3.13) in terms of the characters of the α irreducible representation. Set $j = i$ and sum over i:

$$\sum_i \phi_i^{(\alpha)} = \sum_i \sum_{\mathbf{T}} D_{ii}^{(\alpha)*}(\mathbf{T})\mathbf{T}\phi$$

$$\phi^{(\alpha)} = \sum_{\mathbf{T}} \chi^{(\alpha)*}(\mathbf{T})\mathbf{T}\phi, \tag{3.14}$$

where

$$\phi^{(\alpha)} = \sum_i \phi_i^{(\alpha)}$$

is a basis function for the α representation because any linear combination of α basis functions is also a basis function for α.

If the α irreducible representation is 1-dimensional, Eqs. (3.13) and (3.14) are not essentially different, but if $h > 1$, Eq. (3.13) delivers more information. For row i of $D(\mathbf{T})$, each value of $j = 1, \ldots, h$ delivers a basis function that transforms like row i of the matrix. Basis functions for the same row are called *partners*.

Before proceeding to a proof, here are simple demonstrations of Eq. (3.14) using group **mm2**. With a trial function $\phi = x$, the effect of the group operations is

$$\mathbf{1}x = x \qquad \mathbf{2}_z x = -x \qquad \mathbf{m}_y x = x \qquad \mathbf{m}_x x = -x.$$

Using the characters for irreducible representation B_1 from Table 3.2, Eq. (3.14) delivers

$$\phi^{(\alpha)} = [(1)(1) + (-1)(-1) + (1)(1) + (-1)(-1)]x$$
$$= 4x.$$

This is not surprising, because we showed earlier that B_1 transforms like x. Now use instead the characters for A_2 associated with mass motion rotation using trial function x.

$$\phi^{(\alpha)} = [(1)(1) + (1)(-1) + (-1)(1) + (-1)(-1)]x$$
$$= 0$$

Representation A_2 does not transform like x; the basis function x is not in the space spanned by A_2.

If $\phi_i^{(\alpha)}$ from Eq. (3.13) is truly a basis function, it needs to satisfy this fundamental property of basis functions:

$$\mathbf{T}\phi_j^{(\alpha)} \propto \sum_{i=1}^{h} D_{ij}^{(\alpha)}(\mathbf{T})\phi_i^{(\alpha)}.$$

Begin the proof by multiplying Eq. (3.13) by any group operation $\mathbf{S}$ and then use the Kronecker delta:

$$\mathbf{S}\phi_i^{(\alpha)} = \sum_{\mathbf{T}} D_{ij}^{(\alpha)*}(\mathbf{T})\mathbf{ST}\phi$$
$$= \sum_{k}\sum_{\mathbf{T}} \delta_{ik} D_{kj}^{(\alpha)*}(\mathbf{T})\mathbf{ST}\phi$$
$$= \sum_{n}\sum_{k}\sum_{\mathbf{T}} D_{in}^{(\alpha)*}(\mathbf{S}^{-1}) D_{nk}^{(\alpha)*}(\mathbf{S}) D_{kj}^{(\alpha)*}(\mathbf{T})\mathbf{ST}\phi,$$

where the last step follows because

$$D(\mathbf{S}^{-1})D(\mathbf{S}) = D(\mathbf{E}).$$

Summing over k,

$$\mathbf{S}\phi_i^{(\alpha)} = \sum_{n} D_{in}^{(\alpha)*}(\mathbf{S}^{-1}) \left[\sum_{\mathbf{T}} D_{nj}^{(\alpha)*}(\mathbf{ST})\mathbf{ST}\phi \right]$$
$$= \sum_{n} D_{in}^{(\alpha)*}(\mathbf{S}^{-1}) \left[\sum_{\mathbf{ST}} D_{nj}^{(\alpha)*}(\mathbf{ST})\mathbf{ST}\phi \right],$$

where $\mathbf{ST}$ is a member of the group, so the sum still runs over all group members. The quantity in square brackets is $\phi_n^{(\alpha)}$ by Eq. (3.13):

Table 3.3 Character table for **mm2**

basis functions		**1**	**2$_z$**	**m$_y$**	**m$_x$**
z, x^2, y^2, z^2	A_1	1	1	1	1
xy, L_z	A_2	1	1	-1	-1
x, xz, L_y	B_1	1	-1	1	-1
y, yz, L_x	B_2	1	-1	-1	1

$$S\phi_i^{(\alpha)} = \sum_n D_{in}^{(\alpha)*}(\mathbf{S}^{-1})\phi_n^{(\alpha)}$$

$$= \sum_n D_{ni}^{(\alpha)}(\mathbf{S})\phi_n^{(\alpha)},$$

where the last step follows because D is unitary. The final result proves that $\phi_n^{(\alpha)}$ obeys the definition of a basis function.

An Easy Way to Find Basis Functions

In Section 3.5.1 projection operators and characters were used to find basis functions that transform like a group's irreducible representations. Having seen the method, the lengthy calculation can be avoided by noting that the work has already been done by others with the results tabulated for groups of interest. Character tables in references are typically accompanied by a listing of the basis functions that transform like each irreducible representation. Table 3.3, for example, shows how the character table for **mm2** is displayed in the literature.

For example, Table 3.3 shows that x is a basis function for B_1, y is a basis function for B_2, and z is a basis function for A_1. Put another way, x, y, z transform like B_1, B_2, A_1, respectively, from which the mass motion translation mode is $A_1 + B_1 + B_2$, as we found in Section 3.4 by direct calculation. Similarly, L_z, L_y, L_x transform like A_2, B_1, B_2, so the mass motion rotation mode is $A_2 + B_1 + B_2$.

3.5.2 Visualizing the Water Molecule's Normal Modes

To sketch the water molecule's vibration modes, start by finding basis functions that transform like A_1 and like B_2 with the help of the projection operator in character form Eq. (3.14). Good choices for trial functions are the displacements δy_2 and δz_2 because, from Table 3.3, δy_2 transforms like B_2 and δz_2 transforms like A_1. The group operations acting on these trial functions are

$$1\,\delta y_2 = \delta y_2 \quad 2_z\,\delta y_2 = -\delta y_3 \quad \mathbf{m}_y\,\delta y_2 = -\delta y_3 \quad \mathbf{m}_x\,\delta y_2 = \delta y_2$$
$$1\,\delta z_2 = \delta z_2 \quad 2_z\,\delta z_2 = \delta z_3 \quad \mathbf{m}_y\,\delta z_2 = \delta z_3 \quad \mathbf{m}_x\,\delta z_2 = \delta z_2.$$

With reference to Eq. (3.14) and the character table for **mm2** (Tables 3.2 and 3.3):

$$\phi^{(A_1)} = (1)\delta y_2 - (1)\delta y_3 - (1)\delta y_3 + (1)\delta y_2$$
$$= 2(\delta y_2 - \delta y_3)$$
$$\phi^{(A_1)} = (1)\delta z_2 + (1)\delta z_3 + (1)\delta z_3 + (1)\delta z_2$$
$$= 2(\delta z_2 + \delta z_3)$$
$$\phi^{(B_2)} = (1)\delta y_2 - (-1)\delta y_3 - (-1)\delta y_3 + (1)\delta y_2$$
$$= 2(\delta y_2 + \delta y_3)$$
$$\phi^{(B_2)} = (1)\delta z_2 + (-1)\delta z_3 + (-1)\delta z_3 + (1)\delta z_2$$
$$= 2(\delta z_2 - \delta z_3).$$

The basis functions generated for A_1 have δy_2 and δy_3 opposed and δz_2 parallel to δz_3. The basis function for B_2 has δy_2 parallel to δy_3 and δz_2 opposed to δz_3. The sketches show these results.

The oxygen atom in the sketches is shown with a small amplitude displacement. This is not one of the normal modes, although the oxygen atom does vibrate; it is a constraint due to Newton's laws. Because the molecule is free, its total momentum should be zero and it should have no torques causing it to rotate. The sketch shows that the displacements are of proper direction to satisfy the constraints. The amplitudes are not to scale, but the more massive oxygen atom is shown with only a small amplitude displacement because, for the most abundant isotopes, $M_O/m_H \approx 16$.

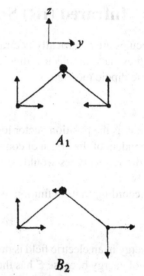

The normal modes in the sketch clearly satisfy the symmetry requirements, but they lack physical meaning. For a given representation, a linear combination of the normal mode basis functions is also a normal mode. The basis functions can be given physical meaning based on mechanics by taking the combinations shown in the sketch, where the modes are given the standard descriptive labels. There are two modes that transform like A_1 and one that transforms like B_2 as required.

Experimentally, the A_1 symmetric bond stretching mode has a small admixture of bond bending, and the A_1 bond bending mode has a small mix of bond stretching. The B_2 mode is a pure asymmetric bond stretching because there is no additional basis function to mix in. Group theory by itself cannot predict the exact way the basis functions combine because this depends on the detailed dynamics and requires lengthy and difficult calculations in quantum mechanics. Nevertheless, using only the tools of simple geometry, elementary algebra, and grade school arithmetic, we showed that the

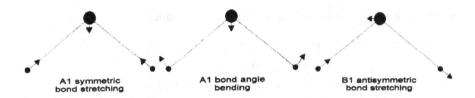

A1 symmetric bond stretching   A1 bond angle bending   B1 antisymmetric bond stretching

water molecule has only three vibration modes, and we made excellent approximations to their physical nature that can be useful guides to advanced calculations. This
is a strength of group theory: symmetry principles alone enable us to make substantial advances in physical situations where the detailed dynamics are either unknown
or difficult to apply.

3.6   Infrared (IR) Spectroscopy

Molecules are electrically neutral, but they can interact with an applied electric field
$\mathcal{E}$ if they have an electric dipole moment $\boldsymbol{p}$. For a system of fixed charges q_i, the
electric dipole moment is

$$\boldsymbol{p} = \sum_i q_1 \mathbf{r}_i,$$

where $\mathbf{r}_i$ is the position vector locating q_i. If the molecule is electrically neutral, $\boldsymbol{p}$ is
independent of the origin of coordinates. In a quantum mechanical treatment, the sum
over discrete charges would be replaced by integration over the molecule's electron
density.

According to its definition, electric dipole moment is a vector:

$$\boldsymbol{p} = p_x \hat{\mathbf{i}} + p_y \hat{\mathbf{j}} + p_z \hat{\mathbf{k}}.$$

Its energy in an electric field is the scalar product $\boldsymbol{p} \cdot \mathcal{E}$. This expression has the dimensions of energy because $\mathcal{E}$ has the dimensions *force/charge* and $\boldsymbol{p}$ has the dimensions
charge $\times$ *distance*. Combining gives *force* $\times$ *distance*, the dimensions of work and
energy.

The oscillating electric field of an electromagnetic wave can interact with a molecule's dipole moment to transfer energy, but only if the dipole moment is changing
with time. Suppose that a molecule has a component of dipole moment p_z changing
harmonically in time with angular frequency ω_v due to molecular vibration. If the
applied electric field of the wave has frequency ω_a, the energy transferred depends on

$$\boldsymbol{p} \cdot \mathcal{E} = p_z \mathcal{E}_0 \sin(\omega_v t) \sin(\omega_a t),$$

where $\mathcal{E}_0$ is the amplitude of the electric field. If ω_a and ω_v are substantially different,
the average energy transferred over time approaches 0; but if $\omega_a \approx \omega_v$, the time
average approaches 1/2.

A molecule can absorb electromagnetic energy only if it has a vibration mode
that causes a dipole moment to change with time; such a mode is said to be *infrared*

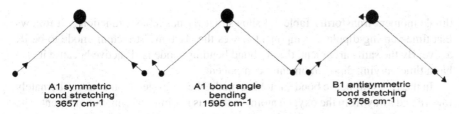

|A1 symmetric bond stretching 3657 cm-1 | A1 bond angle bending 1595 cm-1 | B1 antisymmetric bond stretching 3756 cm-1 |

active. The vibration frequencies of molecules typically fall in the infrared (IR), the region of the electromagnetic spectrum with wavelengths longer than visible red light but shorter than microwaves. The infrared spectrum is usually defined as between wavelengths of 700 or 800 nm up to 1 mm.

Frequencies in IR work are commonly specified as *wave numbers* $\bar{\nu}$, the number of complete wavelength cycles in 1 cm. The units of $\bar{\nu}$ are therefore cm^{-1}. It follows that the wavelength λ in cm equals $1/\bar{\nu}$. If c is the speed of light $\approx 3 \times 10^{10}$ $cm \cdot s^{-1}$, the circular frequency ν (Hz) is

$$c = \nu\lambda$$
$$\nu = \frac{c}{\lambda} = c\bar{\nu}.$$

Researchers often express energies in *electron volts* (eV). An electron passing through a potential difference of 1 volt gains an energy of $e \times 1$ V $\approx 1.6 \times 10^{-19}$ joule (J). Strictly speaking, the electron's charge e is a negative number, but the context shows whether it should be treated as negative or as positive with magnitude $|e|$. The energy E of a photon of frequency ν is $E = h\nu$, where $h \approx 6.63 \times 10^{-34}$ J $\cdot$ s is *Planck's constant*. Hence the energy of a photon can be written

$$E = \frac{h\nu}{e} = \frac{hc}{\lambda e} = \frac{hc}{e}\bar{\nu} \approx 1.2 \times 10^{-4}\,\bar{\nu} \quad eV.$$

E increases as the wave number increases. The spectrum visible to the human eye is only a small portion of the electromagnetic spectrum ranging between red and blue.

deep red $\lambda = 700$ nm, $\bar{\nu} = 1.4 \times 10^4$ cm^{-1}, $E = 1.7$ eV
deep blue $\lambda = 400$ nm, $\bar{\nu} = 2.5 \times 10^4$ cm^{-1}, $E = 3.0$ eV

All three water molecule's vibrational modes are IR active. The sketch shows the modes with their experimentally measured wave numbers. Group theory cannot predict wave numbers, because it does not deal with dynamics.

Note that the A_1 and B_2 bond stretching modes have different symmetries but nearly equal wave numbers because physically they both describe bond stretching oscillations of the H-O bonds.

Whether a mode is IR active or not can be determined either from symmetry or from how the dipole moment vector transforms. Consider the A_1 bond stretching mode. Both H-O bonds stretch in phase, so by symmetry the y component of the dipole moment does not change, but the z component does change as both hydrogen atoms move away and toward the oxygen atom. Looking alternatively at how the dipole moment transforms Table 3.3 shows that A_1 has a basis function z. It follows that time-varying dipole z component causes the A_1 bond stretching mode to be IR active. By the same argument, the A_1 bond bending mode is IR active because it also has a time-varying dipole moment z component.

In the B_2 asymmetric bond stretching mode, the hydrogen atoms move alternately toward and away from the oxygen atom, so there is no time-varying z component. The hydrogen atoms do move back and forth together along the y-axis, so there is a time-varying y component. Group theory also gives this result because B_2 transforms like y, as Table 3.3 shows.

Consider carbon dioxide, CO_2. It is a linear molecule with a central carbon flanked by oxygen atoms as shown. The three atoms give 9 displacement functions. The center of mass can translate in three dimensions x, y, z, but there are only two different mass motion rotations, namely rotation about the two axes x and z perpendicular to the molecule's longitudinal.

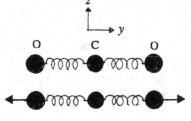

The five constraints leave $9 - 5 = 4$ vibration modes; one is shown in the lower sketch. This vibration mode does not have a changing dipole moment because extension along $+y$ by the right-hand oxygen atom is cancelled by extension along $-y$ by the left-hand atom. Although the oxygen atoms are vibrating, this mode is not IR active.

If a vibration mode is IR active, it can absorb light closely equal in frequency to the vibration frequency. An instrument called a *spectrometer* can record the absorption of IR radiation by a molecular compound versus wave number. IR absorption spectra can be analyzed to give information on the molecule's structure, and it is an important and widely used tool for identifying compounds.

3.6.1   Thermal Radiation

Every material radiates electromagnetic energy that depends on its absolute temperature. The graph, called the *Planck function*, shows two important features of the radiation. First, the intensity increases with higher absolute temperature degrees Kelvin (K) as T^4 (Stefan–Boltzmann law). The curve for 300K $\approx$ 27°C is barely visible on the graph, but the sun, with a surface temperature of 5800K, is too bright to view safely.

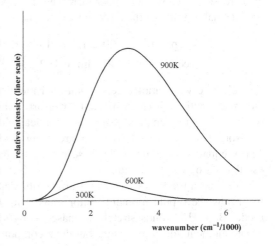

Second, the peak of the intensity moves to larger wave numbers proportional to T (Wien's Displacement law). The coil of an electric heater glows dull red at a low setting, but the much hotter sun appears yellow-white.

In 1900 German physicist Max Planck explained the shape of the radiation curve. His explanation opened the door to the later development of quantum mechanics because he needed to assume that electromagnetic energy is *quantized* – it takes the form of discrete bundles of energy called *photons*. In his model, a photon of frequency v has energy $E = hv$, and a radiating body is a collection of oscillators that emit photons of only certain frequencies and not in a continuous spread of frequencies.

The source of IR radiation in a spectrometer might be a rod of silicon carbide (SiC), a poor conductor with a high melting point. If electric current passes through the rod, the I^2R ohmic dissipation can heat it to 1500K, where the continuous IR radiation covers the spectral range of molecular vibrations. The IR radiation from the source passes through a cell containing a sample in liquid or gas phase that absorbs IR radiation at particular wavelengths. The modified IR passes through a dispersive device such as a grating that can scan across wavelengths. A detector at the output registers the absorption spectrum.

The figure shows the observed absorption spectrum of liquid water in the region of the bond angle bending absorption at 1595 cm^{-1}.

The Earth's atmosphere consists primarily of 78 percent diatomic nitrogen N_2, 21 percent diatomic oxygen O_2, and 1 percent monatomic argon Ar. None of these species is IR active; diatomic molecules of the same atoms cannot have a time-varying dipole moment by symmetry, and an atom cannot have a permanent dipole moment.

The atmosphere also contains many minor chemical species, some of which, like water vapor, are IR active. Water vapor is abundant in the lower atmosphere due to evaporation from oceans, surface water, and vegetation. Because of its ability to absorb IR radiation, water vapor is the most important *greenhouse gas* in keeping the Earth at a habitable average temperature of roughly 20°C. The Earth absorbs a portion of the visible and ultraviolet radiation emitted by the Sun (some is reflected). In turn, the Earth at its much lower temperature emits

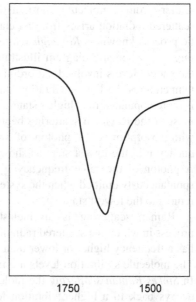

1750　　　　1500

Wavenumbers (cm–1)

U.S. Dept. of Commerce NIST

IR radiation. Absorbed radiation raises a water molecule from a lower to a higher energy level. The radiation spontaneously emitted in the transition from higher to lower energy is radiated in all directions, some back toward the Earth, decreasing the amount of energy lost to space. Without this blanket, the Earth's average global temperature would be $\approx -20°C$.

3.7 Raman Spectroscopy

As we have seen, some vibration modes are IR active and some are not. In 1928 the Indian physicist Chandrasekhara Raman (1888–1970) demonstrated experimentally a new type of spectroscopy now called *Raman scattering* in his honor. Raman scattering can deliver information even on some modes that may not be IR active, and it has practical application in the identification of molecules.

The physical basis of Raman scattering is that an incident electric field $\mathcal{E}$ may be able to induce an electric dipole moment $\boldsymbol{p}_{in} = \alpha\mathcal{E}$ by acting on the molecule's electrons, where α is the *polarizability* of the molecule. Taking a simple case where $\boldsymbol{p}_{in}$ is parallel to $\mathcal{E}$, the energy E is

$$E = \boldsymbol{p}_{in} \cdot \mathcal{E} = \alpha\mathcal{E} \cdot \mathcal{E}.$$

When an electromagnetic wave scatters from a molecule, most of the scattered radiation arises from an elastic process known as *Rayleigh scattering*. The schematic diagram illustrates the energy levels involved. At ordinary temperatures the lowest vibration state is more populated than higher states, so most of the Rayleigh scattering begins with absorption of a photon of frequency ν in the lowest state as shown. A photon of the same frequency ν is spontaneously emitted when the system returns to the lowest state.

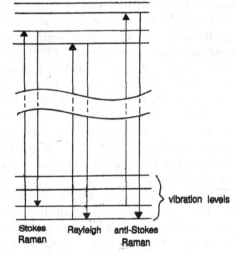

Raman scattering is an inelastic process in which the scattered radiation has a frequency higher or lower than ν. The molecule's vibration levels are involved, as the energy level diagram indicates. In *Stokes Raman scattering* the molecule is excited from the lowest level and then decays back to a higher vibration level, emitting a photon of frequency $<\nu$ (red shifted). In *anti-Stokes Raman scattering* the molecule is excited starting from a higher vibration level and then decays to a lower level, emitting a frequency $>\nu$ (blue shifted).

The energy level diagram demonstrates that incident radiation of any sufficiently short wavelength can generate Raman scattering. The incident radiation does not need to be in the IR region where molecular vibrations occur.

In IR spectroscopy a vibrating molecule can interact with the electric field of incident radiation only if the molecule's physical electric dipole moment varies with time. The same concept applies in Raman scattering except that the dipole moment is induced. Because of its effect on the molecule's charges, all components of an electric field may induce a component of the dipole moment. To describe this effect, the polarizability must be written as a tensor of the second rank with nine components. Here, for example, is the x component of the dipole p_{in} induced by electric field components $\mathcal{E}_x, \mathcal{E}_y, \mathcal{E}_z$:

$$(p_{in})_x = \alpha_{xx}\mathcal{E}_x + \alpha_{xy}\mathcal{E}_y + \alpha_{xz}\mathcal{E}_z.$$

A vibration mode will be Raman active if the vibration causes a change in the molecule's structure, hence a change in the polarizability, hence a change in the induced dipole. As an example, the sketch shows two vibration modes of carbon dioxide. The top pair in the sketch is for a symmetric bond stretching mode and the bottom pair for an asymmetric bond stretching mode. Each pair shows the vibration at maximum amplitude and the same vibration a half period later. The symmetric bond stretching mode changes the molecule's structure because of the symmetrical expanding and shrinking bond lengths; this mode is Raman active but not IR active, as we showed earlier. The asymmetric bond stretching mode does not change the structure of the molecule and is not Raman active, but it is IR active because of its changing dipole moment.

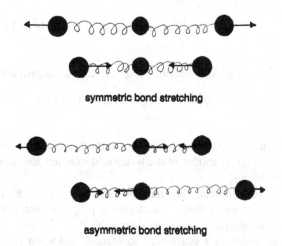

symmetric bond stretching

asymmetric bond stretching

Section 3.6 showed that group theory can predict whether a given vibration is IR active by determining whether any of the functions x, y, z appropriate to a structural electric dipole moment is a basis function for the irreducible representation of the vibration – in other words, to determine whether x, y, z "transform like" the normal mode's irreducible representation. The same principle applies to Raman scattering except that here we are dealing with an induced dipole. Based on the polarizability tensor, possible basis functions for the mode in question could be among the quadratic forms $x^2, xy, xz, yx, \ldots z^2$.

As an example, consider the asymmetric bond stretching mode of water corresponding to irreducible representation B_2. Are x^2 and yz basis functions? The following table shows the effect of the group operations.

To test the suitability of the proposed functions, use the projection operator, Eq. (3.14), and the characters for B_2 from Tables 3.2 and 3.3.

	x	y	z	x^2	yz
1	$x \mapsto x$	$y \mapsto y$	$z \mapsto z$	$(x)(x) = x^2$	$(y)(z) = yz$
$\mathbf{2}_z$	$x \mapsto -x$	$y \mapsto -y$	$z \mapsto z$	$(-x)(-x) = x^2$	$(-y)(z) = -yz$
$\mathbf{m}_y$	$x \mapsto x$	$y \mapsto -y$	$z \mapsto z$	$(x)(x) = x^2$	$(-y)(z) = -yz$
$\mathbf{m}_x$	$x \mapsto -x$	$y \mapsto y$	$z \mapsto z$	$(-x)(-x) = x^2$	$(y)(z) = yz$

$$\phi^{(B_2)} = \sum_{\mathbf{T}} \chi^{(B_2)*}(\mathbf{T})\mathbf{T}\phi$$

Taking $\phi = x^2$ gives

$$\phi^{(B_2)} = (1)(x^2) + (-1)(x^2) + (-1)(x^2) + (1)(x^2)$$
$$= 0.$$

x^2 is not a basis function for B_2; it does not transform like B_2. Now try yz.

$$\phi^{(B_2)} = (1)(yz) + (-1)(-yz) + (-1)(-yz) + (1)(yz)$$
$$= 4yz$$

Hence yz transforms like B_2, so the asymmetric bond stretch B_2 is Raman active. The two A_1 modes of the water molecule are also Raman active; proof is left to the problems.

Raman scattering has low intensity, so the availability of intense laser light sources made Raman scattering a practical research tool. It has been used along with IR spectroscopy to determine the structures of molecules and has become important as a quick way to identify unknown substances. The essential components of a Raman spectrometer are an intense light source focused on the sample with the Raman-scattered light entering a dispersive device that scans over wavelength and then onto a detector. The sample usually requires little or no special preparation and can be liquid, aqueous solution, solid, or even a thin film on a substrate.

Red-shifted Stokes Raman scattering is normally the spectrum recorded because the signal is stronger than for blueshifted anti-Stokes scattering. As the energy diagram implies, the initial energy level for anti-Stokes scattering must be an excited vibration state, but this is

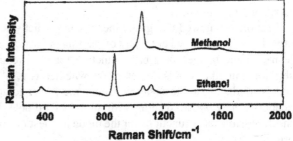

Reprinted figure from l. Boyaci *et al.* J. Raman Spectrosc. **43**, 1171 (2012). Copyright © 2012 John Wiley & Sons, Ltd. Used with permission.

much less populated than the ground level at ordinary temperatures and therefore gives a weaker signal.

A Stokes Raman spectrum is recorded as the wave number shift from the strong Rayleigh peak. The figure shows experimental Stokes Raman scattering spectra for methanol ("wood alcohol") and ethanol ("grain alcohol"). The peaks in the spectra arise from vibrations of various chemical bonds.

The chemical structures of methanol (CH_3OH) and ethanol (CH_3CH_2OH) are very similar. Both have a $-CH_3$ methyl group and a $-OH$ hydroxyl group. Raman scattering allows the two compounds to be readily distinguished to tell if ethanol has been illegally adulterated with toxic methanol.

Methanol

$$\begin{array}{c} H \\ | \\ H-C-OH \\ | \\ H \end{array}$$

Ethanol

$$\begin{array}{c} H \quad H \\ | \quad | \\ H-C-C-OH \\ | \quad | \\ H \quad H \end{array}$$

3.8 Brief Bios

Chandrasekhara Raman (1888–1970) was born in India. In 1917 he became a professor of physics at the University of Calcutta, where he specialized in the scattering of light. He was instrumental in advancing physics in India, helping to establish research institutions and founding technical journals. He was awarded the 1930 Nobel Prize in Physics, cited "for his work on the scattering of light and for the discovery of the effect named after him."

Max Planck (1858–1947) was awarded the 1918 Nobel Prize in Physics, cited "in recognition of the services he has rendered to the advancement of physics by his discovery of energy quanta." He is buried in the Göttingen city cemetery. His tombstone is starkly simple, without even the customary dates of birth and death. Close inspection of the decorative frieze near the bottom reveals the value of the Planck constant written as $h = 6.62 \times 10^{-34}$ W · s^2.

– photo by Robert Kolenkow

Summary of Chapter 3

Chapter 3 demonstrates the application of discrete groups to molecular vibrations.

a) A mass on a spring executes simple harmonic motion, an approximation to the motion of any bound system at small amplitudes.

b) Any motion of a vibrating system of masses and springs can only be a combination of normal modes. A normal mode is simple harmonic motion at one of the particular discrete frequencies characteristic of the system.

c) In Newtonian mechanics, the frequencies of a system's normal modes can be found by diagonalizing the equations of motion to give a new set of coordinates that each obey simple harmonic motion. The method is illustrated by the example of two masses and three springs.

d) In the language of group theory, the normal modes of a system are the basis functions for the irreducible representations of the system's symmetry group.

e) If an irreducible representation of a system's symmetry group has dimension $n > 1$, there will be n normal modes with different motions but the same frequency; these modes are said to be degenerate.

f) A vibrating water molecule involves three masses hence nine displacement coordinates. The molecules' symmetry operations can be determined from a sketch and are identified as the members of the symmetry group **mm2** (Hermann–Mauguin) or $\mathbf{C_{2v}}$ (Schönflies).

g) The nine displacements of the water molecule's atoms are basis functions for a reducible 9×9 representation. The number of times the group's irreducible representations occur in the 9×9 reducible representation can be determined using characters of the irreducible representations. For the water molecule, there are three $\omega = 0$ mass motion normal modes of translation and three $\omega = 0$ rotation modes. Group theory, therefore, shows that the water molecule has three nontrivial normal modes of vibration.

h) To identify the trivial $\omega = 0$ modes, note that translation modes transform like position coordinates (x, y, z) and rotation modes transform like angular momentum (L_x, L_y, L_z).

i) Projection operators can select out basis functions (normal modes) that transform like the nontrivial vibration modes. A simpler method is that character tables in references show basis functions for each of the irreducible representations.

j) The basis functions for a normal mode can give pictures of the modes. It may be necessary to take combinations of modes to give physical meaning to the sketches while maintaining the symmetry.

k) A vibration normal mode is infrared (IR) active able to absorb and emit electromagnetic radiation if it has time-varying basis functions that transform like an electric dipole vector (x, y, z).

l) If a molecule is polarizable, it can have an induced electric dipole moment, allowing it to interact with electromagnetic radiation in Raman scattering. A molecule can undergo Raman scattering if it has time-varying basis functions that transform like the quadratic forms of an induced dipole moment $(x^2, xy, \ldots)$.

Problems and Exercises

3.1 For the example of two masses and three springs in Section 3.2.2 use Eq. (3.5) to calculate the c coefficients and show by direct calculation that $CB = \Omega C$.

3.2 For the example of two masses and three springs in Section 3.2.2 let the springs be equal, $k' = k$. Use Eq. (3.5) to calculate the c coefficients and show by direct calculation that $CB = \Omega C$.

3.3 Boron trifluoride, BF_3, is a planar molecule, with the three fluorines bonded to a central boron. All three bond angles F-B-F are 120°.
(a) What are its symmetry operations (including mirrors/inversions)?
(b) For each symmetry operation, which atoms stay at home?

3.4 Ammonia, NH_3, is tetrahedral, with the three hydrogens in an equilateral triangle base plane and with the nitrogen at the top of the tetrahedron.
(a) What are its symmetry operations (including mirrors/inversions)?
(b) For each symmetry operation, which atoms stay at home?

3.5 Here is the character table for point group **2** (Schönflies **C₂**).

	1	**2**$_z$
A	1	1
B	1	−1

Show by calculation that x is a basis function for irreducible representation B.

3.6 Here is the character table for point group **222** (Schönflies **D₂**). There are three irreducible representations labeled B because any of them can be a principal axis.

	1	**2**$_x$	**2**$_y$	**2**$_z$
A	1	1	1	1
B_1	1	−1	−1	1
B_2	1	−1	1	−1
B_3	1	1	−1	−1

Show by calculation that x is a basis function for irreducible representation B_3.

3.7 Use the character table in Problem 6 to show by calculation that L_z is a basis function for irreducible representation B_1.

3.8 Consider a planar system of three
identical masses and three identical springs. At
equilibrium each mass is at the apex of an equilateral
triangle. A spring links each pair of masses.
(a) How many displacement coordinates are there?
(b) How many normal modes are there?
(c) One of the normal modes is a basis function for the
identity irreducible representation. Sketch this normal mode.

3.9 Consider a planar system of three
identical masses and three identical springs. At
equilibrium each mass is at the apex of an equilateral
triangle. A spring links each pair of masses.
Here is its character table (inversions omitted).

basis functions		1	m_z	3_z	2_y	m_x
$x^2 + y^2, z^2$	A_1	1	1	1	1	1
L_z	A_2	1	1	1	-1	-1
	B_1	1	-1	1	1	-1
z	B_2	1	-1	1	-1	1
$x, y, x^2 - y^2, xy$	E_1	2	2	-1	0	0
xz, yz, L_x, L_y	E_2	2	-2	-1	0	0

Use coordinate displacements as the basis for a reducible representation. In this
representation what are the characters of each of the symmetry operations listed
in the table?

3.10 As in Problem 9, consider a planar system of three identical masses and three
identical springs. At equilibrium each mass is at the apex of an equilateral triangle. A spring links each pair of masses.
Use the character table in Problem 9 and use the coordinate displacements as
the basis for a reducible representation.
(a) Express the reducible representation for the trivial zero-frequency normal
mode translations as a sum of irreducible representations.
(b) Express the reducible representation for the trivial zero-frequency normal
mode rotations as a sum of irreducible representations.

3.11 As in Problem 9, consider a planar system of three identical masses and three
identical springs. At equilibrium each mass is at the apex of an equilateral triangle. A spring links each pair of masses.

The reducible representation based on coordinate displacements can be expressed as the sum of irreducible representations from the character table in Problem 9 as
$A_1 + A_2 + B_2 + 2E_1 + E_2$.
(a) Express the reducible representation of the nontrivial normal modes as a sum of irreducible representations using the character table from Problem 9.
(b) How many different nontrivial frequencies are exhibited by the system's normal modes?

3.12 As in Problem 9, consider a planar system of three masses and three identical springs. At equilibrium each mass is at the apex of an equilateral triangle. A spring links each pair of masses.
Two of the masses have mass m and the third has mass $2m$. What are the symmetry operations of the system?

3.13 Xenon tetrafluoride, XeF_4, is one the few noble gas compounds. Its molecule is planar, assumed to lie in the x-y plane. Each F-Xe-F bond angle is 90°. Model the bonds as four identical springs.
(a) How many displacement coordinates are there? Include possible motion in all three dimensions.
(b) How many nontrivial normal modes are there?
(c) One of the normal modes is a basis function for the identity irreducible representation. Sketch this normal mode.

3.14 Consider the planar XeF_4 molecule described in Problem 13. Assuming motion only in the x-y plane, how many nontrivial normal modes are there?

3.15 The NH_3 molecule is described in Problem 4. Here is its character table.

basis functions		1	3_z	m_x
$z, x^2 + y^2, z^2$	A_1	1	1	1
L_z	A_2	1	1	-1
$x, y, x^2 - y^2, xy, xz, yz, L_x, L_y$	E	2	-1	0

Its nontrivial normal modes are basis functions for the irreducible representations $2A_1 + 2E$.
(a) How many nontrivial normal modes are there?
(b) Which of the normal modes are IR active?
(c) How many different IR frequencies does NH_3 exhibit?

3.16 Boron trifluoride, BF_3, is a planar molecule with F-B-F bond angles of 120°. It has the same symmetry operations as the mass and spring system in Problems 8 and 9, hence has the same character table. Its nontrivial normal modes are basis functions for the irreducible representations $A_1 + B_2 + 2E_1$.

(a) Which of the modes are IR active?

(b) Which of the modes are Raman active?

(c) Which mode is Raman active but not IR active?

3.17 Which of the nontrivial normal modes of H_2O is Raman active?

4

CRYSTALLINE SOLIDS

4.1 Introduction

Crystals have long held a fascination because of their color and symmetry. They are even thought to possess magical and healing powers. From the point of view of this text, their symmetries cry out for the application of group theory. This chapter considers crystal symmetries on the atomic level, a topic that is only a small part of the vast field of solid-state physics that has practical applications in the design of new materials. Solid-state physics is multidisciplinary, engaging physicists, chemists, and materials scientists.

4.2 Bravais Lattices

Every crystalline solid is a *lattice* assembled from a repeated fundamental structure called a *unit cell*. The sketch shows a simple cubic lattice where the filled circles represent the locations of the atomic nuclei; the atoms themselves with their electron clouds would fill most of the lattice volume. A lattice with atoms only at the corners is called *simple*.

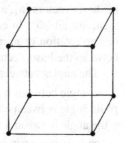

A crystal is held together by the electric forces in chemical bonds. For an *ionic bond*, for example, in sodium chloride, $NaCl$, the single outer valence electron of sodium is transferred to the seven-electron outer shell of chlorine completing the shell and leading to electric attraction between the Na^+ and Cl^- ions. In a *covalent bond* outer electrons are shared between atoms; the various forms of carbon – graphite, diamond, graphene (a sheet one carbon atom thick) – are prime examples. In a *metallic bond* some electrons are shared among some atoms and some move relatively freely through the material and are able to carry electric current when a voltage difference is applied.

For simplicity, consider only *perfect* crystals, where every atom is in its proper location and there are no imperfections. Dealing with effects at the surface is a special field all its own, so we further consider only locations deep inside the crystal far from a surface. This is not a serious limitation because even a small crystal sample consists of a vast number of atoms. For example, a one-carat diamond (1 carat = 0.200 g) contains $\approx 10^{22}$ carbon atoms.

A crystal is a periodic assemblage of identical lattices so that any point inside a lattice experiences the same surroundings as the same point in another lattice. Based on this criterion there are only 14 possible crystal lattices, called the *Bravais lattices*. Two are shown in the sketch.

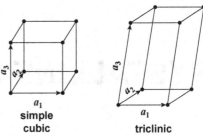

Every lattice has *basis vectors* $\mathbf{a}_i$ along the principal axes. They have dimensions of length numerically equal to the physical length of the axis.

Common units of length for the $\mathbf{a}_i$ are nanometers (nm), picometers (pm), and angstroms (Å). Nanometers and picometers are SI units defined in terms of the meter, and the angstrom is an accepted non-SI unit found mainly in the older literature.

$$1 \text{ nm} = 10^{-9} \text{ m} = 10^{-7} \text{ cm}$$
$$1 \text{ pm} = 10^{-12} \text{ m} = 10^{-10} \text{ cm}$$
$$1 \text{ Å} = 10^{-10} \text{ m} = 10^{-8} \text{ cm}$$
$$1 \text{ nm} = 10 \text{ Å} \qquad 1 \text{ pm} = 0.01 \text{ Å}$$

Crystal axes are typically of the order of a fraction of a nm (a few Å).

Crystallographers usually label the basis vectors $\mathbf{a}, \mathbf{b}, \mathbf{c}$; the corresponding magnitudes are a, b, c. The simple cubic lattice is the most symmetric Bravais lattice because the basis vectors $(\mathbf{a}_1, \mathbf{a}_2, \mathbf{a}_3)$ are equal in length $a = b = c$, and the angles between them are all 90°. In contrast, the triclinic lattice is the least symmetric because it has no rotation or mirror symmetries. Its basis vectors are unequal, and the angles between the basis vectors are unequal with no 90° angle.

The angles between the axes are specified as follows:

α the angle between $\mathbf{b}$ and $\mathbf{c}$
β the angle between $\mathbf{c}$ and $\mathbf{a}$
γ the angle between $\mathbf{a}$ and $\mathbf{b}$.

The 14 Bravais lattices are further divided into seven *crystal families* depending on their fundamental symmetry. The triclinic lattice system has only one member and the cubic system has three as shown in the sketch: simple cubic, body-centered cubic (bcc), and face-centered cubic (fcc).

The lattice for the element polonium (Po) is the only example of a simple cubic system among the elements.

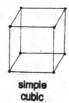
simple
cubic

bcc

fcc

The bcc lattice, a common lattice type in nature, has an atom or ion at the center of the cube (open circle in the sketch). In a crystal volume the bcc lattice is equivalent to two interpenetrating cubes with the filled circles at the corners of one set of cubes and the open circles at the corners of the other. Sodium (Na) and chromium (Cr) are among elements with the bcc lattice. Another example of bcc is the ionic crystal cesium chloride (CsCl). The Cs^+ ions at the corners may appear to outnumber the Cl^- ion in the center, but in a volume of the crystal completely surrounded by neighboring lattices each of the eight Cs^+ ions is shared among eight lattices so that the *stoichiometry* (Greek, measuring the elements) is 1:1 with one Cs^+ ion for every Cl^- ion.

The fcc lattice has an atom at the center of each face of the cube. A common example is sodium chloride (table salt, NaCl). Fcc occurs frequently because it is geometrically equivalent to one way of packing atoms closely in layers. Many elements, including aluminum (Al), copper (Cu), silver (Ag), and solid argon (Ar), have close-packed fcc lattice structures.

4.2.1 Translation Symmetry and Basis Vectors

Translation symmetry is the fundamental symmetry of crystals because a perfect crystal consists of a regular repeated array of identical lattices. The figure shows a lattice with basis vectors a_1, a_2, a_3. They are labeled so that a_3 is directed in a general sense according to the cross product (right-hand rule) $a_1 \times a_2$.

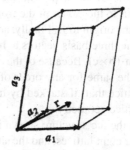

Consider a translation vector $\tau = n_1 a_1 + n_2 a_2 + n_3 a_3$ where the n_i are integers. The translation vector in a simple crystal can move from the location of any atom to any other atom in the crystal. The location of every atom inside a simple crystal is specified by integer multiples $\sum_i n_i a_i$ of the lattice basis vectors.

Let r be a vector to an arbitrary point in a given lattice and let $\phi(r)$ be some function of position in the lattice. The vector sum $r + \tau$ moves to an identical point in another lattice where $\phi(r)$ has the same value as in the original lattice.

Consider the fcc lattice (upper sketch), which has occupied sites (open circles) at the centers of the six faces in addition to sites (filled circles) at the eight corners. Take basis vectors $\mathbf{a}_1$, $\mathbf{a}_2$, $\mathbf{a}_3$ along the principal axes to form a translation $\boldsymbol{\tau}$.

$$\boldsymbol{\tau} = n_1\mathbf{a}_1 + n_2\mathbf{a}_2 + n_3\mathbf{a}_3$$

Because the n_i must be integers for a valid crystal translation, the translation $\boldsymbol{\tau}$ can specify the locations of corner sites but not of face-centered sites. Consider the vectors

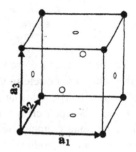

$$\boldsymbol{\sigma}_1 = (n_1 - \tfrac{1}{2})\mathbf{a}_1 + n_2\mathbf{a}_2 + (n_3 - \tfrac{1}{2})\mathbf{a}_3 \quad \text{front, back}$$
$$\boldsymbol{\sigma}_2 = n_1\mathbf{a}_1 + (n_2 - \tfrac{1}{2})\mathbf{a}_2 + (n_3 - \tfrac{1}{2})\mathbf{a}_3 \quad \text{sides}$$
$$\boldsymbol{\sigma}_3 = (n_1 - \tfrac{1}{2})\mathbf{a}_1 + (n_2 - \tfrac{1}{2})\mathbf{a}_2 + n_3\mathbf{a}_3 \quad \text{top, bottom.}$$

The $\boldsymbol{\sigma}_i$ can specify the locations of the sites on the faces, but the coefficients of the basis vectors $\mathbf{a}_i$ are not integers, so the $\boldsymbol{\sigma}_i$ are not valid crystal translations.

Form instead a new set of basis vectors $\mathbf{a}_i'$ by subtracting $\boldsymbol{\sigma}_i$ one at a time from $\boldsymbol{\tau}$.

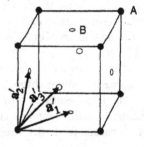

$$\mathbf{a}_1' = \tfrac{1}{2}(\mathbf{a}_1 + \mathbf{a}_2)$$
$$\mathbf{a}_2' = \tfrac{1}{2}(\mathbf{a}_2 + \mathbf{a}_3)$$
$$\mathbf{a}_3' = \tfrac{1}{2}(\mathbf{a}_3 + \mathbf{a}_1)$$

The vector $\boldsymbol{\tau}' = \sum_i m_i\mathbf{a}_i'$, where the m_i are integers, is a valid crystal translation that can specify all of the fcc crystal lattice sites. As an example go from the origin to site A in several steps with the help of the lower sketch. Starting from the origin, $\mathbf{a}_3'$ takes us to the center of the front face. Adding $\mathbf{a}_2'$, the translation $\boldsymbol{\tau}_B' = \mathbf{a}_3' + \mathbf{a}_2'$ goes from the origin to B. Finally, adding $\mathbf{a}_1'$, the translation $\boldsymbol{\tau}_A' = \mathbf{a}_3' + \mathbf{a}_2' + \mathbf{a}_1'$ ends at A.

The three basis vectors $\mathbf{a}, \mathbf{b}, \mathbf{c}$ of any lattice specify a parallelepiped with volume $V = \mathbf{a}\cdot(\mathbf{b}\times\mathbf{c})$. Because of the cyclic vector identity $\mathbf{a}\cdot(\mathbf{b}\times\mathbf{c}) = \mathbf{b}\cdot(\mathbf{c}\times\mathbf{a}) = \mathbf{c}\cdot(\mathbf{a}\times\mathbf{b})$, V is the same for any order of the basis vectors. The parallelepiped is a unit cell of the lattice that, if stacked by translation along the basis vectors, completely defines the lattice geometry.

In the fcc structure shown in the sketches each of the eight corner sites is shared among eight lattices, and the site on each of the six faces is shared among two lattices to give a total of $8 \times \tfrac{1}{8} + 6 \times \tfrac{1}{2} = 4$ sites in the unit cell.

4.3   X-Ray Crystallography

A principal experimental method of determining the structures of crystals is *X-ray diffraction (XRD)*. In 1912 the German physicist Max von Laue (1879–1960, Nobel laureate in physics 1914) realized that the orderly planes of evenly spaced atoms in a perfect crystal could act like the familiar ruled diffraction grating used for spectral analysis, taking into account that a crystal is a 3-dimensional grating.

The physical principle under-
lying XRD is wave interference. The sketch shows incident
electromagnetic waves of wavelength λ reflected elastically
from a model 2-dimensional
crystal. One portion of the incident wave reflects from the first
crystal plane, and another portion reflects from the second
crystal plane. The path lengths
differ by $2x = 2d \sin \theta$, where
d is the spacing between the
planes. If this difference is an
integral number n of wavelengths, the two waves give
strong constructive interference
at θ. In an actual 3-dimensional
crystal, the line of constructive

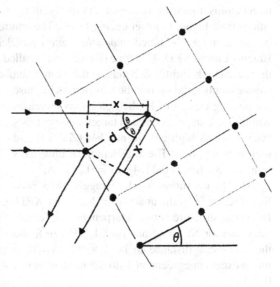

interference traces out a circle, the projection of a cone of included angle 2θ.

The condition for constructive interference can be written

$$n \lambda = 2d \sin \theta, \tag{4.1}$$

a result known as *Bragg's Law*, named after Australian-born British physicist William
Lawrence Bragg (1890–1971), who reported his derivation in 1913 while he was a 22-
year-old student at Trinity College, Cambridge University. Constructive interference
for integer $n = 1$ in Eq. (4.1) is called *first order*. For $n = 2$ it is *second order*, and
so on.

Rewriting Bragg's Law (4.1) as

$$\frac{\lambda}{2d} = \frac{\sin \theta}{n} \leq 1$$

shows that constructive interference from crystals cannot occur unless λ is less than
$2d$. The required radiation was available to Bragg thanks to the 1895 discovery of X-
rays by the German physicist Wilhelm Röntgen (1845–1927). Röntgen named them
X-rays because their nature "X" was at first unknown, but X-rays were later shown to
be electromagnetic radiation of very short wavelength.

X-rays can be produced in a vacuum tube, where electrons from a heated filament
are accelerated by an applied voltage of tens of kilovolts (kV) to crash into a metal
target. The resulting rapid deceleration of the electrons causes them to emit a continuous spectrum of short-wavelength electromagnetic waves called *bremsstrahlung*
(German, braking radiation).

If the voltage applied to an X-ray tube is high enough, collisions of the incoming electrons with the target can eject bound electrons from the inner K shell of the

target atoms. Electrons in the target rapidly fill the vacancy and emit photons in transitions from higher to lower energy levels. The emitted radiation, called K_α radiation, has short and well-defined (*monochromatic*) wavelength ideal for the application of Bragg's Law to XRD. This type of radiation is called *characteristic radiation* because the wavelength emitted depends on the atomic number Z of the target – the higher the atomic number, the shorter the wavelength because the binding energy of K shell electrons increases with Z. Tungsten is a common target in X-ray tubes because of its high atomic number ($Z = 74$) and high melting point. The wavelength of tungsten's principal K_α line is 0.0214 nm = 0.214 Å.

The figure shows W. L. Bragg's 1913 published result from the analysis of $NaCl$ by XRD. Its crystal structure is two interpenetrating fcc lattices, one for Na^+ and one for Cl^-. Bragg found the Na^+-Na^+ distance to be 2×2.8 Å= 0.56 nm in excellent agreement with the modern value, 0.565 nm.

The photo shows Bragg's XRD of zinc sulfide (zinc blende), which has the same fcc crystal structure as $NaCl$. It was published in 1913 in the same paper as his derivation of Bragg's Law. The 4-fold symmetry of the crystal is apparent. The spot in the center is due to the undiffracted beam.

Bragg's X-ray beam produced a continuous spread of wavelengths. Modern equipment uses monochromatic X-rays with the aid of filters, and the crystal is rotated through a wide range of angles during the exposure. Single crystals, as used in Bragg's experiments, can be difficult to fabricate for many materials. Section 4.5.1 discusses the convenient use of powder samples in XRD.

For NaCl

AB = 2·8.10⁻⁸. cm.

Reprinted figure from W. L. Bragg Proc. Roy. Soc. (London) **A89** (610), 248 (1913). The Royal Society (UK).

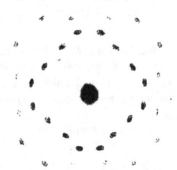

Reprinted figure from W.L. Bragg Proc. Cambridge Phil. Soc. **17**, 43 (1913). Cambridge: Cambridge University Press.

4.4 Fourier Transform

The object of this section is not to delve deeply into the extensive mathematical theory of the Fourier transform but instead to introduce qualitative concepts as a foundation for the upcoming discussion of reciprocal lattices.

The *Fourier transform* is a powerful mathematical tool used throughout solid-state theory, including XRD. It is also used by engineers to analyze signals and images and by scientists to help solve Schrödinger's equation in quantum mechanics.

You may have studied Fourier *series* in math courses. A Fourier series expresses a periodic function as a sum of sines and cosines of multiples of harmonic frequency.

If $\psi(t)$ is a periodic function of time t with period $\mathcal{T}$, its Fourier series can be written in complex form as

$$\psi(t) = \sum_{m=-\infty}^{\infty} c_m e^{2\pi i m t/\mathcal{T}},$$

where m is an integer and the coefficients c_m are found from integrals involving $\psi(t)$.

The Fourier *transform* generalizes the Fourier series and can represent a non-periodic function by letting $\mathcal{T} \to \infty$. Harmonic frequencies for m and $m + 1$ come closer together and in the limit become a continuous function of frequency $m/\mathcal{T} \to \omega$.

Denoting the Fourier transform as $\mathcal{F}(\omega)$:

$$\mathcal{F}(\omega) = \frac{1}{\sqrt{2\pi}} \int_{-\infty}^{\infty} f(t)e^{-i\omega t} dt. \tag{4.2}$$

Note that the Fourier transform of the time-dependent function $f(t)$ is a function of the angular frequency ω. The variables t and ω are said to be *complementary*. The two functions $f(t)$ and $\mathcal{F}(\omega)$ exist in different domains: $f(t)$ is in the time domain, and $\mathcal{F}(\omega)$ is in the frequency domain. The two variables t and ω are inverse to one another as seen from their units; t has units of seconds (s), and ω has units of s^{-1}.

The example in the sketches illustrates Eq. (4.2). The upper sketch shows two rectangular functions in the time domain: a narrow pulse (solid line) and a broader pulse (dashed line). The lower sketch is the relative Fourier transform of each pulse in the frequency domain.

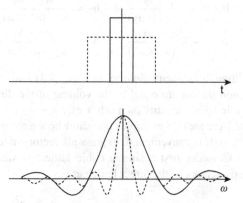

The pulse narrower in time has a wider range in the frequency domain, a fundamental property of complementary variables. It is impossible to have a pulse both narrow in time and narrow in frequency, a classical *uncertainty principle*. This behavior is easy to observe by listening to an AM radio during a thunderstorm. The radio crackles whenever a lightning bolt strikes nearby regardless of the station tuned, because the short pulse generates electromagnetic radiation over a wide range of frequencies.

Like Fourier series, the Fourier transform comes in pairs. The related *inverse transform* retrieves the time function from the frequency transform according to

$$f(t) = \frac{1}{\sqrt{2\pi}} \int_{-\infty}^{\infty} \mathcal{F}(\omega)e^{i\omega t} d\omega. \tag{4.3}$$

For a function of position $\phi(\mathbf{r})$, the variable complementary to $\mathbf{r}$ is the wave vector $\mathbf{k}$, where $|\mathbf{k}| = 2\pi/\lambda$ and λ is wavelength. The units of the complementary variables $\mathbf{r}$ and $\mathbf{k}$ are reciprocal, as expected.

De Broglie's relation $\mathbf{p} = \hbar\mathbf{k}$ between momentum $\mathbf{p}$ and wave vector $\mathbf{k}$, where $\hbar$ ("h bar") $\equiv \frac{h}{2\pi}$, leads to an uncertainty relation between position and momentum. $\mathbf{r}$ and $\mathbf{p}$ are complementary because $\mathbf{r}$ and $\mathbf{k}$ are.

4.5 Reciprocal Lattice

The interference pattern in a 3-D XRD photo is not directly related to diffraction by the atom planes in the physical *direct lattice* crystal sample. Theoretical analysis uses a Fourier transform to calculate scattering from the electron cloud at a single site, then applies it mathematically to the entire periodic lattice. An important result of the analysis is that the observed diffraction pattern actually measures atom spacing in the *reciprocal lattice*, a mathematical construct with no physical existence. The reciprocal lattice nevertheless plays an essential role in solid-state physics.

The calculation has a remarkably simple result. If $\mathbf{a}_1, \mathbf{a}_2, \mathbf{a}_3$ are the basis vectors of the *direct* (physical) lattice, the basis vectors $\mathbf{b}_1, \mathbf{b}_2, \mathbf{b}_3$ of the reciprocal lattice are given by either of the equal cyclic forms:

$$\mathbf{b}_1 = \frac{\mathbf{a}_2 \times \mathbf{a}_3}{\mathbf{a}_1 \cdot (\mathbf{a}_2 \times \mathbf{a}_3)} \qquad \mathbf{b}_2 = \frac{\mathbf{a}_3 \times \mathbf{a}_1}{\mathbf{a}_1 \cdot (\mathbf{a}_2 \times \mathbf{a}_3)} \qquad \mathbf{b}_3 = \frac{\mathbf{a}_1 \times \mathbf{a}_2}{\mathbf{a}_1 \cdot (\mathbf{a}_2 \times \mathbf{a}_3)} \qquad (4.4)$$

$$\mathbf{b}_1 = \frac{\mathbf{a}_2 \times \mathbf{a}_3}{\mathbf{a}_1 \cdot (\mathbf{a}_2 \times \mathbf{a}_3)} \qquad \mathbf{b}_2 = \frac{\mathbf{a}_3 \times \mathbf{a}_1}{\mathbf{a}_2 \cdot (\mathbf{a}_3 \times \mathbf{a}_1)} \qquad \mathbf{b}_3 = \frac{\mathbf{a}_1 \times \mathbf{a}_2}{\mathbf{a}_3 \cdot (\mathbf{a}_1 \times \mathbf{a}_2)}. \qquad (4.5)$$

Although the denominators in Eqs. (4.4) and (4.5) are different in appearance, all denominators are equal to the volume of the direct lattice's unit cell because of the cyclic vector identity on $\mathbf{a} \cdot (\mathbf{b} \times \mathbf{c})$.

Here are two examples that show how a reciprocal lattice is obtained from a direct lattice. It is convenient to express all vectors in terms of Cartesian unit vectors.

Consider first a simple cubic lattice of side a. The basis vectors of the direct lattice are

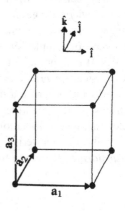

$$\mathbf{a}_1 = a\hat{\mathbf{i}}$$
$$\mathbf{a}_2 = a\hat{\mathbf{j}}$$
$$\mathbf{a}_3 = a\hat{\mathbf{k}}.$$

Equation (4.4) gives

$$\mathbf{b}_1 = \frac{1}{a}\hat{\mathbf{i}}$$
$$\mathbf{b}_2 = \frac{1}{a}\hat{\mathbf{j}}$$
$$\mathbf{b}_3 = \frac{1}{a}\hat{\mathbf{k}}.$$

Here $\mathbf{b}_i$ is parallel to $\mathbf{a}_i$, so the Fourier transform of a simple cubic is also a simple cubic but with reciprocal spacing $\frac{1}{a}$.

As seen in this example, the basis vectors of a reciprocal lattice have dimensions of reciprocal length (length^{-1}) as expected for complementary variables. An interpretation is that if the atoms in a direct lattice are closely spaced, the diffraction pattern is spread out and the atoms in the reciprocal lattice are correspondingly far apart, analogous to the Fourier transform example in Section 4.4.

Consider next a fcc lattice of side a. We found its basis vectors in Section 4.2.1:

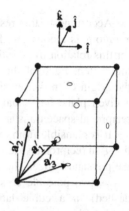

$$\mathbf{a}_1' = \tfrac{a}{2}(\hat{\mathbf{i}} + \hat{\mathbf{j}})$$
$$\mathbf{a}_2' = \tfrac{a}{2}(\hat{\mathbf{j}} + \hat{\mathbf{k}})$$
$$\mathbf{a}_3' = \tfrac{a}{2}(\hat{\mathbf{k}} + \hat{\mathbf{i}}).$$

Using Eq. (4.4) gives

$$\mathbf{b}_1 = \tfrac{1}{a}(\hat{\mathbf{i}} + \hat{\mathbf{j}} - \hat{\mathbf{k}})$$
$$\mathbf{b}_2 = \tfrac{1}{a}(-\hat{\mathbf{i}} + \hat{\mathbf{j}} + \hat{\mathbf{k}})$$
$$\mathbf{b}_3 = \tfrac{1}{a}(\hat{\mathbf{i}} - \hat{\mathbf{j}} + \hat{\mathbf{k}}).$$

Because of the minus signs, adjacent cells must be included to show geometrically that translations of these basis vectors can reach every site.

Without proof, the $\mathbf{b}_i$ are the basis vectors for a bcc lattice, showing that a bcc lattice is the reciprocal of a fcc lattice and vice versa. This result is not altogether surprising because any space-filling lattice must be one of the Bravais lattices. The reciprocal lattice of any Bravais lattice is therefore also a Bravais lattice. Further, the reciprocal lattice of a reciprocal lattice is the original direct lattice.

To conclude this section, here are some properties of the reciprocal lattice. Using Eq.(4.4) and vector identities, the volume V_r of a cell in reciprocal space is $V_r = \mathbf{b}_1 \cdot (\mathbf{b}_2 \times \mathbf{b}_3) = \frac{1}{V_d}$, where $V_d = \mathbf{a}_1 \cdot (\mathbf{a}_2 \times \mathbf{a}_3)$ is the cell's volume in physical space.

Another property is orthogonality of the basis vectors $\mathbf{a}_i$ of a physical lattice and the basis vectors $\mathbf{b}_i$ of its reciprocal lattice. Using Eq. (4.4),

$$\mathbf{b}_1 \cdot \mathbf{a}_1 = \frac{\mathbf{a}_1 \cdot (\mathbf{a}_2 \times \mathbf{a}_3)}{\mathbf{a}_1 \cdot (\mathbf{a}_2 \times \mathbf{a}_3)} = 1$$
$$\mathbf{b}_1 \cdot \mathbf{a}_2 = \frac{\mathbf{a}_2 \cdot (\mathbf{a}_2 \times \mathbf{a}_3)}{\mathbf{a}_1 \cdot (\mathbf{a}_2 \times \mathbf{a}_3)} = 0.$$

The result $\mathbf{b}_1 \cdot \mathbf{a}_2 = 0$ follows because $\mathbf{a}_2$ is perpendicular to $(\mathbf{a}_2 \times \mathbf{a}_3)$ by the properties of the cross product. Generalizing,

$$\mathbf{b}_i \cdot \mathbf{a}_j = \delta_{ij}. \tag{4.6}$$

4.5.1 Miller Indices

Consider a translation $\boldsymbol{\tau} = n_1\mathbf{a}_1 + n_2\mathbf{a}_2 + n_3\mathbf{a}_3$ in physical space. A translation $\boldsymbol{\xi}$ in the corresponding reciprocal space is $\boldsymbol{\xi} = m_1\mathbf{b}_1 + m_2\mathbf{b}_2 + m_3\mathbf{b}_3$ where the n_i and m_i are all integers. Using Eq. (4.6),

$$\boldsymbol{\xi} \cdot \boldsymbol{\tau} = m_1n_1 + m_2n_2 + m_3n_3$$
$$= \text{an integer } N.$$

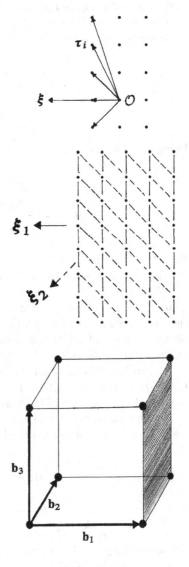

According to this result, the projection of $\boldsymbol{\tau}$ onto $\boldsymbol{\xi}$ is a constant. There are many ways for this relation to be satisfied, as indicated in the sketch. Each of the translations $\boldsymbol{\tau}_i$ from the origin $\mathcal{O}$ in the physical lattice to sites on a given plane has equal projections onto the reciprocal-space translation $\boldsymbol{\xi}$.

It is plausible from the sketch that translation $\boldsymbol{\xi}$ in reciprocal space is normal to planes of the physical lattice. The 2-dimensional sketch shows two possible crystal planes (solid and dashed) in a cubic lattice with the reciprocal lattice vectors $\boldsymbol{\xi}_1$, $\boldsymbol{\xi}_2$ normal to each plane.

Crystal planes in a lattice are classified in terms of a set of three indices called *Miller indices*. From the standpoint of group theory, the Miller index consists of the components $(h\,k\,\ell)$ of a vector that is normal to a crystal plane of the reciprocal lattice, reduced to lowest terms.

To illustrate, consider a simple cubic lattice of side a. The basis vectors of its reciprocal lattice are $\mathbf{b}_1 = \frac{1}{a}\hat{\mathbf{i}}$, $\mathbf{b}_2 = \frac{1}{a}\hat{\mathbf{j}}$, $\mathbf{b}_3 = \frac{1}{a}\hat{\mathbf{k}}$ as found in Section 4.5. In the sketch, the normal to the shaded right-hand crystal plane is $\frac{1}{a}[(1)\mathbf{b}_1 + (0)\mathbf{b}_2 + (0)\mathbf{b}_3]$, so for the shaded face its Miller indices in lowest terms is 100.

For a simple cubic crystal of side a, the spacing d between adjacent crystal planes is

$$d = \frac{a}{\sqrt{h^2 + k^2 + \ell^2}}.$$

Only two crystal planes contribute in this case, so $a = d$. Table 4.1 for an fcc crystal lists d/a for a few Miller indices.

Table 4.1 Indices and d/a for fcc crystal (n = 1)

index	d/a
100	$\frac{1}{2\sqrt{1}}$
110	$\frac{1}{2\sqrt{2}}$
111	$\frac{1}{\sqrt{3}}$
210	$\frac{1}{2\sqrt{5}}$
311	$\frac{1}{\sqrt{11}}$

Powder XRD

In the years 1915–17, Albert Hull (1880–1966) in the United States and researchers in Germany independently demonstrated XRD with powder samples. Powder samples are easy to fabricate and only a small sample is needed, both of which are advantages over single crystal samples.

Interference patterns are achieved with powder XRD because even a small grain has a large representative lattice structure. Bragg's law can be satisfied by many different portions of the powder that each have grains oriented the same way. Thus, a single powder XRD experiment can generate more data than from a single crystal in fixed orientation.

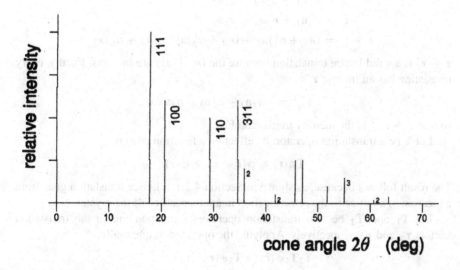

Credit: Data for the figure are taken from Table VlI, A. Hull Phys. Hev. 10 661 (1917). Copyright by the American Physical Society 1917.

The figure shows the measured positions and estimated intensities from Hull's experiment with powdered aluminum metal. Most of the interferences are first order, but several (labeled 2 or 3) are higher order. The first few are labeled with Hull's assignment of Miller indices.

By applying Bragg's law for various Bravais lattices to his data, Hull got the best fit by taking aluminum to have fcc structure. Consider, for example, the 100 case (Bragg angle $\theta = 10.3°$). Hull's X-ray tube used a molybdenum target ($\lambda = 0.0714$ nm), so Bragg's law gives

$$d = \frac{\lambda}{2 \sin \theta} = \frac{0.0714}{0.358}$$
$$= 0.200 \text{ nm.}$$

From Table 4.1, $a = 2d = 0.400$ nm at room temperature, in good agreement with the modern value, 0.405 nm.

4.6   Lattice Translation Group

The translations τ of a given crystal lattice form a group with elements

$$\tau = n_1 \mathbf{a}_1 + n_2 \mathbf{a}_2 + n_3 \mathbf{a}_3,$$

where the n_i are integers. The combination rule for the translation group elements is addition.

First, there is an identity translation $\mathbf{E}$ defined by $n_1 = n_2 = n_3 = 0$. Second, the sum of two translations is also a translation.

$$\tau = n_1 \mathbf{a}_1 + n_2 \mathbf{a}_2 + n_3 \mathbf{a}_3$$
$$\tau' = n_1' \mathbf{a}_1 + n_2' \mathbf{a}_2 + n_3' \mathbf{a}_3$$
$$\tau + \tau' = (n_1 + n_1') \mathbf{a}_1 + (n_2 + n_2') \mathbf{a}_2 + (n_3 + n_3') \mathbf{a}_3$$

$\tau + \tau'$ is a valid lattice translation because the $(n_i + n_i')$ are integers. Finally, every translation has an inverse τ^{-1},

$$\tau^{-1} = -n_1 \mathbf{a}_1 - n_2 \mathbf{a}_2 - n_3 \mathbf{a}_3,$$

so that $\tau + \tau^{-1}$ is the identity translation $\mathbf{E}$.

Let $\mathbf{T}$ be a translation operator. Its effect on a function $\phi(\mathbf{r})$ is

$$\mathbf{T}\phi(\mathbf{r}) = \phi(\mathbf{r} + \tau) = \phi(\mathbf{r}).$$

The result follows because, as shown in Section 4.2.1, a lattice translation goes from a location in a lattice to the same location in another lattice of the crystal.

Let $\mathbf{T}_1$ and $\mathbf{T}_2$ be two translation operators corresponding to the translation vectors τ_1 and τ_2, respectively. Applying the operators sequentially,

$$\mathbf{T}_2 \mathbf{T}_1 \phi(\mathbf{r}) = \mathbf{T}_2 \phi(\mathbf{r} + \tau_1)$$
$$= \phi(\mathbf{r} + \tau_1 + \tau_2),$$

and in the reverse order,

$$\mathbf{T}_1\mathbf{T}_2\phi(\mathbf{r}) = \mathbf{T}_1\phi(\mathbf{r} + \boldsymbol{\tau}_2)$$
$$= \phi(\mathbf{r} + \boldsymbol{\tau}_2 + \boldsymbol{\tau}_1).$$

By the rule for vector addition, $\boldsymbol{\tau}_1 + \boldsymbol{\tau}_2 = \boldsymbol{\tau}_2 + \boldsymbol{\tau}_1$ so that $\mathbf{T}_2\mathbf{T}_1 = \mathbf{T}_1\mathbf{T}_2$. Translation operators commute, and the group of crystal translations is therefore an Abelian group. Theorem 9 in Section 2.7.4 states that the irreducible representations of an Abelian group are all 1-dimensional. A second result from Theorem 9 is that every member of an Abelian group is a class by itself. The number of irreducible representations of a group is equal to the number of its classes, so there is an irreducible representation for every possible translation $\boldsymbol{\tau}$. This is a very large number for even a small crystal sample. If the crystal sample has length ℓ (perhaps a few cm) along a lattice axis with length a (typically a fraction of a nm), the number of irreducible representations $n_0 \approx \frac{\ell}{a}$ is in the tens of millions. The translation group can be viewed as a continuous group.

Because the reciprocal lattice is just another lattice, its properties are analogous to the direct lattice. For instance, it has a translation group. A valid translation $\boldsymbol{\rho}$ in the reciprocal lattice is

$$\boldsymbol{\rho} = n_1\mathbf{b}_1 + n_2\mathbf{b}_2 + n_3\mathbf{b}_3,$$

with integers n_i.

Generalizing from Eq. (4.6), translation $\boldsymbol{\tau}$ in the direct lattice is related to translation $\boldsymbol{\rho}$ in the reciprocal lattice by

$$\boldsymbol{\rho} \cdot \boldsymbol{\tau} = 1. \tag{4.7}$$

4.6.1 Bloch's Theorem

The crystal lattice translation group is a *cyclic group* $\{\mathbf{T}_1, \mathbf{T}_2 = \mathbf{T}_1^2, \ldots, \mathbf{T}_{n_0} = \mathbf{T}_1^{n_0}\}$ where the group members are sequential powers of one group member $\mathbf{T}_1$. Proof is left to the problems.

With repeated translations, the maximum translation $\mathbf{T}_{n_0}$ reaches a surface of the crystal where no further translations are possible. To avoid the problem of dealing with surfaces, a common approach is to imagine the crystal slab bent into a cylinder so that the left-hand and right-hand surfaces coincide: $\mathbf{T}_{n_0} = \mathbf{T}_0$, so $\mathbf{T}_1^{n_0} = \mathbf{E}$. Another approach is to imagine the right-hand surface butted up against the left-hand surface of an identical crystal slab and so on ad infinitum. Both methods are examples of *periodic boundary conditions* in the way the structure is made to repeat.

Because the irreducible representations are all 1-dimensional,

$$D(\mathbf{E}) = (1)$$
$$D(\mathbf{T}_1^{n_0}) = (1)$$
$$(D(\mathbf{T}_1))^{n_0} = (1).$$

To simplify the notation, let $\xi \equiv D(\mathbf{T}_1)$ so that the condition becomes $\xi^{n_0} = 1$. The solution is $\xi = e^{i2\pi m}$, as follows from de Moivre's theorem: $e^{i2\pi m} = \cos(2\pi m) + i \sin(2\pi m) = 1 + 0 = 1$, where m is an integer. In standard notation, $k \equiv 2\pi m/a$, so $D(\mathbf{T}_1) = e^{ika}$.

Generalizing to three dimensions,

$$\phi(\mathbf{r}) = e^{i\mathbf{k}\cdot\mathbf{r}} u_k(\mathbf{r}), \tag{4.8}$$

where the function $u_k(\mathbf{r})$ obeys translation symmetry $u_k(\mathbf{r}+\boldsymbol{\tau}) = u_k(\mathbf{r})$. Equation (4.7) is known as *Bloch's theorem*. In Bloch's theorem, $\mathbf{k}$, the *wave vector*, serves to label the irreducible representations of the translation group. As pointed out, the translation group of a crystal has a very large number of members. In principle it is a discrete group, but the members are so closely spaced that it can be taken to be a continuous group labeled as a continuous function of $\mathbf{k}$.

Equation (4.8) has a quantum-mechanical interpretation. Consider a plane wave $\psi(x, t)$ traveling along the x-axis:

$$\psi(x,t) \propto e^{i(kx-\omega t)},$$

where $k = 2\pi/\lambda$. This is a *plane* wave because it does not depend on the transverse coordinates y, z. More generally, a plane wave moving along the direction of a vector $\mathbf{k}$ is

$$\psi(\mathbf{r}, t) \propto e^{i(\mathbf{k}\cdot\mathbf{r}-\omega t)}.$$

The space part, $e^{i\mathbf{k}\cdot\mathbf{r}}$, has the same form as the first factor in Eq. (4.8). In quantum mechanics it is a solution of the time-independent Schrödinger equation and represents the de Broglie matter wave of an electron moving freely in the absence of forces.

Let $\mathbf{r} \mapsto \mathbf{r}+\boldsymbol{\tau}$ in Eq. (4.7), where $\boldsymbol{\tau}$ is a valid translation. Then:

$$\phi(\mathbf{r}+\boldsymbol{\tau}) = e^{i\mathbf{k}\cdot(\mathbf{r}+\boldsymbol{\tau})} u_k(\mathbf{r}).$$

By the periodicity of the crystal lattice, this becomes

$$\phi(\mathbf{r}+\boldsymbol{\tau}) = e^{i\mathbf{k}\cdot\boldsymbol{\tau}} e^{i\mathbf{k}\cdot\mathbf{r}} u_k(\mathbf{r}+\boldsymbol{\tau})$$

$$\phi(\mathbf{r}+\boldsymbol{\tau}) = e^{i\mathbf{k}\cdot\boldsymbol{\tau}} \phi(\mathbf{r}).$$

The function $u_k(\mathbf{r})$ is invariant under translation in a crystal lattice, so Eq. (4.8) can be interpreted as the quantum-mechanical wave function of an electron moving in the periodic potential of a crystal lattice, with only a *phase factor* $e^{i\mathbf{k}\cdot\boldsymbol{\tau}}$ from one lattice point to an equivalent point.

4.6.2   Quantum Mechanics of Crystals

Consider a mass m in a periodic 1-dimensional potential $V(x)$ that repeats every a units so that $V(x + a) = V(x)$. The 1-dimensional Schrödinger equation with eigenfunction $\psi(x)$ is

$$-\frac{\hbar^2}{2m}\frac{d^2\psi}{dx^2} + V(x)\psi = E\psi. \tag{4.9}$$

Expand V in a Fourier series in terms of reciprocal lattice vectors b and expand the eigenfunctions ψ in terms of k.

$$V(x) = \sum_b V_b e^{ibx}$$

$$\psi(x) = \sum_k c_k e^{ikx}$$

Substituting these expressions in Eq. (4.9) gives

$$\frac{\hbar^2}{2m} \sum_k k^2 c_k e^{ikx} + \sum_b \sum_k V_b c_k e^{i(b+k)x} = E_k \sum_k c_k e^{ikx}.$$

In the second term on the left, take the values $k \to k - b$ from the sum over k to make all terms proportional to e^{ikx}.

$$\left(\frac{\hbar^2}{2m} k^2 c_k - E_k c_k + \sum_b V_b c_{(k-b)} \right) e^{ikx} = 0$$

The coefficient of e^{ikx} must equal 0.

$$\left(\frac{\hbar^2 k^2}{2m} - E_k \right) c_k + \sum_b V_b c_{(k-b)} = 0 \tag{4.10}$$

Equation (4.10) involves an infinite sum $b = 0, \pm b, \pm 2b, \ldots$, but in calculations the coefficients $c_{(k-b)}$ tend to drop off rapidly with increasing $|b|$. The expansion for ψ can be written

$$\psi_k = \sum_b c_{(k-b)} e^{i(k-b)x}$$

$$= e^{ikx} \left(\sum_b c_{(k-b)} e^{-ibx} \right),$$

which is a statement of Bloch's theorem, with

$$u_k(x) = \sum_b c_{(k-b)} e^{-ibx}.$$

u_k is periodic, as required:

$$u_k(x + a) = \sum_b c_{(k-b)} e^{-ib(x+a)}$$

$$= \sum_b c_{(k-b)} e^{-ibx} e^{-iba}$$

$$= \sum_b c_{(k-b)} e^{-ibx}$$

$$= u_k,$$

because the 1-dimensional form of $\rho \cdot \tau = 1$ is $ba = 1$.

4.7 Crystallographic Point Groups and Rotation Symmetry

Rotation symmetry and inversion symmetry operations of a lattice also form a group. Such groups are called *crystallographic point groups* if the rotation axes all pass through the same point, leaving one point in the lattice invariant. There are 32 crystallographic point groups.

Point groups also apply to symmetries of molecules. Section 3.4 showed that the symmetry operations of a water molecule form a group called **mm2** or C_{2v}. This is a point group because all the group operations leave the position of the oxygen atom unchanged.

A rotation symmetry operation of a point group can be either a physical *proper rotation* about an axis or an *improper rotation* – a proper rotation followed by either a mirror operation (Hermann–Mauguin) or an inversion operation: $x, y, z \mapsto -x, -y, -z$ (Schönflies). The two systems use different names for point groups, as we saw for the water molecule, but both give the same physical results.

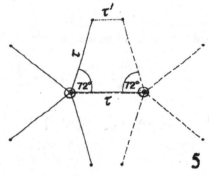

A 2-dimensional lattice model shows that the only proper rotations possible for the point group of a lattice are 1-fold, 2-fold, 3-fold, 4-fold, and 6-fold, where an *n-fold* rotation is rotation by $\frac{360°}{n}$ or equivalently by $\frac{2\pi}{n}$ radians.

The rotation axes in the sketches are perpendicular to the page and pass through the circled lattice points. The translation between lattice points along the line between the circled points is τ.

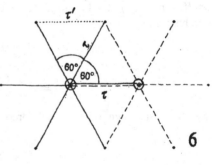

In the sketch for 5-fold rotation $\tau' = \tau - 2\tau \cos{(72°)}$. Valid translation symmetry requires that, on a given axis, the distance between lattice points must be $n\tau$ where n is some integer $n = 0, \pm 1, \pm 2, \ldots$. This condition cannot be satisfied for the proposed 5-fold rotation, hence a perfect crystal cannot have 5-fold rotation symmetry (although a molecule might).

The sketch shows a 6-fold rotation. Here, $\tau' = \tau$ satisfies the requirement of translation symmetry, so 6-fold is an allowed rotation symmetry.

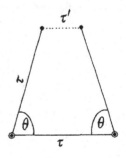

Table 4.2 All possible rotation symmetries of a lattice

n	0	-1	+1	+2	+3
θ (deg)	60°	0°	90°	120°	180°
θ (rad)	$\pi/3$	0	$\pi/2$	$2\pi/3$	π
symmetry	6-fold	1-fold	4-fold	3-fold	2-fold

All possible rotation symmetries of a lattice can be found with the help of the sketch, using the condition $\tau' = n\,\tau$.

$$\tau' = (1 - 2\cos\theta)\tau = n\tau$$

$$1 - 2\cos\theta = n$$

Table 4.2 lists all real solutions for n = integer. In summary, crystal point groups cannot have a 5-fold proper rotation or a proper rotation greater than 6-fold.

Here is an alternate proof using group representation theory. The matrix for rotation by θ about a z-axis perpendicular to the page is

$$\begin{pmatrix} \cos\theta & -\sin\theta & 0 \\ \sin\theta & \cos\theta & 0 \\ 0 & 0 & 1 \end{pmatrix}.$$

Consider the product of two successive rotations, first by θ_1 then by θ_2:

$$\begin{pmatrix} \cos\theta_2 & -\sin\theta_2 & 0 \\ \sin\theta_2 & \cos\theta_2 & 0 \\ 0 & 0 & 1 \end{pmatrix} \begin{pmatrix} \cos\theta_1 & -\sin\theta_1 & 0 \\ \sin\theta_1 & \cos\theta_1 & 0 \\ 0 & 0 & 1 \end{pmatrix} = \begin{pmatrix} \cos\theta_2\cos\theta_1 - \sin\theta_2\sin\theta_1 & -\cos\theta_2\sin\theta_1 - \sin\theta_2\cos\theta_1 & 0 \\ \sin\theta_2\cos\theta_1 + \cos\theta_2\sin\theta_1 & -\sin\theta_2\sin\theta_1 + \cos\theta_2\cos\theta_1 & 0 \\ 0 & 0 & 1 \end{pmatrix}$$

$$= \begin{pmatrix} \cos(\theta_1+\theta_2) & -\sin(\theta_1+\theta_2) & 0 \\ \sin(\theta_1+\theta_2) & \cos(\theta_1+\theta_2) & 0 \\ 0 & 0 & 1 \end{pmatrix}.$$

Using $\theta_1 + \theta_2 = \theta_2 + \theta_1$ shows that rotations about the z-axis commute and consequently form an Abelian group. The irreducible representations are 1-dimensional, so the reduced matrix has entries only on the main diagonal.

$$\begin{pmatrix} m_1 & 0 & 0 \\ 0 & m_2 & 0 \\ 0 & 0 & m_3 \end{pmatrix}$$

The m_i are all integers because the location of every lattice point is specified by integer multiples of the lattice basis vectors. The character of the reduced matrix is therefore $\sum_i m_i = n'$ = an integer. The character $1 + 2\cos\theta$ of the original rotation matrix is unchanged by the similarity transformation that reduces it, giving the result:

$$1 + 2\cos\theta = n'.$$

This relation leads to the same values for allowed rotation symmetries of a lattice as listed in Table 4.2.

4.7.1   Plane Diagrams of Group Symmetry Operations

There are two principal methods for depicting a group's symmetry operations in a plane diagram. One method uses characteristic symbols for the axes of allowed rotations:

2-fold: ❙   3-fold: ▲   4-fold: ◆   6-fold: ⬤

The sketch shows the *diagram of symmetry elements* for the **32** group. The triangle ▲ at the center symbolizes 3-fold rotation about the axis perpendicular to the paper. Each of the three flip axes is terminated with the symbol ❙ for the 180° 2-fold flips.

The other type of diagram is called a *stereogram*. It is based on the principle of *stereographic projection* illustrated in the sketch. Consider a point **P** on the surface of a sphere. A line drawn from **P** to the South pole **S** establishes the projected point **P′** where the line intersects the equatorial plane. If **P** is in the southern hemisphere, the line is drawn from the North pole.

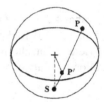

The sketch is the stereogram for the **32** group. To see how it is constructed, consider circular markers on a sphere (filled in the northern hemisphere, open in the southern). First apply the group's rotations about the polar axis. A circle on the sphere's surface projects to a circle in the stereogram, so the placement of the three filled circles illustrates the rotations by 0°, 120°, 240°. Next consider flips about the linear axes. Circles in the northern hemisphere are flipped to circles in the southern, denoting 180° flips.

4.8   Crystallographic Space Groups and the Seitz Operator

The group consisting of point group operations and translation operations is called the *space group* of the lattice. Crystallographers have identified 230 space groups and have devised compact notations to label them. However, only about half a dozen space groups characterize the majority of crystal lattices, mainly because nature prefers close packing. Details can be found in the literature, but here only some group theoretical aspects are discussed.

The dual notation $\{\mathbf{R}|\mathbf{T}\}$, called the *Seitz operator*, is commonly used to describe space groups. Here **T** is a lattice translation operator and **R** represents other symmetries such as proper rotations. Special cases:

pure translation: $\{\mathbf{E}|\mathbf{T}\}$
pure rotation: $\{\mathbf{R}|\mathbf{0}\}$

Given a rotation operator $\mathbf{R}$ and a translation $\boldsymbol{\tau}$, the Seitz operator acts on a vector $\mathbf{r}$ to yield a vector $\mathbf{r}'$.

$$\mathbf{r}' = \mathbf{R}\mathbf{r} + \boldsymbol{\tau}$$

Applying this result twice gives the rule for the product of Seitz operators.

$$\begin{aligned}
\{\mathbf{R}'|\mathbf{T}'\}\{\mathbf{R}|\mathbf{T}\}\mathbf{r} &= \{\mathbf{R}'|\mathbf{T}'\}(\mathbf{R}\mathbf{r} + \boldsymbol{\tau}) \\
&= \mathbf{R}'(\mathbf{R}\mathbf{r} + \boldsymbol{\tau}) + \boldsymbol{\tau}' \\
&= \mathbf{R}'\mathbf{R}\mathbf{r} + \mathbf{R}'\boldsymbol{\tau} + \boldsymbol{\tau}'
\end{aligned} \tag{4.11}$$

This result can be written as the Seitz operator:

$$\{\mathbf{R}'\mathbf{R}|\mathbf{R}'\boldsymbol{\tau} + \boldsymbol{\tau}'\}.$$

Interchanging the unprimed and primed operators gives

$$\begin{aligned}
\{\mathbf{R}|\mathbf{T}\}\{\mathbf{R}'|\mathbf{T}'\}\mathbf{r} &= \{\mathbf{R}|\mathbf{T}\}(\mathbf{R}'\mathbf{r} + \boldsymbol{\tau}') \\
&= \mathbf{R}\mathbf{R}'\mathbf{r} + \mathbf{R}\boldsymbol{\tau}' + \boldsymbol{\tau}.
\end{aligned} \tag{4.12}$$

Equations (4.11) and (4.12) are not equal. Seitz operators do not commute, because rotations and translations do not, in general, commute.

The translation $\boldsymbol{\tau} = \tau_x \hat{\mathbf{i}} + \tau_y \hat{\mathbf{j}} + \tau_z \hat{\mathbf{k}}$ of a point (x, y, z) can be expressed in matrix form as

$$\begin{pmatrix} 1 & 0 & 0 & \tau_x \\ 0 & 1 & 0 & \tau_y \\ 0 & 0 & 1 & \tau_z \\ 0 & 0 & 0 & 1 \end{pmatrix} \begin{pmatrix} x \\ y \\ z \\ 1 \end{pmatrix} = \begin{pmatrix} x + \tau_x \\ y + \tau_y \\ z + \tau_z \\ 1 \end{pmatrix}.$$

Note that a 4×4 matrix with a dummy last line is needed to describe a translation in three dimensions.

To see how rotations and translations are combined, consider a rotation by θ about the z-axis followed by a translation $\boldsymbol{\tau}$. The rotation matrix must be enlarged to 4×4 for compatibility with the translation matrix.

$$\begin{pmatrix} 1 & 0 & 0 & \tau_x \\ 0 & 1 & 0 & \tau_y \\ 0 & 0 & 1 & \tau_z \\ 0 & 0 & 0 & 1 \end{pmatrix} \begin{pmatrix} \cos\theta & -\sin\theta & 0 & 0 \\ \sin\theta & \cos\theta & 0 & 0 \\ 0 & 0 & 1 & 0 \\ 0 & 0 & 0 & 1 \end{pmatrix} \begin{pmatrix} x \\ y \\ z \\ 1 \end{pmatrix} = \begin{pmatrix} \cos\theta & -\sin\theta & 0 & \tau_x \\ \sin\theta & \cos\theta & 0 & \tau_y \\ 0 & 0 & 1 & \tau_z \\ 0 & 0 & 0 & 1 \end{pmatrix} \begin{pmatrix} x \\ y \\ z \\ 1 \end{pmatrix}$$

4.9 Crystal Symmetry Operations

Chapter 3 showed how to use group theory to determine the vibration modes of molecules and their corresponding optical activity. The principles are the same for a crystal lattice. As a refresher, here is a summary of the steps.

1) Determine the group of the lattice.

2) Look up the character table for the irreducible representations of the group.

3) Irreducible representations that transform like an electric dipole moment (x, y, z) are IR active. Irreducible representations that transform, like the quadratic forms of the polarizability tensor (x^2, xy, yz, ...), are Raman active. Luckily, mathematicians have done the necessary work, using projection operators so that published character tables list basis functions for the irreducible representations.

4.9.1 Notation

Considering only proper rotations (no inversions), the operation C_n (Schönflies notation) represents rotation by $2\pi/n$. The cubic family has the following 24 symmetry operations:

1: the identity; make no changes (1 member)

C_2: rotation by π (180°) about an axis from the midpoint of one edge to the midpoint of the opposite edge (6 members)

C_3: rotation by $2\pi/3$ (120°) about an axis from one vertex to the far opposite vertex (8 members)

C_4: rotation by $\pi/2$ (90°) or by $3\pi/2$ (270°) about an axis from the center of one face to the center of the opposite face (6 members)

C_4^2: rotation by π (180°) about an axis from the center of one face to the center of the opposite face (3 members)

A further word about notation: C_n^m is a rotation by $2\pi m/n$, so it would appear that the symmetry operation C_4^2 should be included with C_2. However, C_4^2 is a class of its own and therefore corresponds to a separate, distinct, irreducible representation. Recall that the number of a group's irreducible representations is equal to the number of its classes.

The sketch shows each of the specified axes. Note that every axis passes through one point of the cube (the center) so that the center remains stationary under all rotations, as expected for a crystallographic point group. Put another way, there is a center of inversion so that the cube structure is invariant under inversion. To help visualize the rotations, play with a physical cube or check the Internet for graphics (some animated) that show symmetries of the cube.

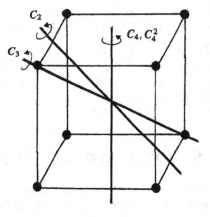

Table 4.3 Character table for **O**

basis functions		**1**	**6C$_2$**	**8C$_3$**	**6C$_4$**	**3C$_4^2$**
r^2	A_1	1	1	1	1	1
	A_2	-1	1	1	-1	1
$x^2 - y^2, 3z^2 - r^2$	E	2	0	0	-1	2
x, y, z	T_1	3	-1	0	1	-1
xy, yz, zx	T_2	3	1	0	-1	-1

4.9.2 Cubic Lattice Character Table

As an illustration, consider the point group for all three members of the cubic family (simple cubic, bcc, fcc). Considering only proper rotations, the group has 24 members and is called **432** (Hermann–Mauguin) or **O** (Schönflies). Table 4.3 shows the character table for **O** using Schönflies notation for the operations.

In character tables that use Schönflies notation, the number in front of the operation symbol is the number of such operations. The operation symbol itself specifies the nature of the operation. Thus, there are six 2-fold operations **C$_2$** and a total of 24 group elements.

There are five classes, hence five irreducible representations: two 1-dimensional A_1, A_2; one 2-dimensional E; and two 3-dimensional T_1, T_2. The symbol **1** is used for the identity operation to avoid confusion with representation E. Note that Theorem 8 (Section 2.7.4) is satisfied:

$$24 = (1)^2 + (1)^2 + (2)^2 + (3)^2 + (3)^2 \quad \checkmark$$

Improper rotations are also symmetry operations of the cubic family. An improper rotation is a proper rotation followed by an inversion $\{x, y, z\} \mapsto \{-x, -y, -z\}$. The **O** group of 24 members is expanded to a group of 48 members by applying an inversion to each element of the **O** group. A general approach to expansion is to form the direct product of the original group with the 2-dimensional inversion group. The inversion operation is symbolized $\bar{1}$ (Hermann–Mauguin) and **i** (Schönflies). The expanded group is called **m3m** (Hermann–Mauguin) or **O$_h$** (Schönflies).

$$\bar{1} \text{ or } \mathbf{i} = \begin{pmatrix} 1 & 1 \\ 1 & -1 \end{pmatrix}$$

$$\mathbf{m3m} = \mathbf{432} \otimes \bar{1}$$

$$\mathbf{O}_h = \mathbf{O} \otimes \mathbf{i}$$

Additional information can be gleaned from the character table. If an inversion operation does not change the sign of the basis function for an irreducible representation, the irreducible representation is termed *even* (German *gerade*) and is denoted by the subscript g. If it changes sign, it is *odd* (German *ungerade*) and has subscript u.

Table 4.4 Segment of character table $\mathbf{O}_h$

basis functions		**1**	**6C$_2$**	...
r^2	A_{1+}	1	1	...
	A_{2+}	-1	1	...
$x^2 - y^2, 3z^2 - r^2$	E_+	2	0	...
	T_{1+}	3	-1	...
xy, yz, zx	T_{2+}	3	1	...

Table 4.5 Segment of character table $\mathbf{O}_h$

basis functions		**1**	**6C$_2$**	...
	A_{1-}	1	1	...
	A_{2-}	-1	1	...
	E_-	2	0	...
x, y, z	T_{1-}	3	-1	...
	T_{2-}	3	1	...

In modern work the subscript g is replaced by $+$ or $(+)$ for *positive parity* and u by $-$ or $(-)$ for *negative parity*. Parity corresponds to the eigenvalues ± 1 of the inversion operator.

As an example, T_{2+} is *even* with subscript $+$ because xy does not change sign upon inversion. These segments of the character table for $\mathbf{O}_h$ indicate that A_{1+}, E_+, and T_{2+} are Raman active (Table 4.4) and T_{1-} is IR active (Table 4.5).

4.10   Lattice Vibrations

The vibrations of atoms in a crystal are involved in a wide range of important properties, such as specific heat, sound waves, and transport properties such as heat conduction.

4.10.1   Diatomic Linear Chain and Dispersion

The discussion of molecular vibrations in Chapter 3 introduced a Newtonian mechanical model consisting of masses and springs, and showed that a small molecule has only a relatively small number of fundamental vibration modes (the normal modes of the system). A masses-and-springs model for a crystal lattice becomes swamped by the complexity of the large 3-dimensional structure. For pedagogical purposes we use instead a 1-dimensional model called the *diatomic linear chain*. The diatomic chain model is treated in many texts to introduce the subject of lattice vibrations because useful solutions are readily found and because it gives qualitative understanding of phenomena observed experimentally.

As shown in the sketch, the model consists of two possibly different masses m and M arranged alternately in a very long 1-dimensional chain of springs. The springs are all taken to have the same spring constant C because interatomic forces depend almost entirely on electronic structure independent of nuclear mass. Let a be the length of the smallest unit that repeats in constituting the chain. Thus, the distance between a mass m and a neighboring mass M is $\frac{a}{2}$.

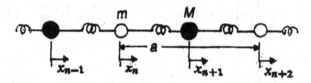

Each mass is acted on only by the spring forces to its left and to its right. By Newton's second law, the equations of motion for mass m (coordinate x_n) and mass M (coordinate x_{n+1}) are therefore

$$m\frac{d^2x_n}{dt^2} = -C(x_n - x_{n+1}) - C(x_n - x_{n-1})$$

$$= -C(2x_n - x_{n-1} - x_{n+1}) \qquad (4.13)$$

$$M\frac{d^2x_{n+1}}{dt^2} = -C(-x_{n+2} + x_{n+1}) - C(x_{n+1} - x_n)$$

$$= -C(2x_{n+1} - x_n - x_{n+2}). \qquad (4.14)$$

Now look for wavelike solutions in the form $Ae^{i(\frac{2\pi x}{\lambda} - \omega t)}$. As the sketch indicates, consider mass m located at coordinate $x_n = na$ so that mass M next up the chain is located at $x_{n+1} = (n + \frac{1}{2})a$. Let $q = \frac{2\pi}{\lambda}$ be the wave vector. The solutions are then

$$x_{n-1} = A_{n-1}\, e^{i(q(n-\frac{1}{2})a - \omega t)}$$

$$x_n = A_n\, e^{i(qna - \omega t)}$$

$$x_{n+1} = A_{n+1}\, e^{i(q(n+\frac{1}{2})a - \omega t)}$$

$$x_{n+2} = A_{n+2}\, e^{i(q(n+1)a - \omega t)}.$$

The amplitude must be the same at every mass m in the chain so that $A_{n+2} = A_n$, and similarly the amplitude is the same at every mass M so that $A_{n-1} = A_{n+1}$. Substituting in Eqs. (4.13) and (4.14) using these conditions and cancelling the common factor $e^{i(qna - \omega t)}$ gives the result in matrix form:

$$\begin{pmatrix} m\omega^2 - 2C & 2C\cos\left(\frac{qa}{2}\right) \\ 2C\cos\left(\frac{qa}{2}\right) & M\omega^2 - 2C \end{pmatrix} \begin{pmatrix} A_n \\ A_{n+1} \end{pmatrix} = \begin{pmatrix} 0 \\ 0 \end{pmatrix}.$$

These equations have the useless solution $A_n = A_{n+1} = 0$, so to obtain a nontrivial result set the determinant of coefficients equal to zero, the same approach used in Section 3.2.2 to find the normal modes of a system of two masses and three springs. The condition here becomes a biquadratic equation in ω^2:

$$mM\omega^4 - 2C(m + M)\omega^2 + 4C^2 \sin^2\left(\tfrac{qa}{2}\right) = 0.$$

Using the trigonometric identity $\sin^2(\eta) = \tfrac{1}{2}(1 - \cos(2\eta))$, the two roots are

optical branch:

$$\omega_+^2 = \left(\frac{C}{mM}\right)\left(m + M + \sqrt{m^2 + M^2 + 2mM\cos(qa)}\right) \qquad (4.15)$$

acoustic branch:

$$\omega_-^2 = \left(\frac{C}{mM}\right)\left(m + M - \sqrt{m^2 + M^2 + 2mM\cos(qa)}\right). \qquad (4.16)$$

Taking the case $M = 3m$ as a numerical example, the figure shows ω versus q for the two roots. The roots are periodic in the inverse wavelength q, so q in the range $\pm\left(\frac{\pi}{a}\right)$ exemplifies the behavior that is found anywhere along the infinite chain. The range $\pm\left(\frac{\pi}{a}\right)$ is called the *first Brillouin zone*.

Unlike the small number of discrete vibrations typically found for molecular vibrations, the angular frequency ω for lattice vibrations is a continuous function of the inverse wavelength q. In addition, the diatomic chain exhibits two separate continuous functions called the *optical branch* and the *acoustic branch*.

To understand the names of the branches, consider the ratio of the vibration amplitude of mass M compared to the vibration amplitude of its neighboring mass m. Taking $q = 0$ for simplicity, $A_{n+1}/A_n = -m/M$ for the optical branch; adjacent masses move opposite to one another. In an ionic crystal, this motion could give rise to a time-varying electric dipole moment,

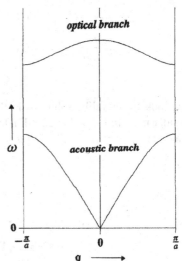

making the vibration optically active. For the acoustic branch at $q = 0$, $A_{n+1}/A_n = 1$; the masses move together in phase, characteristic of a sound wave propagating along the chain.

Some have called the chain model a "toy" because it is only 1-dimensional and does not include transverse oscillations that occur in a real 3-dimensional crystal. Nevertheless, the chain model accounts qualitatively for some observed phenomena.

The figure shows optical and acoustic branches for half the Brillouin zone in crystalline silicon as measured by neutron spectroscopy, to be described later. The experimental data exhibit four branches: longitudinal optical (LO) and longitudinal acoustic (LA), and transverse optical (TO) and transverse acoustic (TA). The longitudinal branches follow the behavior predicted by the linear chain model, which does not include transverse modes. Note the absence of a gap between the LA and LO modes at $q = q_{max}$, which is predicted by the chain model for $M = m$. The optical branches in silicon are not optically active, because there cannot be a time-varying electric dipole moment in a monatomic crystal.

As shown by Taylor's series expansion of Eq. (4.16), ω for the acoustic branch is to a good approximation linearly proportional to q for small q. Let v be the circular frequency of the oscillation:

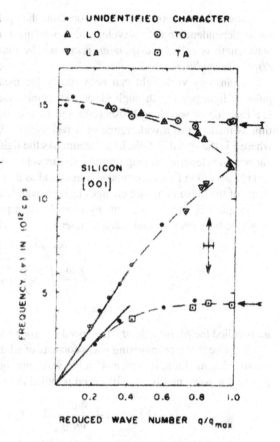

Reprinted figure from B. N. Brockhouse Phys. Rev. Lett. **2**, 256 (1959). Copyright 1959 by the American Physical Society.

$$\omega \propto q$$

$$2\pi v = \mathrm{v}\frac{2\pi}{\lambda}$$

$$\mathrm{v} = v\lambda,$$

so that v for small q is the speed of a propagating wave, identified as a sound wave. The heavy solid lines at the lower left in the data for silicon represent the experimentally measured speed of sound in silicon. The lines are seen to closely approximate the slopes of the acoustic branches for small q. For the linear chain model, the maximum value of ω for the acoustic branch is $\omega \leq \sqrt{2C/M}$ for $M > m$. Sound waves of higher frequency cannot propagate along the chain.

For small q, the acoustic branch $d\omega/dq = \mathrm{v} \approx$ constant. For larger q, $d\omega/dq$ is not constant, as seen from the figures for the chain model and for the silicon experimental data. For $d\omega/dq \neq$ constant, the speed of a sound wave has a non-linear dependence on its wavelength. A nonlinear relation between frequency and wavelength is called a *dispersion relation*. The diatomic chain is an example of a *dispersive medium*.

An analogy with light can help clarify the meaning of dispersion. Consider a pulse of light passing through glass, a dispersive medium whose refractive index n is a function of wavelength. According to Fourier decomposition, a pulse narrow in time is made up of a wide range of wavelengths. The speed of light in glass is c/n, where c is the speed of light in a vacuum. As the light pulse travels through glass, the various wavelengths making up the pulse travel at different speeds so the amplitudes and phases of the Fourier terms no longer add as they did to give the initial pulse. The shape of the pulse therefore changes as it propagates.

To quantify dispersion, start by considering a propagating wave ϕ of amplitude A, single frequency ω, and wave number k.

$$\phi = A e^{i(kx - \omega t)}$$

$$\frac{\omega}{k} = \left(\frac{\omega}{2\pi}\right)\left(\frac{2\pi}{k}\right) = \nu\lambda$$

$$\equiv u_p$$

u_p is called the *phase velocity*, the speed of propagation of a single wavelength.

Suppose that a propagating wave consists of a bundle (*packet*) of wavelengths. In a dispersive medium, ω depends on the wave number. Let the packet be narrow and centered at wave number k_0. Its speed is called the *group velocity u_g*.

$$\omega(k) \approx \omega_0 + \left.\frac{\partial\omega}{\partial k}\right|_{k_0} (k - k_0)$$

$$\phi = A e^{i(kx - \omega_0 t - \frac{\partial\omega}{\partial k}(k - k_0)t)}$$

$$= A e^{i(k_0 x - \omega_0 t)} e^{i((k - k_0)x - \frac{\partial\omega}{\partial k}(k - k_0)t)}$$

so that

$$u_g = \frac{\frac{\partial\omega}{\partial k}(k - k_0)}{(k - k_0)} = \frac{\partial\omega}{\partial k}. \tag{4.17}$$

When ω depends linearly on k, u_g is constant as in the experimental data for silicon.

4.10.2 Visualizing Lattice Vibrations

In Chapter 3 the vibration modes of a water molecule were visualized by finding the basis functions for the chosen irreducible representation. The tool was the projection operator described in Section 3.5.1:

$$\phi^{(\alpha)} = \sum_{\mathbf{T}} \chi^{(\alpha)*}(\mathbf{T})\mathbf{T}\phi, \tag{4.18}$$

where on the right-hand side the sum is over the symmetry operations $\mathbf{T}$ of the group, $\mathbf{T}\phi$ is the action of $\mathbf{T}$ on a trial function ϕ, and $\chi^{(\alpha)*}(\mathbf{T})$ is the character of $\mathbf{T}$ in irreducible representation α. On the left-hand side, the result is a basis function $\phi^{(\alpha)}$ that transforms like irreducible representation α and hence represents a normal mode.

For an isolated molecule in free space, trivial zero-frequency representations for translations and rotations have to be subtracted. This is not necessary for a lattice because lattice sites are fixed in space.

In Chapter 3 the normal modes for molecular vibrations were generated by a Cartesian triad of virtual displacements $\delta x, \delta y, \delta z$ of each atom. In a lattice all sites reachable by a lattice translation are identical, so it is only necessary to consider the displacements of one of the equivalent sites.

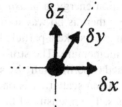

An ionic crystal with a bcc lattice (symmetry group $\mathbf{O}$) has equivalent corner lattice sites. As an example consider its IR active irreducible representation T_{1-}. The table shows the effect of group operations on a trial function $\phi = \delta z$. Because of the simple geometry of a cube, the table entries can be found by inspection or with the help of sketches. For lattices with more complicated geometry, references on the Internet show how to calculate rotations about an arbitrary axis.

$\chi^{(T_{1-})}(\mathbf{C}_3) = 0$ and does not contribute to Eq. (4.15).

Inserting the values from the table into Eq. (4.15) generates the basis function for $\phi^{(T_{1-})} = 8\delta z$. The corner ions are equivalent and therefore all move with the same amplitude.

$\mathbf{T}$	1	$\mathbf{C}_2$	$\mathbf{C}_3$	$\mathbf{C}_4$	$\mathbf{C}_4^2$
$\chi^{(T_{1-})}$	3	-1	0	1	-1
$0°$	δz				
$90°_x$				$-\delta y$	
$270°_x$				δy	
$90°_y$				δx	
$270°_y$				$-\delta x$	
$90°_z$				δz	
$270°_z$				δz	
$180°_x$					$-\delta z$
$180°_y$					$-\delta z$
$180°_z$					δz
$180°$		$-\delta y$			
$180°$		δy			
$180°$		δx			
$180°$		$-\delta x$			
$180°$		$-\delta z$			
$180°$		$-\delta z$			

The sketch shows this vibration mode for T_{1-}, with all the corner ions moving upward. If the lattice is at rest with no external forces acting, the body-centered ion must move downward for conservation of momentum. One half-cycle later, the directions of motion are all reversed. The oscillation of the ions causes a time-varying electric dipole moment as expected for an IR mode.

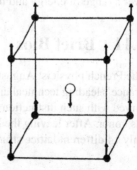

4.10.3   Phonons

When lattice vibrations are treated according to quantum mechanics, the harmonic oscillator is a good model for vibrations of small amplitude. In quantum mechanics, the energy levels E_n of a harmonic oscillator are discrete and evenly spaced according to $E_n = (n + \frac{1}{2})\hbar\omega$, where $n = 0, 1, 2, \ldots$ and $\hbar \equiv \frac{h}{2\pi}$.

The energy increment $h\nu$ in vibration energy is interpreted as a quantum of vibration energy, a *quasiparticle* called a *phonon*. A phonon is not a physical particle – there is no way to make a beam of phonons. However, in lattice vibrations a phonon acts like a particle because it obeys mechanical laws and can transfer energy and momentum. The structure of the branches in lattice vibrations can therefore be studied experimentally by scattering physical particles off phonons.

Raman scattering is one method used to study lattice vibrations, but it has limitations. The intensity of the scattered light is weak, requiring powerful laser sources. A photon has momentum $\frac{h}{\lambda}$, so a photon of long wavelength has small momentum. The change of the photon's momentum as it scatters from a phonon is therefore very small except near the center of the acoustic branch where the phonon also has small momentum $q \approx 0$.

Neutrons are the principal tool for studying the full extent of both the acoustic and optical branches. Neutrons from a reactor can be slowed by collisions with atoms of low mass as the neutrons pass through a moderator such as water or graphite. By the de Broglie relation, the momentum of a particle is $p = \frac{h}{\lambda}$, and for slow neutrons λ can be comparable to the lattice spacing of a crystal. They are called *thermal neutrons* because their energy is comparable to the average energy of a gas molecule at ordinary temperatures. A neutron with energy 0.04 eV (comparable to a gas molecule at 300K) travels at 2,700 m/s, and its de Broglie wavelength is 0.146 nm = 1.46 Å, in the range of lattice spacings.

If an incident neutron of energy E_i undergoes inelastic scattering by phonons in a crystal, conservation of energy gives

$$E_i = E_s + n\hbar\omega \quad (n \text{ phonons created})$$
$$E_i = E_s - n\hbar\omega \quad (n \text{ phonons annihilated}),$$

where E_s is the energy of the scattered neutron. Like photons, the number of phonons is not conserved. The branches are mapped out by detecting the scattered neutrons over a range of energy and momentum.

4.11   Brief Bios

The French physicist Auguste Bravais (1811–63) studied at the École Polytechnique, France's leading technical university. He joined the French Navy upon graduation and served with such distinction in Scandinavia that a mountain in Norway was named in his honor. After leaving the Navy, he taught at the École, and in 1849 he showed that only 14 different lattice structures can occur.

German engineer and physicist Wilhelm Röntgen (1845–1923) discovered X-rays when he was studying electric discharges in gases, noticing that a nearby crystal glowed when the discharge was on. The X-ray photograph of the bones in his wife's hand complete with ring was one of the first X-ray photos made and created a world-wide sensation. Röntgen was awarded the 1901 Nobel Prize in Physics, cited "in recognition of the extraordinary services he has rendered by the discovery of the remarkable rays subsequently named after him." His Nobel Prize in Physics was the first ever awarded.

W. L. Bragg often used apparatus built by his physicist father William Henry Bragg (1862–1942). The father-and-son team shared the 1915 Nobel Prize in Physics, cited "for their services in the analysis of crystal structure by means of X-rays." At age 25 W. L. Bragg was the youngest person (as of 2020) to be a Nobel laureate in physics.

XRD has become a powerful tool for providing insight into the structure of biological molecules. A famous example, mired in controversy, is "Photo 51," an XRD photograph of DNA fibers taken by Rosalind Franklin at Cambridge University. Without authorization, a worker showed the photo to James Watson and Francis Crick. The photo supported their hypothesized double helix model for DNA and they shared the 1962 Nobel Prize in Physiology for this fundamental discovery. Franklin died in 1958 and was not eligible because Nobel Prizes are awarded only to the living.

Miller indices were developed in the 1830s by British mineralogists W. Whewell (1794–1866) and W. H. Miller (1801–80). Whewell was a person of broad learning in science, theology, and poetry who coined the terms "scientist" and "physicist."

Albert Hull (1880–1966) majored in Greek in college, taking only one course in physics. After teaching language for several years, he went to Yale and earned a Ph.D. in physics. He spent his professional career at the General Electric Research Laboratory in Schenectady, New York. Among his many accomplishments, he developed the powder XRD method and invented the magnetron vacuum tube, which in modern versions powers radar installations and microwave ovens.

Felix Bloch (1905–83) was born in Switzerland and took physics courses with Schrödinger in Zurich. He left Germany in the 1930s to accept a post at Stanford University. He shared the 1952 Nobel Prize in Physics with Harvard physicist Edward Purcell (1912–97) for their development of nuclear magnetic resonance (NMR). They were cited "for their development of new methods for nuclear magnetic precision measurements and discoveries in connection therewith."

Summary of Chapter 4

Chapter 4 deals with the application of group theory to the symmetry of crystalline solids.

  a) A crystal is an assembly of identical lattices, so that an atom or ion at a given lattice site experiences the same surroundings anywhere in the crystal (assumed far from a surface).

b) Every crystal must consist of one of the only 14 possible lattices known as Bravais lattices.

c) Position in a lattice is characterized by basis vectors $\mathbf{a_i}$ along each of the lattice's principal axes. Each basis vector is assigned the physical length of its axis.

d) The translation vector $\sum m_i \mathbf{a}_i$, where the m_i are integers, is a displacement from a point in a lattice to the identical point in another lattice of the given crystal.

e) X-rays can have wavelengths comparable to the lattice site spacings in a crystal. Crystal structures are found experimentally by the diffraction of X-rays (XRD) from lattice planes, using either a single crystal sample or a powder sample.

f) The Fourier transform relates functions in a space to functions in a corresponding reciprocal space. For example, time t and angular frequency ω are reciprocal variables, and position r and inverse wavelength (wave vector) k are reciprocal variables.

g) The Fourier transform analysis of an XRD pattern involves a lattice reciprocal to the physical lattice. Simple vector formulas show how to convert the basis vectors of the physical lattice to basis vectors for the reciprocal lattice. A reciprocal lattice is always a Bravais lattice, because any space-filling lattice must be a Bravais lattice.

h) A Miller index consists of the components of a vector normal to the crystal plane of the reciprocal lattice, reduced to lowest terms.

i) A crystallographic point group is the group of symmetries that leaves a point in the lattice invariant. The translation vectors of a crystal form a cyclic Abelian group, with irreducible representations $e^{i\mathbf{k}\cdot\boldsymbol{\tau}}$.

j) An improper rotation symmetry is a proper (physical) rotation followed by an inversion ($x \mapsto -x, y \mapsto -y, z \mapsto -z$). Rotational symmetry of a lattice, consisting of proper and improper rotations, can only be 1-fold, 2-fold, 3-fold, 4-fold, or 6-fold.

k) Crystallographic space groups consist of combined rotational and translational symmetries; these symmetry operations do not commute, in general.

l) A crystal can be IR optically active if its lattice symmetry group has an irreducible representation that transforms like a time-varying vector (x, y, z), or Raman optically active if it transforms like a time-varying quadratic polarizability $(x^2, xy, yz, \dots)$.

m) The longitudinal vibrations of a crystal lattice can be modeled by a linear chain consisting of a chain of particles (either homonuclear with equal masses or heteronuclear with different masses) connected by equal springs. The vibrational frequencies ω resolve into two continuous branches, the optical branch and the acoustic branch, with ω a function of inverse wavelength $q = \frac{2\pi}{\lambda}$.

n) In a dispersive medium $\omega = \omega(k)$ and the group velocity of an oscillation packet is $u_g = \frac{\partial \omega}{\partial k}$. There is no dispersion if ω depends linearly on k.

o) Lattice vibrations act like quasi-particles called phonons that can transfer energy and momentum. The structure of the branches can be studied experimentally by the scattering of low-energy (thermal) neutrons, which can have de Broglie wavelengths of the order of lattice spacings.

Problems and Exercises

4.1 Copper is often the target in an X-ray tube because of its high thermal conductivity. The target is typically mechanically rotated at several thousand revolutions per minute to average the heating effect of the electron beam. Copper's principal K_α line has a wavelength of 0.154 nm. What is the shortest lattice spacing that can be observed by XRD using a copper target?

4.2 Turquoise is an attractive bluish gemstone often fashioned into jewelry by Native Americans. In one XRD experiment, a turquoise sample was shown to have a triclinic Bravais lattice. In a Cartesian frame the measured basis vectors are

$$\mathbf{a} = 0.741\hat{\mathbf{i}}$$
$$\mathbf{b} = 0.322\hat{\mathbf{i}} + 0.692\hat{\mathbf{j}}$$
$$\mathbf{c} = 0.344\hat{\mathbf{i}} + 0.241\hat{\mathbf{j}} + 0.896\hat{\mathbf{k}},$$

where basis vector $\mathbf{a}$ is taken to lie along the x-axis and lengths are in nm.

(a) What is the angle between basis vectors $\mathbf{b}$ and $\mathbf{c}$?
(b) What is the volume of the triclinic Bravais lattice?

4.3 The limestone White Cliffs of Dover, cave stalactites, and pearls are among naturally occurring forms of calcium carbonate ($CaCO_3$). Aragonite is one of the minerals found in limestone. It has an orthorhombic lattice with all angles 90° and with measured axes

$$a = 0.50 \text{ nm}$$
$$b = 0.80 \text{ nm}$$
$$c = 0.57 \text{ nm}.$$

What are the axes of its reciprocal lattice?

4.4 The *primitive cell* of a Bravais lattice is the smallest volume that can translate to fill its Bravais lattice completely. A primitive cell has only one lattice site (counting the sites shared with neighboring lattices).
(a) What is a primitive cell for the simple cubic lattice of NaCl?
(b) What is the volume of the primitive cell?

4.5 There are often several choices for the basis vectors of a primitive cell (defined in Problem 4). Let a be the length of one edge of a bcc Bravais lattice.

(a) One set of possible basis vectors for the bcc primitive cell is

$$\mathbf{a} = a\hat{\mathbf{i}}$$
$$\mathbf{b} = a\hat{\mathbf{j}}$$
$$\mathbf{c} = \left(\frac{a}{2}\right)(\hat{\mathbf{i}} + \hat{\mathbf{j}} + \hat{\mathbf{k}}).$$

What is the volume of the primitive cell described by these basis vectors?
(b) Another possible set of basis vectors for the bcc primitive cell is

$$\mathbf{a} = \left(\frac{a}{2}\right)(-\hat{\mathbf{i}} + \hat{\mathbf{j}} + \hat{\mathbf{k}})$$
$$\mathbf{b} = \left(\frac{a}{2}\right)(\hat{\mathbf{i}} - \hat{\mathbf{j}} + \hat{\mathbf{k}})$$
$$\mathbf{c} = \left(\frac{a}{2}\right)(\hat{\mathbf{i}} + \hat{\mathbf{j}} - \hat{\mathbf{k}}).$$

What is the volume of the primitive cell described by these basis vectors?

4.6 Problem 2 lists experimentally determined basis vectors for a triclinic lattice. Calculate the basis vectors of its reciprocal lattice.

4.7 Show that the translation group is cyclic by proving $\mathbf{T}_2 = \mathbf{T}_1^2$.

4.8 Sketch a simple cubic lattice and shade a plane with Miller index (001).

4.9 Sketch a simple cubic lattice and shade a plane with Miller index (110).

4.10 From the XRD of aluminum powder Hull determined the lattice to be fcc, with lattice cell size $a = 0.405$ nm (modern value). He observed strong interference at Bragg angle $\theta = 8.90°$ using a molybdenum X-ray tube ($\lambda = 0.0714$ nm). He assigned Miller index 111. Show that this measurement is consistent with a.

4.11 From the XRD of aluminum powder Hull determined the lattice to be fcc, with lattice cell size $a = 0.405$ nm (modern value). He observed strong interference at Bragg angle $\theta = 17.25°$ using a molybdenum X-ray tube ($\lambda = 0.0714$ nm). He assigned Miller index 311. Show that this measurement is consistent with a.

4.12 *Avogadro's number* N_A is a universal constant equal to the number of particles (such as atoms, molecules) in one *mole* of a substance. One mole of aluminum has a mass of 26.98 g. The density of aluminum is 2.70 g·cm^{-3}. In XRD experiments with aluminum powder, where he determined the lattice structure to be fcc, Hull observed strong interference at Bragg angle $\theta = 14.63°$ using a molybdenum X-ray tube ($\lambda = 0.0714$ nm). He assigned Miller index 110. From these data, what is the predicted value of N_A?

4.13 The icosahedral group **I** (Schönflies) or **532** is large, with 60 members and five irreducible representations of orders $A = 1$, $F_1 = 3$, $F_2 = 3$, $G = 4$, and $H = 5$. Table 4.6 shows its character table (omitting inversion) and where $\eta = \frac{1+\sqrt{5}}{2}$.

Table 4.6 Character table for **I**

basis functions		1	$12\mathbf{C}_5$	$12\mathbf{C}_5^2$	$20\mathbf{C}_3$	$15\mathbf{C}_2$
r^2	A	1	1	1	1	1
x, y, z, L_x, L_y, L_z	F_1	3	η	$1 - \eta$	0	-1
	F_2	3	$1 - \eta$	η	0	-1
	G	4	-1	-1	1	0
$3z^2 - r^2, x^2 - y^2, xy, yz, zx$	H	5	0	0	-1	1

(a) Why is **I** not a crystallographic point group?

(b) For the group $\mathbf{I}_h = \mathbf{I} \otimes \mathbf{i}$ what are the basis functions for *gerade* (+) and *ungerade* (−) representations?

4.14 The identity is the only symmetry operation for a triclinic lattice. Sketch its stereogram.

4.15 Sketch the diagram of symmetry elements for a square, including 4-fold rotations and flips.

4.16 Sketch the stereogram for a square, including 4-fold rotations and flips.

4.17 The figure is a stereogram for an orthorhombic lattice ($\alpha = \beta = \gamma = 90°, a \neq b \neq c$). Including rotations and flips, what symmetry operations does the stereogram show?

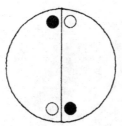

4.18 The table gives the basis functions for the group **32**. What are the *gerade* (+) and *ungerade* (−) irreducible representations of the group $\overline{\mathbf{3}}\mathbf{m} = \mathbf{32} \otimes \overline{\mathbf{1}}$?

basis functions	
$z^2, x^2 + y^2$	A_1
z, L_z	A_2
$x, y, x^2 - y^2, xy, xz, yz, L_x, L_y$	E_1

4.19 The table gives the basis functions for the group **622**. What are the *gerade* $(+)$ and *ungerade* $(-)$ irreducible representations of the group $\mathbf{6/mmm} = \mathbf{622} \otimes \mathbf{\bar{1}}$?

basis functions	
$z^2, x^2 + y^2$	A_1
z, L_z	A_2
	B_1
	B_2
x, y, xz, yz, L_x, L_y	E_1
$x^2 - y^2, xy$	E_2

4.20 Show that for a monatomic crystal $m = M$ the linear chain model predicts that the acoustic branch and the optical branch give the same value of ω at $q = q_{max}$.

4.21 A sound wave is propagating along a linear diatomic chain $(M > n)$.
(a) Calculate its maximum possible frequency.
(b) What is the speed (phase velocity) of the sound wave traveling with its maximum possible frequency?

5

BOHR'S QUANTUM THEORY AND MATRIX MECHANICS

5.1 Introduction

The first part of the twentieth century was a period of intense activity among theoretical physicists as they attempted to formulate a theory to account for unexplained atomic phenomena. Notable theorists were the Danish physicist Niels Bohr (1885–1962) and the German physicists Werner Heisenberg (1901–76), Max Born (1882–1970), and Erwin Schrödinger (1887–1961). All later were Nobel laureates in physics.

As the title suggests, this chapter has two main topics. The first is a review of Bohr's quantum theory. His model, proposed in 1913, had great success in explaining the spectral lines of atomic hydrogen but could not be extended further to multi-electron atoms. The second main topic is matrix mechanics. In 1925 Heisenberg and Born used matrix algebra to develop *matrix mechanics*, the first complete quantum theory. It was followed a few months later by Schrödinger's *wave mechanics*, equivalent to matrix mechanics but a very different approach. Wave mechanics is treated in Chapter 6.

5.2 Bohr's Model

In Bohr's model (now called the *old quantum theory*) atomic electrons orbit the nucleus with discrete energies; but, although they are accelerated, they are assumed not to radiate, contrary to classical electromagnetism. A spectral line of frequency ν_{ji} is emitted in a transition from a higher energy state j to a lower state i: $E_j - E_i = h\nu_{ji}$.

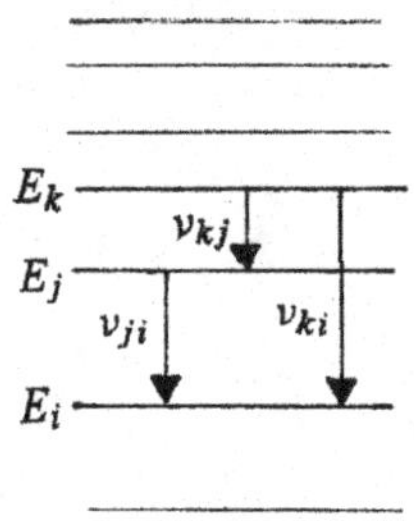

One experimental proof for the existence of discrete states came from experimental measurements of wavelengths λ of spectral lines. In some cases the frequencies $\nu = \frac{c}{\lambda}$ of three particular lines satisfy the relation

$$\nu_{kj} + \nu_{ji} = \nu_{ki},$$

known as the *Ritz combination principle*. The schematic diagram shows how Bohr's model accounts for the Ritz principle.

In the old quantum theory, Bohr's *quantization condition* set the mechanical quantity $\oint p\,dq$, called the *action* in classical physics, equal to an integer multiple of Planck's constant.

$$\oint p\,dq = nh \tag{5.1}$$

To model an electron in orbit around an atom's nucleus, Bohr applied the quantization condition to a mass m moving with speed v in a circular orbit of radius r.

$$p = mvr$$
$$dq = d\theta$$
$$\oint p\,dq = \int_0^{2\pi} mvr\,d\theta$$
$$= 2\pi mvr = nh$$
$$mvr = n\left(\frac{h}{2\pi}\right) \equiv n\hbar$$

This is interpreted as quantization of angular momentum in units of $\hbar$. Using this quantization condition, Bohr derived the discrete energy levels E_n of a hydrogen atom. E_n and the discrete orbital radii r_n, calculated from Bohr's model, are

$$E_n = -\left(\frac{\mu_0^2 c^4 m_e e^3}{8h^2}\right)\left(\frac{1}{n^2}\right) = -\left(\frac{\mathcal{R}_\infty hc}{e}\right)\left(\frac{1}{n^2}\right) \quad \text{eV} \tag{5.2}$$

$$r_n = \left(\frac{\epsilon_0 h^2}{\pi m_e e^2}\right)n^2 = a_0 n^2 \quad \text{m}, \tag{5.3}$$

where

$$\mathcal{R}_\infty = \text{Rydberg constant for nuclear mass} \to \infty$$
$$= \left(\frac{\mu_0^2 c^3 m_e e^4}{8h^3}\right) \approx 1.097 \times 10^7 \, \text{m}^{-1}$$
$$\left(\frac{\mathcal{R}_\infty hc}{e}\right) \approx 13.61 \, \text{eV} = 2.18 \times 10^{-18} \, \text{J}$$

a_0 = Bohr radius $\approx 5.29 \times 10^{-11}$ m $= 0.529$ Å

μ_0 = permeability of vacuum defined as $4\pi \times 10^{-7}$ SI units

ϵ_0 = permittivity of vacuum $\approx 8.854 \times 10^{-12}$ SI units

$$c = \text{speed of light} = \frac{1}{\sqrt{\epsilon_0 \mu_0}} \approx 3 \times 10^8 \text{ m} \cdot \text{s}^{-1}.$$

n is known as the *principal quantum number* because it specifies the major energy levels even though Bohr introduced n as quantization of angular momentum.

In his quantum theory Bohr assumed that the frequency ν of a spectral line emitted in a transition from an excited state with principal quantum number m to a lower state with quantum number n is

$$h\nu = E_m - E_n. \tag{5.4}$$

Equation (5.4) is called the *Bohr frequency condition*.

From Eq. (5.2) the wavelength λ of a spectral line emitted in a transition from an excited state with quantum number m to the *ground state* of lowest energy, assigned quantum number n, is therefore

$$\frac{1}{\lambda} = \mathscr{R}_\infty \left(\frac{1}{n^2} - \frac{1}{m^2} \right), \tag{5.5}$$

where $m > n$.

For a given n, Eq. (5.5) defines a *series* of spectral lines in the hydrogen atom spectrum. Each series has an infinite number of spectral lines. The shortest wavelength (highest frequency) $\lambda = \frac{n^2}{\mathscr{R}}$ marks the *series limit* $m \to \infty$. In each series the spectral line with the longest wavelength occurs for the transition from the first excited state $m = n + 1$ to the ground state n. Bohr's model gained support because it agreed accurately with Eq. (5.5), called the Rydberg formula, which had been established empirically in 1888 by Swedish physicist Johannes Rydberg (1854–1919).

The first few series are named after their discoverers; for example, the Lyman series $n = 1$, the Balmer series $n = 2$, and the Paschen series $n = 3$. In the Lyman series the longest wavelength is 121.6 nm = 1216 Å. It is in the ultraviolet and is called the Lyman-α line L_α. The Sun is a strong emitter of L_α radiation, which would be harmful to life on Earth except that the ozone (O_3) layer in the upper atmosphere strongly absorbs much of it.

If an electron with energy greater than an atom's ionization potential interacts with an atom and becomes bound, the emitted spectrum is not discrete but is instead a continuous spectrum with unquantized energies. The Bohr model does not have a method for calculating the continuous spectrum.

The 1927 photo shows the hydrogen Balmer emission spectrum with discrete lines verging into a continuous spectrum on the right.

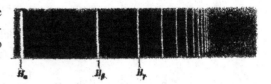

Reprinted figure from G. Herzberg Ann. Physik **84**, 585 (1927) with permission from John Wiley & Sons.

Equation (5.3) shows that the orbit radius increases as n^2. An atom excited to an orbit with large principal quantum number such as $n \approx 100$ is called a *Rydberg atom*. According to Eq. (5.3), a Rydberg atom can be macroscopic in size with r_n a fraction of a millimeter. Rydberg atoms have been detected in outer space, where the gas pressure is very low, and they have been created in the laboratory in gases at low pressure using tunable dye lasers. Low pressure is required because, according to Eq. (5.2), an electron with large n is nearly at the ionization limit and is easily released in a collision.

Bohr's quantization condition Eq. (5.1) was remarkably successful in accounting for the line spectra of *hydrogenic* atoms – atoms so heavily ionized as to possess only one electron. One problem with Bohr's orbit model is that zero angular momentum $n = 0$ was arbitrarily excluded because it implies an electron on a straight-line path heading toward collision with the nucleus. More important, Bohr quantization is unable to deal with multi-electron atoms such as helium (He) because Bohr's model has no way to include the Coulomb force between atomic electrons.

5.2.1   Bohr's Correspondence Principle

Bohr's Correspondence Principle was an important guide for theorists in the development of quantum mechanics. It states that in the limit of large quantum numbers, a quantum theory should give the same result as classical mechanics. Max Planck had earlier used a correspondence principle $h \rightarrow 0$ to relate the Planck function for thermal radiation to classical approximations.

Applying Bohr's correspondence principle to a hydrogen atom, consider the frequency emitted in a transition between a state with quantum number $n + 1$ and the next lower state n. From Eq. (5.5):

$$h\nu = E_{n+1} - E_n$$

$$\nu = -\mathscr{R}_\infty c \left(\frac{1}{(n+1)^2} - \frac{1}{n^2} \right)$$

$$\lim_{n \gg 1} \nu = \left(\frac{m_e e^4}{4h^3 \epsilon_0^2} \right) \frac{1}{n^3}. \tag{5.6}$$

From Kepler's third law of planetary motion, the period $\mathcal{T}$ of an electron in a circular orbit in hydrogen is related to the radius r as

$$\mathcal{T}^2 = \left(\frac{16\pi^3 m_e \epsilon_0}{e^2}\right) r^3 = \left(\frac{16\pi^3 m_e \epsilon_0}{e^2}\right) a_0^3 n^6$$

$$\mathcal{T} = \sqrt{\left(\frac{16\pi^3 m_e \epsilon_0}{e^2}\right)\left(\frac{\epsilon_0 h^2}{\pi m_e e^2}\right)^3 n^6} = \left(\frac{4h^3 \epsilon_0{}^2}{m_e e^4}\right) n^3. \tag{5.7}$$

Comparing Eqs. (5.6) and (5.7),

$$v\mathcal{T} \approx \left(\frac{1}{n^3}\right) n^3 = 1,$$

which is the classical result that the frequency emitted by a circulating charge equals its frequency of revolution.

The Planck and Bohr Correspondence Principles, although historically useful, do not apply to all physical systems.

5.3 Matrix Mechanics

5.3.1 Some Concepts

Some background in classical physics, possibly familiar from courses in mathematical physics, is needed to understand how matrix mechanics developed.

The Hamiltonian

In 1833 Irish mathematician and physicist William Rowan Hamilton (1805–65) reformulated Newtonian mechanics to eliminate the anthropomorphic concept of force. He introduced the *Hamiltonian function* $\mathcal{H}(p_i, q_i)$, where in standard notation q_i is a generalized coordinate and p_i is its *conjugate momentum* $p_i = m\dot{q}_i$.

In this text we take the Hamiltonian to be the total energy of a system. Texts on classical mechanics give more general definitions.

$$\mathcal{H} = \text{total energy}$$

$$= \text{kinetic energy} + \text{potential energy}$$

The total energy of a system is a useful concept because it is conserved if no work is done on the system, although components of the total energy may change from one form to another, as in the exchange of kinetic energy and gravitational potential energy by a falling stone.

The Hamilton equations of motion replace Newton's second law $F = ma$ by

$$\dot{q} = \frac{\partial \mathcal{H}}{\partial p} \qquad \dot{p} = -\frac{\partial \mathcal{H}}{\partial q}, \tag{5.8}$$

where time derivatives use Newton's dot notation $\dot{q} \equiv \frac{\partial q}{\partial t}$ and so on.

The harmonic oscillator is a simple application of Eq. (5.8).

$$\mathcal{H} = \left(\frac{1}{2m}\right)p^2 + \left(\frac{k}{2}\right)q^2$$

$$\dot{q} = \frac{\partial \mathcal{H}}{\partial p} = \frac{p}{m}$$

$$p = m\dot{q} \implies \dot{p} = m\ddot{q}$$

From Eq. (5.8),

$$\dot{p} = -\frac{\partial \mathcal{H}}{\partial q} = -kq$$

$$m\ddot{q} + kq = 0,$$

the familiar equation for simple harmonic motion.

Fourier Series

A periodic function of time $y(t)$ can be expressed by Fourier series:

$$y(t) = \sum_{n=-\infty}^{\infty} \phi_n e^{in\omega t}$$

$$= \ldots \phi_{-1} e^{-i\omega t} + \phi_0 + \phi_1 e^{i\omega t} + \ldots,$$

where n is an integer $-\infty \leq n \leq \infty$ and ω is angular frequency. The coefficients ϕ_n are determined by an integral condition.

If $y(t)$ is real, then ϕ_0 is real and independent of time. The other coefficients ϕ_n can be real or complex, but for $y(t)$ real $\phi_{-n} = \phi_n^*$ must be satisfied. Proof is left to the problems.

5.3.2 The Beginnings of Matrix Mechanics – From Continuous to Discrete

In classical physics quantities such as acceleration and electric fields are represented by continuous functions. Newton's invention of calculus was specifically designed to deal with continuous functions, as shown by the derivative:

$$\frac{dv}{dt} = \lim_{\Delta t \to 0} \left(\frac{v(t + \Delta t) - v(t)}{\Delta t}\right).$$

Quantum mechanics called for a new kind of mathematics. Heisenberg said that there was no hope of observing the orbits in Bohr's planetary atomic model. Instead, quantum mechanics should be based only on observable quantities and relations between them. Heisenberg took the example of discrete transitions that account for the observed frequencies and intensities of spectral lines and proposed that the continuous quantities of classical physics should be replaced by differences.

Heisenberg resorted to representation by Fourier series because it was otherwise unclear how to express a function in a new quantum mechanics. This example from Heisenberg's groundbreaking 1925 paper uses Fourier series to evaluate the product of two functions $x(t)$ and $y(t)$, first in classical physics and then in the new quantum algebra where quantities such as $\omega_{nn'}$ are discrete differences $\omega_n - \omega_{n'}$.

In modern notation, where n, α, and β are integers,

classical:

$$x(t) = \sum_{\alpha=-\infty}^{\infty} \mathfrak{A}_\alpha e^{i\alpha\omega t}$$

$$x(t)y(t) = \sum_{\alpha=-\infty}^{\infty} \mathfrak{A}_\alpha \mathfrak{B}_{\beta-\alpha} e^{i(\alpha+\beta-\alpha)\omega t}$$

quantum:

$$x(t) = \sum_{\alpha=-\infty}^{\infty} \mathfrak{A}_{n\,n-\alpha} e^{i\omega_{n\,n-\alpha}t}$$

$$x(t)y(t) = \sum_{\alpha=-\infty}^{\infty} \mathfrak{A}_{n\,n-\alpha} e^{i\omega_{n\,n-\alpha}t} \mathfrak{B}_{n-\alpha\,n-\beta} e^{i\omega_{n-\alpha\,n-\beta}t} \tag{5.9}$$

$$= \sum_{\alpha=-\infty}^{\infty} \mathfrak{A}_{n\,n-\alpha} \mathfrak{B}_{n-\alpha\,n-\beta} e^{i\omega_{n\,n-\beta}t}. \tag{5.10}$$

Born was the first to recognize that the sum over α in Eq. (5.10) is the rule for multiplication of matrices. Equation (5.10) is evidence that quantities should be represented by matrices. Matrix elements are by their nature discrete and well suited to Heisenberg's emphasis on differences.

For the time dependence to be the same in Eqs. (5.9) and (5.10),

$$\omega_{n\,n-\beta} = \omega_{n\,n-\alpha} + \omega_{n-\alpha\,n-\beta}. \tag{5.11}$$

Equation (5.11) is just the Ritz combination principle for frequencies. In classical Fourier series, frequencies combine as integers:

$$e^{in\omega t} = e^{in_1\omega t} e^{in_2\omega t}$$
$$n = n_1 + n_2.$$

In contrast, Eq. (5.11) shows that the frequencies of a Fourier series in matrix mechanics are not integer multiples of the same frequency. The frequencies in matrix mechanics combine according to the Ritz principle in a modified Fourier series.

Note that because of the Ritz combination principle, all terms have the same time dependence, a general result that holds for any quantum quantities and for any functions of the quantities. Having the same time dependence ensures that calculated frequencies of a system's spectral lines all vary the same way with time. The time dependence of a system's terms is therefore a common factor usually dropped from equations.

5.4 Matrix Mechanics Quantization

Heisenberg's thoughts in his 1925 paper on the new calculation techniques are difficult to follow even by accomplished theorists. A few months later, a paper by Born and his assistant Pascual Jordan clarified how Heisenberg's ideas could be expressed in matrix form.

5.4.1 Matrix Mechanics Dynamics

A principal feature of the new mechanics is that classical dynamics such as Hamilton's equations of motion are applied unchanged, but kinematic quantities like momentum are to be represented by matrices, in this chapter printed in bold. If $\mathcal{H}$ is the matrix for the Hamiltonian, with $\mathbf{q}$ and $\mathbf{p}$ the matrices for a coordinate and its conjugate momentum, the equations of motion in matrix mechanics are written

$$\dot{\mathbf{q}} = \frac{\partial \mathcal{H}}{\partial \mathbf{p}} \qquad \dot{\mathbf{p}} = -\frac{\partial \mathcal{H}}{\partial \mathbf{q}}. \tag{5.12}$$

Equation (5.12) is identical in algebraic form to the classical Hamilton's equations, Eq. (5.8), except that here $\mathcal{H}$, $\mathbf{q}$, and $\mathbf{p}$ are matrices and not continuous functions.

Equation (5.12) contains time derivatives of matrices. To define the time derivative of a matrix, note that matrices $\mathbf{q}$ and $\mathbf{p}$ and any matrix $\mathbf{a}$ that is a function of $\mathbf{q}$ and $\mathbf{p}$ all have the same time dependence $e^{2\pi i \nu_{nm} t}$. The elements of some matrix $\mathbf{a}$ are then

$$a_{nm} e^{2\pi i \nu_{nm} t}$$

and the elements of its time derivative $\dot{\mathbf{a}}$ are

$$\dot{a}_{nm} = (2\pi i \nu_{nm}) a_{nm} e^{2\pi i \nu_{nm} t}$$

for states with quantum numbers n and m. If $\nu_{nm} \neq 0$ for $n \neq m$, then $\dot{\mathbf{a}} = 0$ implies that $a_{nm} = 0$ for $n \neq m$. In other words, a matrix with zero time derivative is therefore a diagonal matrix with nonzero elements only along the main diagonal.

5.4.2 Quantization Condition

Heisenberg was at first puzzled that products in his calculations did not necessarily commute. Unlike most physicists of his time, Born had studied matrix algebra and realized that Heisenberg's equations represented matrix multiplication where dynamical quantities might not commute.

In matrix mechanics, the Heisenberg–Born–Jordan quantization condition is

$$\mathbf{pq} - \mathbf{qp} = \frac{h}{2\pi i}\mathbf{1} \qquad (5.13)$$
$$= -i\hbar\mathbf{1},$$

where $\mathbf{1}$ is the identity matrix. Equation (5.13) is a fundamental postulate of matrix mechanics. It clearly exhibits the essential role of noncommutativity in matrix mechanics. A plausible derivation following Born–Jordan is given in Appendix B.

The diagonal elements of a Fourier series matrix are independent of time, and the nondiagonal elements depend on time through an exponential function of complex argument. The right-hand side of Eq. (5.13) is a diagonal matrix, showing that the quantization condition is independent of time.

The matrices in Eq. (5.13) have infinite dimension. For a proof by contradiction, suppose that the matrices are finite $N \times N$ and take the trace of both sides. Because the trace of a matrix product $\mathbf{AB}$ equals the trace of $\mathbf{BA}$, the trace of the left-hand side equals zero, but on the right-hand side the trace of $\mathbf{1} = N$. Hence the matrices cannot be finite. The matrices must be infinite because their elements include all possible values of their operators.

The fundamental postulates of matrix mechanics contain Planck's constant only in the quantization condition. According to Planck's correspondence principle, $\hbar \to 0$,

$$\lim_{\hbar \to 0}(\mathbf{pq} - \mathbf{qp}) = \lim_{\hbar \to 0}(-i\hbar\mathbf{1})$$
$$= \mathbf{0},$$

where $\mathbf{0}$ is the matrix whose elements are all 0. In the classical limit, $\mathbf{q}$ and $\mathbf{p}$ in Eq. (5.13) become ordinary continuous functions that commute.

It follows from Eq. (5.13) that the matrices $\mathbf{p}$ and $\mathbf{q}$ cannot both be diagonal because they would then commute. Physically, this means that $\mathbf{q}_i$ and its conjugate $\mathbf{p}_i$ cannot be measured simultaneously, a result that reflects the Heisenberg uncertainty principle.

If there are more than one $\mathbf{q}$ and $\mathbf{p}$, the $\mathbf{q}_i$ commute among themselves and the $\mathbf{p}_i$ commute among themselves.

$$\mathbf{q}_j\mathbf{q}_k - \mathbf{q}_k\mathbf{q}_j = 0 \qquad (5.14)$$
$$\mathbf{p}_j\mathbf{p}_k - \mathbf{p}_k\mathbf{p}_j = 0 \qquad (5.15)$$

A coordinate $\mathbf{q}_i$ is noncommutative only with its conjugate $\mathbf{p}_i$, so Eq. (5.13) can be written

$$\mathbf{p}_j\mathbf{q}_k - \mathbf{q}_k\mathbf{p}_j = \frac{h}{2\pi i}\delta_{jk},$$

where δ_{jk} equals the matrix $\mathbf{1}$ for $j = k$ and equals the matrix $\mathbf{0}$ for $j \neq k$.

A coordinate commutes with a function only of coordinates $\mathbf{f(q)}$, and similarly a momentum commutes with a function only of momentum $\mathbf{g(p)}$. Stated without proof, a general function $\mathbf{f(p, q)}$ satisfies the commutation relations

$$\mathbf{fq} - \mathbf{qf} = -i\hbar\frac{\partial \mathbf{f}}{\partial \mathbf{p}} \tag{5.16}$$

$$\mathbf{fp} - \mathbf{pf} = i\hbar\frac{\partial \mathbf{f}}{\partial \mathbf{q}}. \tag{5.17}$$

This chapter is on matrix mechanics but has yet to show an explicit matrix or at least a portion of one – they do have infinite dimension, after all. Operations in group theory are defined by their algebraic relations, as in the quantization condition Eq. (5.13). It is a distinguishing feature of group representations that explicit forms are not needed, although they might come in handy for calculations.

$$\begin{pmatrix} 0 & q_{01} & 0 & 0 & 0 & \dots \\ q_{10} & 0 & q_{12} & 0 & 0 & \dots \\ 0 & q_{21} & 0 & q_{23} & 0 & \dots \\ \vdots & \vdots & \vdots & \vdots & \vdots & \end{pmatrix}$$

The figure shows a portion of a harmonic oscillator's coordinate matrix $\mathbf{q}$. It has infinite dimension because it contains all possible values that could be assumed by the coordinate. The first row and first column are assigned index 0. Heisenberg obtained this form by reasoning that transitions occur only between adjacent states $n \to n \pm 1$ so that the diagonal elements $q_{nn} = 0$.

5.4.3   Time and the Hamiltonian

For any matrix $\mathbf{a}$, its commutator with the Hamiltonian matrix $\mathcal{H}$ generates the time derivative of $\mathbf{a}$.

$$\mathcal{H}\mathbf{a} - \mathbf{a}\mathcal{H} = -i\hbar\dot{\mathbf{a}} \tag{5.18}$$

Equation (5.18) can be proved to hold for any Hamiltonian $\mathcal{H}(\mathbf{p}, \mathbf{q})$.

For a simplified proof, assume that $\mathcal{H}$ can be expressed as a power series in $\mathbf{p}$.

$$\mathcal{H} = \sum_k c_k \mathbf{p}^k \tag{5.19}$$

Multiply the quantization condition Eq. (5.13) by $\mathbf{p}$ from the left.

$$\mathbf{p}^2\mathbf{q} - \mathbf{pqp} = -i\hbar\mathbf{p} \tag{5.20}$$

Substitute $\mathbf{pq} = \mathbf{qp} - i\hbar\mathbf{1}$ in Eq. (5.20).

$$\mathbf{p}^2\mathbf{q} - \mathbf{qp}^2 = -2i\hbar\mathbf{p}$$

Continue the iteration to give

$$\mathbf{p}^k\mathbf{q} - \mathbf{qp}^k = -ki\hbar\mathbf{p}^{k-1}. \tag{5.21}$$

Take the derivative of Eq. (5.19) with respect to **p**.

$$\frac{\partial \mathcal{H}}{\partial \mathbf{p}} = \sum_k c_k k \mathbf{p}^{k-1} \tag{5.22}$$

Multiply Eq. (5.21) by c_k and sum over k.

$$\sum_k c_k \mathbf{p}^k \mathbf{q} - \mathbf{q} \sum_k c_k \mathbf{p}^k = -i\hbar \sum_k c_k k \mathbf{p}^{k-1}$$

Use Eqs. (5.19) and (5.22) to give

$$\mathcal{H}\mathbf{q} - \mathbf{q}\mathcal{H} = -i\hbar \frac{\partial \mathcal{H}}{\partial \mathbf{p}}.$$

From the Hamilton equations of motion Eq. (5.12),

$$\frac{\partial \mathcal{H}}{\partial \mathbf{p}} = \dot{\mathbf{q}}$$

$$\mathcal{H}\mathbf{q} - \mathbf{q}\mathcal{H} = -i\hbar\dot{\mathbf{q}},$$

or, in general, for any matrix **a**,

$$\mathcal{H}\mathbf{a} - \mathbf{a}\mathcal{H} = -i\hbar\dot{\mathbf{a}}.$$

The solution of Eq. (5.18) for **a** is a similarity transformation that transforms **a** in abstract space, taking it from $t = 0$ to a new time t.

$$\mathbf{a}(t) = e^{i\frac{\mathcal{H}}{\hbar}t}\mathbf{a}(0)e^{-i\frac{\mathcal{H}}{\hbar}t} \tag{5.23}$$

5.5 Consequences of Matrix Mechanics

A complete and consistent theory must account for all physical results without the help of additional assumptions. Any guesswork additions or assumptions grafted onto a theory or calculation are described by an *Ansatz* (German, method for calculation). In Bohr's model, $\Delta E = h\nu$ is an *Ansatz*. Because matrix mechanics is a complete theory of quantum phenomena, all physical laws can be derived from its postulates without the need for any *Ansatz*. A few examples follow.

5.5.1 Conservation of Energy

Replace **a** by $\mathcal{H}$ in Eq. (5.18).

$$-i\hbar\dot{\mathcal{H}} = \mathcal{H}\mathcal{H} - \mathcal{H}\mathcal{H}$$

$$= 0$$

The Hamiltonian $\mathcal{H}$ is, therefore, independent of time unless it depends explicitly on time. As a corollary, $\mathcal{H}$ must be a diagonal matrix. Its matrix elements are zero everywhere except along the main diagonal, where the elements are real numbers independent of time that are interpreted as the energies of the system's quantum states. Hence the condition $\dot{\mathcal{H}} = 0$ is equivalent to the statement that energy is conserved – it does not change with time.

The matrices $\mathbf{q}, \mathbf{p}$, and $\mathcal{H}$ are all Hermitian matrices equal to their adjoints. Recall that the adjoint $\mathcal{A}^{\dagger}$ of a matrix $\mathcal{A}$ is its transpose complex conjugate. For two quantum states n and m:

$$\mathcal{H} = \mathcal{H}^{\dagger}$$
$$\mathcal{H}_{nm} = \mathcal{H}_{mn}^{*}.$$

The diagonal matrix elements $m = n$ are therefore real $\mathcal{H}_{nn} = \mathcal{H}_{nn}^{*}$.

5.5.2   Bohr Frequency Condition

The Bohr frequency condition $h\nu = E_n - E_m$ is an *Ansatz* in the old quantum theory, but in matrix mechanics it follows from the fundamental principles.

The Fourier series for $\mathbf{q}$ and $\dot{\mathbf{q}}$ are

$$\mathbf{q}_{nm} = q_{nm}\, e^{2\pi i \nu_{nm} t} \tag{5.24}$$

$$\dot{\mathbf{q}}_{nm} = (2\pi i\, \nu_{nm})\, q_{nm}\, e^{2\pi i \nu_{nm} t}. \tag{5.25}$$

From Eq. (5.25):

$$\dot{\mathbf{q}} = \frac{i}{\hbar}(\mathcal{H}\mathbf{q} - \mathbf{q}\mathcal{H}) = \frac{2\pi i}{h}(\mathcal{H}\mathbf{q} - \mathbf{q}\mathcal{H})$$

$$(2\pi i\, \nu_{nm})\, q_{nm}\, e^{2\pi i \nu_{nm} t} = \frac{2\pi i}{h} \sum_{k} \left(\mathcal{H}_{nk} q_{km}\, e^{2\pi i \nu_{nk} t} - q_{nk}\, e^{2\pi i \nu_{nk} t}\, \mathcal{H}_{km} \right).$$

$\mathcal{H}$ is diagonal $\mathcal{H}_{ij} = \delta_{ij}$, so the sum over k gives

$$(2\pi i\, \nu_{nm})\, q_{nm}\, e^{2\pi i \nu_{nm} t} = \frac{2\pi i}{h} \left(\mathcal{H}_{nn} q_{nm}\, e^{2\pi i \nu_{nm} t} - q_{nm}\, e^{2\pi i \nu_{nm} t}\, \mathcal{H}_{mm} \right).$$

Common terms cancel, leaving

$$h\nu_{nm} = \mathcal{H}_{nn} - \mathcal{H}_{mm}$$
$$= E_n - E_m,$$

proving the Bohr frequency condition.

5.5.3 Conservation of Angular Momentum

A Hamiltonian is called *separable* if it can be expressed as the sum of two functions, one that depends only on momentum and a second that depends only on coordinates. $\mathcal{H}$ for central force motion is an example of a separable Hamiltonian.

$$\mathcal{H} = \text{kinetic energy} + \text{potential energy}$$

$$\mathcal{H} = \left(\frac{1}{2m}\right)p^2 + U(q)$$

Symbolically,

$$\mathcal{H} = \mathcal{H}_1(\mathbf{p}) + \mathcal{H}_2(\mathbf{q}).$$

It is convenient to use the notation

$$\mathcal{H} \equiv \mathcal{P}(\mathbf{p}) + \mathcal{Q}(\mathbf{q}).$$

In classical mechanics, the angular momentum of a particle is $\mathbf{M} = \mathbf{r} \times \mathbf{p}$, where $\mathbf{r}$ is the position vector of a particle and $\mathbf{p}$ is its momentum.

Using Cartesian coordinates, $\mathbf{M}$ can be expressed in determinant form as

$$\mathbf{M} = \mathbf{r} \times \mathbf{p}$$

$$= \begin{vmatrix} \hat{\mathbf{i}} & \hat{\mathbf{j}} & \hat{\mathbf{k}} \\ q_x & q_y & q_z \\ p_x & p_y & p_z \end{vmatrix}$$

$$= (q_y p_z - p_y q_z)\hat{\mathbf{i}} + (p_x q_z - q_x p_z)\hat{\mathbf{j}} + (q_x p_y - p_x q_y)\hat{\mathbf{k}}.$$

In matrix mechanics, the expression for the angular momentum of a particle is the same as the classical form except that all quantities are matrices. The order in the $\mathbf{p}_i \mathbf{q}_j$ products is unimportant because $\mathbf{p}_i \mathbf{q}_j - \mathbf{q}_j \mathbf{p}_i = 0$ if $i \neq j$. Consider $\mathbf{M}_z$:

$$\mathbf{M}_z = (\mathbf{q}_x \mathbf{p}_y - \mathbf{p}_x \mathbf{q}_y).$$

Taking the time derivative:

$$\dot{\mathbf{M}}_z = \dot{\mathbf{q}}_x \mathbf{p}_y + \mathbf{q}_x \dot{\mathbf{p}}_y - \dot{\mathbf{p}}_x \mathbf{q}_y - \mathbf{p}_x \dot{\mathbf{q}}_y. \tag{5.26}$$

In Newtonian mechanics, torque $\boldsymbol{\tau}$ is

$$\boldsymbol{\tau} = \mathbf{r} \times \left(\frac{d\mathbf{p}}{dt}\right).$$

In matrix mechanics, $\boldsymbol{\tau}_z$ is

$$\boldsymbol{\tau}_z = \mathbf{q}_x \dot{\mathbf{p}}_y - \dot{\mathbf{p}}_x \mathbf{q}_y. \tag{5.27}$$

Comparing Eqs. (5.26) and (5.27):

$$\dot{\mathbf{M}}_z = \dot{\mathbf{q}}_x \mathbf{p}_y - \mathbf{p}_x \dot{\mathbf{q}}_y + \boldsymbol{\tau}_z. \tag{5.28}$$

We now show that the first two terms on the right-hand side of Eq. (5.28) cancel. Assuming $\mathcal{H} = \mathcal{P} + \mathcal{Q}$ Eq. (5.18) gives

$$
\begin{aligned}
-i\hbar\dot{\mathbf{q}} &= (\mathcal{P} + \mathcal{Q})\mathbf{q} - \mathbf{q}(\mathcal{P} + \mathcal{Q}) \\
&= (\mathcal{P}\mathbf{q} - \mathbf{q}\mathcal{P}) + (\mathcal{Q}\mathbf{q} - \mathbf{q}\mathcal{Q}) \\
&= (\mathcal{P}\mathbf{q} - \mathbf{q}\mathcal{P}) + 0.
\end{aligned}
$$

Inserting this result in Eq. (5.28) gives

$$
\dot{\mathbf{M}}_z = \left(\frac{i}{\hbar}\right)[(\mathcal{P}\mathbf{q}_x - \mathbf{q}_x\mathcal{P})\mathbf{p}_y - \mathbf{p}_x(\mathcal{P}\mathbf{q}_y - \mathbf{q}_y\mathcal{P})] + \tau_z.
$$

Applying Eq. (5.16):

$$
\dot{\mathbf{M}}_z = \left(\frac{d\mathcal{P}}{d\mathbf{p}_x}\right)\mathbf{p}_y - \mathbf{p}_x\left(\frac{d\mathcal{P}}{d\mathbf{p}_y}\right) + \tau_z. \tag{5.29}
$$

$\mathcal{P}$ is the kinetic energy part of the Hamiltonian, so

$$
\mathcal{P} = \left(\frac{1}{2m}\right)(\mathbf{p}_x^2 + \mathbf{p}_y^2)
$$

$$
\frac{d\mathcal{P}}{d\mathbf{p}_x} = \left(\frac{1}{m}\right)\mathbf{p}_x \qquad \frac{d\mathcal{P}}{d\mathbf{p}_y} = \left(\frac{1}{m}\right)\mathbf{p}_y
$$

and Eq. (5.29) becomes

$$
\begin{aligned}
\dot{\mathbf{M}}_z &= \left(\frac{1}{m}\right)(\mathbf{p}_y\mathbf{p}_x - \mathbf{p}_x\mathbf{p}_y) + \tau_z \\
&= 0 + \tau_z
\end{aligned}
$$

because $\mathbf{p}_x$ and $\mathbf{p}_y$ commute, Eq. (5.15). In matrix mechanics, the total angular momentum of a particle is conserved, $\dot{\mathbf{M}}_z = 0$, if no torque acts, just as in classical mechanics.

5.6 Heisenberg Uncertainty Relation

The matrix mechanics quantization condition shows that $\mathbf{p}$ and $\mathbf{q}$ do not commute and hence cannot both be diagonal matrices with definite sharp eigenvalues. Heisenberg expressed this quantitatively in his *uncertainty relation*, $\Delta x \Delta p_x \geq \frac{1}{2}\hbar$, that demonstrates a limit to the accuracy of simultaneously measuring both a position x and its conjugate momentum p_x. His relation has generated a vast number of publications by both physicists and philosophers who have struggled to understand the deeper meaning of the uncertainty relation and where many (including Einstein) tried unsuccessfully to find cases it does not hold. Suppose that the momentum of a particle is known precisely so that its de Broglie wavelength is a sharply defined value. The uncertainty relation implies that x could be anywhere in the range $\pm\infty$. If now the particle's position is measured with good accuracy, the momentum suffers an uncertain change bounded by the uncertainty relation. The act of measurement destroys some of what was known about the system.

To estimate the limit of $\Delta x \Delta p_x$ Heisenberg used thought experiments based on accepted physical principles. His thought experiments are possible in principle but might not be possible in practice due to limitations of technology.

Suppose that a beam of particles traveling along the y axis is normally incident upon a slit of width d. The position uncertainty in the x direction is therefore $\Delta x = d$. Before reaching the slit, momentum in the x direction is $p_x = 0$. The particles have de Broglie wavelength $\lambda = \frac{h}{p_y}$, so on passing through the slit diffraction spreads the beam through $\approx \theta$:

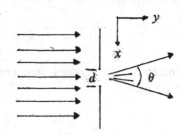

$$\theta \approx \frac{\lambda}{d} = \frac{\lambda}{\Delta x}.$$

Due to diffraction, the particles now have momentum in the x direction:

$$\Delta p_x \approx p_y \theta$$

$$\approx \frac{h}{\lambda}\theta = \left(\frac{h}{\lambda}\right)\left(\frac{\lambda}{\Delta x}\right)$$

$$\Delta x \Delta p_x \approx h.$$

In another thought experiment, Heisenberg used the *Compton effect*, where a photon of known frequency ν scatters from an electron, causing its frequency to decrease slightly to ν'. Imagine a photon with energy $h\nu$ and momentum $\frac{h\nu}{c}$ colliding with an electron (mass m_e) traveling with speed $V \ll c$. Assume that initially the photon and the electron are both moving along the **y** axis as shown. Suppose that after the collision the electron is scattered through some angle with speed V' and the scattered photon is observed along the **x** axis.

Initial:

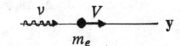

Final:

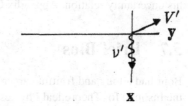

There are no external forces, so energy and momentum are conserved.

$$\text{energy: } h\nu + \frac{p_y^2}{2m_e} = h\nu' + \frac{(p'^2_x + p'^2_y)}{2m_e}$$

$$y \text{ momentum: } \frac{h\nu}{c} + p_y = p'_y$$

$$x \text{ momentum: } \frac{h\nu'}{c} = p'_x$$

There are three equations and three unknowns p'_x, p'_y, ν'. Eliminate p'_x and use the good approximation $\nu'^2 + \nu^2 \approx 2\nu^2$ to give

$$p_y = -\frac{m_e c \nu'}{\nu} + m_e c - \frac{h\nu}{c}.$$

Inaccuracy in p'_y is due to inaccuracy in ν' because ν is accurately known and $m_e c$ is a constant.

$$\Delta p_y \approx \left(\frac{m_e c}{\nu}\right) |\Delta\nu'|$$

To determine ν' accurately, it must be measured for at least a time T where $T\Delta\nu' \approx T\Delta\nu = 1$. During time T, the electron's position along y is uncertain by

$$\Delta y \approx T \left| \frac{p_y - p'_y}{m_e} \right|.$$

From the conservation of y momentum, $p'_y - p_y = \frac{h\nu}{c}$, so

$$\Delta y \approx T \left(\frac{h\nu}{m_e c}\right).$$

Hence:

$$\Delta y \Delta p_y \approx \left(\frac{m_e c \Delta\nu'}{\nu}\right)\left(\frac{h\nu}{m_e c \Delta\nu'}\right)$$
$$\approx h.$$

Heisenberg developed a rigorous and sophisticated quantum-mechanical proof of his uncertainty relation. Appendix C presents a simplified version due to Weyl.

5.7   Brief Bios

Bohr had a long and fruitful career in both physics and public policy. He established the Institute for Theoretical Physics in Copenhagen, still a leader in research. In 1943 the British military secretly flew him from Sweden to England, a perilous trip due to a poorly fitting oxygen mask. In England he joined the Allied atomic bomb project, and after the war worked for international understanding. Bohr was awarded the 1922 Nobel Prize in Physics, cited "for his services in the investigation of the structure of atoms and of the radiation emanating from them."

Werner Heisenberg (1901–76) was awarded the 1932 Nobel Prize in Physics, cited "for the creation of quantum mechanics, the application of which has, inter alia, led to the discovery of the allotropic forms of hydrogen." In the 1930s the

German government passed over Heisenberg for a professorship even though he was a Nobel laureate and the founder of quantum mechanics. The government considered the indeterminacy of quantum mechanics to be "non-Aryan" science.

Max Born (1882–1970, born in Poland) shared the 1954 Nobel Prize in Physics with Walther Bothe. Born was cited "for his fundamental research in quantum mechanics, especially for his statistical interpretation of the wave function." Born's Nobel award was greatly delayed because Born was forced to leave Germany in the 1930s, spending the remainder of his active career in England, India, and Scotland. He returned to Germany after the war, and he and his wife are buried in the Göttingen city cemetery where Planck is also buried. The quantization condition $\mathbf{pq} - \mathbf{qp} = \frac{h}{2\pi i}$ is on Born's tombstone.

– photo courtesy of Julian Herzog/Wikipedia (CC BY 4.0)

Summary of Chapter 5

Chapter 5 reviews Bohr's early quantum theory and lays out the principles and consequences of matrix mechanics, the first complete quantum theory.

a) Before matrix mechanics, Bohr developed a model that introduced the ideas of quantization and discrete energy levels. In Bohr's theory a particle's angular momentum is taken to be an integer multiple of Planck's constant, $mvr = n\hbar$. Using only classical physics and his quantization postulate, Bohr quantitatively accounted for the main features of the hydrogen atom's discrete spectrum.

b) Bohr's frequency condition, $h\nu_{mn} = E_m - E_n$, relates the frequency of a spectral line to the energy difference of the transition.

c) Bohr's model was a dead end because it could not be extended to multi-electron atoms and because its picture of electrons in planetary orbits was too literal.

d) Correspondence principles were helpful guides to the early development of quantum mechanics. Bohr's correspondence principle states that in the limit of large quantum numbers a quantum theory should give the same results as classical physics. Planck's correspondence principle states that classical results should be obtained in the limit $h \to 0$. These principles, though useful, do not hold for all physical systems.

e) Matrix mechanics was the first complete quantum theory. The founders, Heisenberg, Born, and Jordan, were steeped in classical physics. They used its concepts – Hamiltonian, Fourier series – to make the transition from continuous classical to discrete quantum.

f) Heisenberg considered Bohr orbits to be unobservable and considered that a true quantum theory should be based on discrete differences. His early calculations presented puzzling features until Born realized that Heisenberg's equations were a statement of matrix multiplication.

g) In matrix mechanics, physical quantities such as a coordinate $\mathbf{q}$ and its conjugate momentum $\mathbf{p}$ are represented by infinite matrices that contain all possible values of the quantities. Matrix mechanics has a single quantization condition that applies to all physical systems, written as the commutator $\mathbf{pq} - \mathbf{qp} = -i\hbar\mathbf{1}$. A coordinate and its conjugate momentum do not commute. Coordinates, momenta, and a coordinate with a nonconjugate momentum all commute.

h) The commutator of the Hamiltonian matrix $\mathcal{H}$ with any matrix $\mathbf{a}$ gives the time derivative of $\mathbf{a}$ $\mathcal{H}\mathbf{a} - \mathbf{a}\mathcal{H} = -i\hbar\dot{\mathbf{a}}$.

i) A complete quantum theory must account for all physical principles. As an example, the commutator of $\mathcal{H}$ with itself is obviously 0, so its time derivative is 0 unless $\mathcal{H}$ depends explicitly on time, showing that energy does not change with time – in other words, energy is conserved. All other principles such as conservation of linear and angular momentum, Bohr's frequency condition, and so forth can be derived from matrix mechanics.

j) In lectures Heisenberg showed that his famous uncertainty principle follows from idealized thought experiments based on known physical principles such as de Broglie's matter waves. He and others also derived the principle rigorously from quantum mechanics.

Problems and Exercises

5.1   What is the longest wavelength of the Paschen series spectrum? Would it be visible to the human eye?

5.2   For the Lyman series, what is the wavelength of the line with the second longest wavelength? Would it be visible to the human eye?

5.3 In 1924 French theorist Louis de Broglie (1892–1987, Nobel laureate in physics 1929) proposed that matter has a wavelike nature expressed by $\lambda = \frac{h}{mv}$. Show that his relation agrees with Bohr's model, assuming that an electron in a circular orbit has an integral number of matter wavelengths along the circumference.

5.4 This problem presents an example of the Correspondence Principle. Suppose that the electron in hydrogen is traveling in a circular orbit with quantum number n and makes a transition to $n+1$, where $n \gg 1$. Show that the Bohr theory result for the energy change agrees closely with classical physics.

5.5 Write the Hamiltonian for the electron in a hydrogenic atom. Assume that the nucleus is fixed in position.

5.6 Write the Hamiltonian for the two electrons (labeled 1 and 2) in a helium atom. Assume that the nucleus is fixed in position.

5.7 A function $y(t)$ is expressed as a Fourier series.

$$y(t) = \sum_{n=-\infty}^{\infty} \phi_n e^{in\omega t}$$
$$= \cdots + \phi_{-1} e^{-i\omega t} + \phi_0 + \phi_1 e^{i\omega t} + \ldots$$

Show that if $y(t)$ is real, $\phi_{-n} = \phi_n^*$.

5.8 Use the quantization condition Eq. (5.13) to show that $\mathbf{pq}^2 - \mathbf{q}^2\mathbf{p} = -2i\hbar\mathbf{q}$.

5.9 Consider a particle moving freely in the absence of applied forces. Show that according to the principles of matrix mechanics its linear momentum is conserved.

5.10 Consider the electron in a hydrogen atom traveling in a circular orbit. Show that according to the principles of matrix mechanics the electron's angular momentum commutes with the Hamiltonian.
What is the physical meaning of this result?

5.11 Consider the Hamiltonian for a 1-dimensional harmonic oscillator. According to the quantization condition, Eq. (5.13), $\mathbf{pq} - \mathbf{qp}$ is a diagonal matrix. Show that this is a necessary condition by proving that the matrix $\mathbf{d} = \mathbf{pq} - \mathbf{qp}$ must be diagonal. Use the Hamilton equations of motion, but do not use Eq. (5.13).

6

WAVE MECHANICS, MEASUREMENT, AND ENTANGLEMENT

6.1 Introduction

In late 1925, Heisenberg, Born, and Jordan published their "three men's paper" that laid out the principles of matrix mechanics. Only a few months later, Austrian physicist Erwin Schrödinger, who was fluent in German and English, published papers in both languages that presented a different approach to quantum mechanics called *wave mechanics*.

Matrix mechanics and wave mechanics could hardly be more different in appearance. As we saw in Chapter 5, matrix mechanics is based on the discrete nature of atomic transitions and uses the algebra of infinite matrices. Wave mechanics arose from de Broglie's momentum-wavelength relation and from an analogy with light waves; it uses the calculus of partial differential equations and the mathematical properties of special functions.

Nevertheless, both theories give the same experimentally verified results such as the Balmer formula. In the late 1920s, Hungarian-born American mathematical physicist and ex-child prodigy John von Neumann showed that both theories are isomorphic to the same mathematical structure and hence are fundamentally the same.

6.2 Schrödinger's Wave Mechanics

In 1925, Schrödinger read de Broglie's 1924 thesis on the wave nature of matter as expressed by the relation $p = \frac{h}{\lambda}$ between momentum and wavelength. Schrödinger was struck by these ideas and argued that "particles" are only waves and that what is called a "particle" is actually a bundle or *packet* of waves. He drew an analogy with optics. If the wavelength of light is small compared to the dimensions of measuring equipment, it is accurate to treat the light as traveling in straight lines or rays, leading to consequences such as Snell's Law. If, however, light enters an aperture with dimensions comparable to the wavelength, spreading of the beam by diffraction reveals the wave nature of light.

According to Schrödinger's analogy, the wave nature of matter would be apparent on the atomic scale. Consider de Broglie's relation.

$$\lambda = \frac{h}{p} = \frac{h}{mv}$$

Let r be of atomic dimensions.

$$\frac{\lambda}{r} = \frac{h}{mvr}$$

Bohr's quantization condition is $mvr = n\hbar$,

$$\frac{\lambda}{r} = \frac{h}{n\hbar} = \frac{2\pi}{n},$$

so that λ and r are comparable in magnitude for small n. For a Rydberg atom, n is large and λ is small compared to the atom's size. Then classical mechanics holds more accurately, as expected from Bohr's Correspondence Principle.

6.2.1 Wave Packets

To illustrate his concept of wave packets, Schrödinger calculated a wave packet for a 1-dimensional harmonic oscillator having amplitude A and natural angular frequency ω_0 so that the position of the packet is $x = A\cos(\omega_0 t)$. The sketch shows calculated wave packets centered at $x_0 = 0.0$ and at 0.9 A during the course of the harmonic motion. The envelopes of the packets are identical Gaussian (normal) distributions, corrugated by de Broglie wave oscillations. The de Broglie wavelength $\lambda = \frac{h}{p}$ is shortest at the center $x_0 = 0.0$, where the classical particle is moving fastest, and longer near at $x_0 = 0.9$ A, near a turning point where the particle is moving slower.

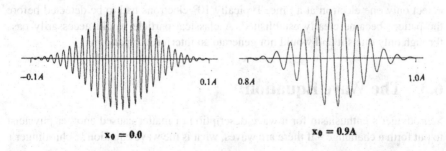

$x_0 = 0.0$ $x_0 = 0.9A$

If a light pulse constructed from Fourier series passes into a medium with varying refractive index, the different frequencies of the Fourier series propagate with different speeds and eventually the pulse disperses, broadening and decreasing in amplitude. The wave packets in Schrödinger's calculation are constructed from a series of harmonic oscillator wave functions (Hermite polynomials) and do not disperse; the wave packets act coherently and do not change overall shape throughout the motion, acting like a particle.

The tic marks in the sketch are drawn outside the main body of the packet for clarity. In work with pulses it is customary to express the width of a pulse as its *FWHM*: *F*ull *W*idth at *H*alf the *M*aximum pulse height. The packets in the sketch both have FWHM ≈ 0.03 A. They are well localized and indeed act like a particle.

Schrödinger's harmonic oscillator model illustrates his concept of the wave nature of matter, but it does not apply to physical systems in general. For the harmonic oscillator, the position of the oscillating particle is known from classical mechanics, but in the scattering of particles the position of the scattered particle is unknown because the wave spreads out from the scattering center. Wave mechanics can only predict the probability that the particle has been scattered into a particular range of angles.

6.2.2   Two-Slit Interference

Two-slit interference (*Young's experiment*) is the classic experiment for demonstrating the wave nature of light. A collimated beam illuminates two narrow slits, as shown in the sketch. The spacing between the slits (not to scale) is comparable to the  light's wavelength. The intensity pattern cast on a distant screen displays interference fringes because of constructive and destructive interference between the diffracted beams.

Technical advances made two-slit experiments with electrons possible. Using nanotechnology, experimenters fabricated slits in a thin gold foil 95 nm wide spaced by 430 nm. A two-slit interference pattern was seen, even when using fast detectors to detect only one electron at a time. Typically 10^5 electrons had to be detected before the pattern became clearly established. A classical particle would necessarily pass through only one slit and would not generate an interference pattern.

6.3   The Wave Equation

Schrödinger's enthusiasm for a wave description of matter caused another physicist to put forth a challenge – if there are waves, what is the wave equation? Schrödinger's response to the challenge was his famous Wave Equation. His initial derivation used mathematical techniques beyond the scope of this text. The approach here follows his later simpler but nonrigorous derivation.

Consider a 1-dimensional wave equation from classical physics,

$$\frac{\partial^2 \psi}{\partial x^2} - \frac{1}{u_p^2} \frac{\partial^2 \psi}{\partial t^2} = 0,$$

where ψ is the wave function and $u_p = v\lambda$ is the *phase velocity* of a wave with a single angular frequency $\omega = 2\pi v$ and wave number $k = 2\pi/\lambda$. A solution of the wave equation for a wave of amplitude C is

$$\psi = Ce^{i(kx-\omega t)},$$

where

$$\frac{\omega}{k} = (2\pi v)\left(\frac{\lambda}{2\pi}\right)$$

$$= v\lambda = u_p$$

$$\frac{\partial^2 \psi}{\partial t^2} = (-i\omega)^2 Ce^{i(kx-\omega t)}$$

$$= -\omega^2 \psi,$$

giving the *time-independent equation*

$$\frac{\partial^2 \psi}{\partial x^2} + \frac{\omega^2}{u_p^2} \psi = 0. \tag{6.1}$$

Here is the central issue: if the wave packet is to represent a particle, the group velocity u_g of the wave packet must equal the velocity v of the particle. This can be achieved if there is dispersion (Section 4.10.1) to make u_g variable:

$$u_g = \frac{d\omega}{dk} \tag{6.2}$$

Let $E = T + V(x)$ be the total energy of the particle where T is kinetic energy, and $V(x)$ is potential energy.

$$T = \frac{1}{2}mv^2$$

$$v = \sqrt{\frac{2T}{m}}$$

$$= \frac{1}{m}\sqrt{2m(E - V)}$$

From Eq. (6.2),

$$\begin{aligned}\frac{1}{u_g} &= \frac{dk}{d\omega}\\ &= \frac{d}{d\nu}\left(\frac{1}{\lambda}\right) = \frac{d}{d\nu}\left(\frac{\nu}{u_p}\right)\\ &= \frac{d}{dE}\left(\frac{E}{u_p}\right),\end{aligned}$$

using $h\nu = E$. For a wave packet to represent a particle, the condition $u_g = v$ must be satisfied:

$$\begin{aligned}\frac{1}{u_g} &= \frac{1}{v}\\ \frac{d}{dE}\left(\frac{E}{u_p}\right) &= \frac{m}{\sqrt{2m(E-V)}}.\end{aligned}$$

Integrating,

$$\begin{aligned}\frac{E}{u_p} &= \int \frac{m}{\sqrt{2m(E-V)}}dE\\ &= \sqrt{2m(E-V)}\\ u_p &= \frac{E}{\sqrt{2m(E-V)}}.\end{aligned} \tag{6.3}$$

Inserting Eq. (6.3) in Eq. (6.1)

$$\frac{\partial^2\psi}{\partial x^2} + \frac{2m\omega^2(E-V)}{E^2}\psi = 0.$$

In the denominator of the second term, use $E = h\nu = \hbar\omega$:

$$\begin{aligned}\frac{\partial^2\psi}{\partial x^2} + \frac{2m(E-V)}{\hbar^2}\psi &= 0\\ \frac{\hbar^2}{2m}\frac{\partial^2\psi}{\partial x^2} + (E-V)\psi &= 0\\ -\frac{\hbar^2}{2m}\frac{\partial^2\psi}{\partial x^2} + V\psi &= E\psi,\end{aligned}$$

the standard form of the time-independent Schrödinger wave equation. For a two-body problem, m would be the reduced mass, as in classical physics. Extending to three dimensions, $\psi = \psi(x, y, z)$ and $V = V(x, y, z)$,

$$-\frac{\hbar^2}{2m}\nabla^2\psi + V\psi = E\psi. \tag{6.4}$$

In Eq. (6.4) ψ is called the *wave function*. Because of de Broglie's relation, the term $-\frac{\hbar^2}{2m}\nabla^2$ is the quantum-mechanical equivalent of kinetic energy $\frac{p^2}{2m}$. The operator on the left-hand side of Eq. (6.4) is therefore the Hamiltonian $\mathcal{H} = kinetic$ *energy* $+$ *potential energy.*

$$\mathcal{H}\psi = E\psi \qquad (6.5)$$

In Eq. (6.5) $\mathcal{H}$ is the matrix of a Hamiltonian operator, and E is the matrix of energy values that satisfy the wave equation.

The Hamiltonian in Schrödinger's wave equation, Eq. (6.4), makes no reference to spin variables; it is an equation only for the spatial (physical coordinate) dependence of the wave function.

Matter waves undergo dispersion as they pass through a region where the potential $V(x, y, z)$ is a function of position, analogous to the dispersion of a light wave as it passes through a medium of varying refractive index. The wave equation is seen to be a description of matter wave dispersion ensuring that wave packets act like particles. As Schrödinger put it, his wave equation contains in itself the law of dispersion.

To derive the time-dependent wave equation, note that ψ depends on time:

$$\psi \sim e^{-i\omega t} = e^{-i2\pi\nu t} = e^{-\frac{iEt}{\hbar}}$$

$$\dot{\psi} = -\frac{iE}{\hbar}\psi$$

$$E\psi = i\hbar\dot{\psi}$$

$$\mathcal{H}\psi = i\hbar\frac{\partial\psi}{\partial t}. \qquad (6.6)$$

6.3.1 Eigenvalues, Eigenvectors, and Matrix Elements

Finite matrices play an important role in wave mechanics. One reason is that the squared magnitude of a matrix element gives the probability that a system makes a transition from one quantum state to another.

From Eq. (6.5) for a state of $\mathcal{H}$ with quantum label n:

$$\mathcal{H}\psi_n = E_n\psi_n. \qquad (6.7)$$

Multiply Eq. (6.7) from the left by the wave function ψ_m^* for state m of $\mathcal{H}$ and integrate.

$$\int \psi_m^* \mathcal{H}\psi_n \, d\tau = E_n \int \psi_m^* \psi_n \, d\tau \qquad (6.8)$$

The quantity on the left-hand side is called the *matrix element* $\mathcal{H}_{mn}$ of $\mathcal{H}$ between states m and n.

$$\mathcal{H}_{mn} \equiv \int \psi_m^* \mathcal{H} \psi_n \, d\tau$$

Suppose that $\mathcal{H}$ is diagonal. If the wave functions are normalized to 1, Eq. (6.8) becomes

$$\int \psi_m^* \mathcal{H} \psi_n \, d\tau = \mathcal{H}_{mn} \, \delta_{mn} = E_n \qquad (6.9)$$

$$\mathcal{H}_{mn} = 0 \qquad (m \neq n).$$

In wave mechanics, the quantized values E_n correspond to the diagonal elements of a diagonal matrix $\mathcal{H}$ and are called the *eigenvalues* of $\mathcal{H}$ (German *eigen*, special). The wave function ψ_n is the *eiqenfunction* corresponding to eigenvalue E_n.

The same nomenclature is used for any operator that satisfies a relation like Eq. (6.7).

$$\mathcal{F} \phi_n = F_n \phi_n$$

$\int \phi_m^* \mathcal{F} \phi_n \, d\tau$ is the matrix element of $\mathcal{F}$ between states m and n, F_n is the eigenvalue for state n, and ϕ_n is the eigenfunction of state n. If $\mathcal{F}$ is a diagonal matrix, the eigenvalues F are the diagonal matrix elements of $\mathcal{F}$.

6.3.2   Dirac's Bra-Ket Bracket Notation

Dirac's *bra-ket* bracket notation replaces the left-hand wave function in a matrix element by a *bra* vector $\langle \; |$, where complex conjugation of the bra vector is assumed. The right-hand wave function is replaced by a *ket* vector, $| \; \rangle$. A bra vector is the adjoint of the corresponding ket. In *bra-ket* notation, a matrix element of **T** is written

$$\int \psi^*(\alpha, j, m) \mathbf{T} \psi(\alpha', j', m') \, d\tau \equiv \langle \alpha, j, m | \mathbf{T} | \alpha', j', m' \rangle,$$

where the $\alpha, j, \ldots$ are various quantum numbers. A bracket does not specify the particular mathematical form of the matrix element, which we have been showing as an integral.

Taking **T** to be the identity operator, $\langle \psi_1 | \psi_2 \rangle$ is the *inner product* of wave functions ψ_1 and ψ_2, analogous to the scalar product familiar from vector algebra. For example, taking $\langle \xi |$ as a 2-dimensional row vector and $| \phi \rangle$ as a 2-dimensional column vector, the inner product $\langle \xi | \phi \rangle$ is a scalar.

$$\begin{pmatrix} \xi_1 & \xi_2 \end{pmatrix} \begin{pmatrix} \phi_1 \\ \phi_2 \end{pmatrix} = \xi_1^* \phi_1 + \xi_2^* \phi_2$$

The *outer product* $| \psi_1 \rangle \langle \psi_2 |$ is a matrix. As an example, $| \phi \rangle \langle \xi |$ is

$$\begin{pmatrix} \phi_1 \\ \phi_2 \end{pmatrix} \begin{pmatrix} \xi_1 & \xi_2 \end{pmatrix} = \begin{pmatrix} \phi_1 \xi_1^* & \phi_1 \xi_2^* \\ \phi_2 \xi_1^* & \phi_2 \xi_2^* \end{pmatrix}.$$

Here is a useful theorem that applies to matrix elements. Consider vector functions χ and ϕ each with two components. Let an operator $\mathbf{S}$ represented by a 2×2 matrix act upon ϕ in a scalar product with χ.

$$
\begin{aligned}
\langle \chi | \mathbf{S}\phi \rangle &= \sum_i \chi_i^* (\mathbf{S}\phi)_i \\
&= \sum_i \sum_j \chi_i^* S_{ij} \phi_j \\
&= \chi_1^* S_{11}\phi_1 + \chi_1^* S_{12}\phi_2 + \chi_2^* S_{21}\phi_1 + \chi_2^* S_{22}\phi_2 \quad (6.10)
\end{aligned}
$$

Now consider

$$
\begin{aligned}
\langle \mathbf{S}^\dagger \chi | \phi \rangle &= \sum_i (\mathbf{S}^\dagger \chi)_i^* \phi_i \\
&= \sum_i \sum_j (S_{ij}^\dagger \chi_j)^* \phi_i \\
&= \sum_i \sum_j (S_{ji}^* \chi_j)^* \phi_i \\
&= \sum_i \sum_j S_{ji} \chi_j^* \phi_i \\
&= S_{11}\chi_1^*\phi_1 + S_{12}\chi_1^*\phi_2 + S_{21}\chi_2^*\phi_1 + S_{22}\chi_2^*\phi_2. \quad (6.11)
\end{aligned}
$$

The vector components and the matrix elements are numbers, so the order of multiplication is unimportant. Equations (6.10) and (6.11) are seen to be equal.

Theorem 10 *The value of the matrix element of an operator is unchanged if the operator acts on one wave function or, alternatively, if its adjoint acts on the other.*

$$
\langle \chi | \mathbf{S}\phi \rangle = \langle \mathbf{S}^\dagger \chi | \phi \rangle \quad \textit{bra-ket form}
$$
$$
\int \chi^* \mathbf{S}\phi \, d\tau = \int (\mathbf{S}^\dagger \chi)^* \phi \, d\tau \quad \textit{integral form} \quad (6.12)
$$

6.4 Quantization Conditions in Wave Mechanics

In matrix mechanics, a single quantization condition Eq. (5.13) in Section 5.4.2 applies to every physical system. What is the quantization condition in wave mechanics? In contrast to matrix mechanics, the wave mechanics solution for every physical system carries within itself its own quantization conditions.

Mathematically, the wave equation Eq. (6.4) has a multitude of possible solutions. Schrödinger showed that choosing only wave functions that are *well behaved* (*physically admissible*) not only greatly restricts the number of possible solutions for a given system but also generates the system's eigenvalues.

The conditions for a wave function to be well behaved are as follows:

- ψ is everywhere continuous, single valued, finite, and differentiable to second order.

- ψ remains finite for $r \to 0$.

- ψ decreases at least as fast as $1/r$ for $r \to \infty$.

- $\int \psi^* \psi d\tau$ integrated over all space is finite.

The last condition implies that $\psi^* \psi$ must satisfy a *normalization condition* such as $\int \psi^* \psi d\tau = 1$.

For a hydrogen atom where $V = -\frac{e^2}{4\pi\epsilon_0 r}$, the wave functions for $E > 0$ are all well behaved and energy levels for $E > 0$ are therefore not quantized. They correspond to hyperbolic orbits in Bohr's model and account for the continuous spectrum. The wave functions for $E > 0$ decrease only slowly as r increases and are therefore significant over distances greater than the size of an atom.

For $E < 0$ (Bohr's elliptic orbits), well-behaved wave functions are possible only for certain values of energy, giving rise to discrete quantized energy levels. These wave functions also extend over all space, but they decrease exponentially with distance, becoming very small beyond atomic dimensions.

6.4.1   Example: Rigid Rotor (Dumbbell)

Quantization of a dumbbell rotating about a fixed axis is a simple problem in wave mechanics, but it nevertheless illustrates the essential features of applying Schrödinger's equation to a physical problem.

The dumbbell consists of two equal masses M linked by a rigid massless rod of total length ρ. The dumbbell rotates in the x-y plane about a fixed axis along z through the center of mass. The potential function is $V = 0$.

The reduced mass μ converts this two-body problem to an equivalent one-body problem, a procedure familiar from classical mechanics.

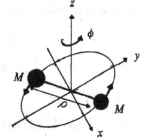

$$\mu = \frac{MM}{M + M} = \frac{M}{2}$$

The Laplacian in polar coordinates is

$$\nabla^2 = \frac{1}{r^2}\frac{\partial}{\partial r}\left(r^2\frac{\partial}{\partial r}\right) + \frac{1}{r^2 \sin\theta}\frac{\partial}{\partial \theta}\left(\sin\theta \frac{\partial}{\partial \theta}\right) + \frac{1}{r^2 \sin^2\theta}\frac{\partial^2}{\partial \phi^2},$$

where θ is the polar angle measured from the z axis, and ϕ is the azimuthal angle in the x-y plane. Rotation is in the x-y plane so that $\theta = \frac{\pi}{2}$. In the one-body problem, $r = \rho$. Because r and θ are constant,

$$\frac{\partial \psi}{\partial r} = 0 \qquad \frac{\partial \psi}{\partial \theta} = 0,$$

leaving only dependence on the angle of rotation ϕ. The wave equation reduces to

$$-\left(\frac{\hbar^2}{2\mu\rho^2}\right)\frac{d^2\psi}{d\phi^2} = E\psi$$

$$\left(\frac{\hbar^2}{M\rho^2}\right)\frac{d^2\psi}{d\phi^2} + E\psi = 0. \tag{6.13}$$

The moment of inertia I about the center of mass is

$$I = M\left(\frac{\rho}{2}\right)^2 + M\left(\frac{\rho}{2}\right)^2 = \left(\frac{M}{2}\right)\rho^2$$

$$M\rho^2 = 2I.$$

Using $M\rho^2 = 2I$ in Eq. (6.13), the wave equation becomes

$$\frac{d^2\psi}{d\phi^2} + \left(\frac{2EI}{\hbar^2}\right)\psi = 0. \tag{6.14}$$

The solution of Eq. (6.14) is

$$\psi(\phi) = Ae^{i\alpha\phi},$$

where A is determined by an initial condition, and α is determined by a quantization condition. Substituting the assumed solution in Eq. (6.14) gives

$$\left(\frac{2EI}{\hbar^2}\right) = \alpha^2. \tag{6.15}$$

For ψ to be single-valued and continuous, the quantization condition is $\alpha = n$, where n is an integer $\ldots, -1, 0, +1, \ldots$ because ψ must remain the same if ϕ is increased by 2π.

$$Ae^{in\phi} = Ae^{in(\phi+2\pi)}$$

$$= Ae^{in\phi}$$

From Eq. (6.15) the energy levels E_n are

$$\frac{2E_n I}{\hbar^2} = n^2$$

$$E_n = \frac{n^2\hbar^2}{2I}. \tag{6.16}$$

In classical mechanics the energy E of a rotating dumbbell is

$$E = \frac{L^2}{2I},$$

where L is the angular momentum of the dumbbell. The classical result agrees with the quantum result Eq. (6.16) for $L = n\hbar$, which is just Bohr's condition for the quantization of angular momentum.

6.4.2   Example: Rigid Rotor (Diatomic Molecule)

The sketch shows two atoms of masses M_1 and M_2 bound in a diatomic molecule. Let M_1 be a distance ρ_1 from the center of mass and M_2 a distance ρ_2. The molecule's bond length is $\rho = \rho_1 + \rho_2$. The molecule rotates about its center of mass (*cm*), but for a molecule in a gas the axis of rotation is not fixed. Hence θ and ϕ in this quantization problem are both variables. Neglect vibration of the bond length so that $V = 0$ and ψ is independent of r.

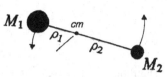

The reduced mass μ of the equivalent one-body problem is

$$\mu = \frac{M_1 M_2}{M_1 + M_2}.$$

With $\frac{\partial \psi}{\partial r} = 0$ the wave equation is

$$\frac{\hbar^2}{2\mu\rho^2} \left(\frac{1}{\sin\theta} \frac{\partial}{\partial\theta} \left(\sin\theta \frac{\partial\psi}{\partial\theta} \right) + \frac{1}{\sin^2\theta} \frac{\partial^2\psi}{\partial\phi^2} \right) + E\psi = 0.$$

The moment of inertia I about the center of mass is

$$I = M_1\rho_1^2 + M_2\rho_2^2.$$

Use the definition of center of mass to eliminate ρ_1 and ρ_2 in favor of ρ.

$$M_1\rho_1 = M_2\rho_2$$

$$\rho_2 = \rho - \rho_1$$

$$\rho_1 = \frac{M_2\rho}{M_1 + M_2}$$

$$\rho_2 = \frac{M_1\rho}{M_1 + M_2}$$

$$I = M_1 \left(\frac{M_2}{M_1 + M_2} \right)^2 \rho^2 + M_2 \left(\frac{M_1}{M_1 + M_2} \right)^2 \rho^2$$

$$= \mu\rho^2$$

The wave equation is then

$$\frac{\hbar^2}{2I} \left(\frac{1}{\sin\theta} \frac{\partial}{\partial\theta} \left(\sin\theta \frac{\partial\psi}{\partial\theta} \right) + \frac{1}{\sin^2\theta} \frac{\partial^2\psi}{\partial\phi^2} \right) + E\psi = 0. \qquad (6.17)$$

Equation (6.17) can be reduced to two equations in θ and in ϕ by separation of variables $\psi(\theta, \phi) = \Theta(\theta)\Phi(\phi)$. Substituting, multiplying by $\sin^2\theta$, and dividing by $\Theta\Phi$ gives

$$\frac{1}{\Theta}\left(\sin\theta \frac{d}{d\theta}\left(\sin\theta \frac{d\Theta}{d\theta}\right)\right) + \left(\frac{2EI}{\hbar^2}\right)\sin^2\theta + \frac{1}{\Phi}\frac{d^2\Phi}{d\phi^2} = 0.$$

The last term on the left-hand side is a function only of ϕ, and the other terms are functions only of θ. Each set of terms must therefore equal the same constant, leading to separate equations for Φ and Θ.

$$\frac{1}{\Phi}\frac{d^2\Phi}{d\phi^2} = \beta \tag{6.18}$$

This equation is also satisfied by the rotating dumbbell in Section 6.4.1. As shown there, its well-behaved solution is

$$\Phi = Ae^{im\phi},$$

where m is an integer $\ldots, -1, 0, +1, \ldots$. Substituting this solution in Eq. (6.18), it follows that $\beta = -m^2$. (Using m in place of n here agrees with standard notation.)
The equation for Θ is

$$\frac{\sin\theta}{\Theta}\frac{d}{d\theta}\left(\sin\theta \frac{d\Theta}{d\theta}\right) + \left(\frac{2EI}{\hbar^2}\right)\sin^2\theta + \beta = 0.$$

Substituting $\beta = -n^2$ and dividing by $\sin^2\theta$ puts the equation in standard form.

$$\frac{1}{\sin\theta}\frac{d}{d\theta}\left(\sin\theta \frac{d\Theta}{d\theta}\right) + \left(\frac{2EI}{\hbar^2} - \frac{m^2}{\sin^2\theta}\right)\Theta = 0 \tag{6.19}$$

The solutions for Θ are the associated Legendre functions $P_\ell^m(\cos\theta)$.
The remaining quantization condition comes from the properties of the associated Legendre functions. They are well-behaved only if the term $\propto EI$ in Eq. (6.19) is equal to $\ell(\ell + 1)$, where ℓ is an integer that satisfies the condition $-m \le \ell \le m$. Hence

$$\frac{2EI}{\hbar^2} = \ell(\ell + 1)$$

$$E = \frac{\hbar^2\ell(\ell + 1)}{2I}. \tag{6.20}$$

ℓ is identified as an angular momentum quantum number and m as a magnetic quantum number. The product functions $\Theta\Phi$ are the *spherical harmonics* $Y_\ell^m(\theta, \phi)$, Section 8.8.2.

$$Y_\ell^m(\theta, \phi) \propto P_\ell^m(\cos\theta)e^{im\phi}$$

Spherical harmonics appear in the solution, because as we shall show in Chapter 8, spherical harmonics are basis functions for the group SO(3) of physical rotations and this problem involves rotation in physical space.

For large ℓ, Eq. (6.20) becomes $E \propto \frac{\ell^2}{2I}$, as expected from Bohr's Correspondence Principle, because in classical mechanics the rotational energy E_r of a rotating body with angular momentum $L = I\omega$ is

$$E_r = \frac{1}{2}I\omega^2 = \frac{L^2}{2I}.$$

6.4.3   Rotational Spectra of Diatomic Molecules

Energy levels of molecules can be qualitatively ordered in a hierarchy that depends on the interaction. Electronic transitions have the largest energy differences, with photons emitted in the visible range of the spectrum. Next are vibrational levels with spectra in the IR. Rotational energies are the smallest, with energy differences in the microwave region. This section discusses the rotational spectra of diatomic molecules, drawing upon Eq. (6.20) for the quantized energy of a rotating diatomic molecule.

Equation (6.20) can be expressed as

$$E = BL(L + 1) \qquad\qquad (6.21)$$

$$B \equiv \frac{\hbar^2}{2I} \quad (B \text{ in energy units}),$$

where B is called the *rotational constant*, L is orbital angular momentum, and E is in units of energy. In molecular spectroscopy the energies Eq. (6.21) are expressed in wave numbers $\bar{\nu}$, so the rotational constant is then

$$E = h\nu = hc\bar{\nu}$$

$$\bar{\nu} = \left(\frac{1}{hc}\right)\left(\frac{\hbar^2}{2I}\right)L(L + 1) = BL(L + 1)$$

$$B \equiv \frac{\hbar}{4\pi c I} \quad (B \text{ in wave number units}).$$

The selection rule for rotational transitions is $\Delta L = \pm 1$, so the frequency of an emitted photon in a transition from level L to the neighboring level $L - 1$ is

$$h\nu = \Delta E = BL(L + 1) - B(L - 1)(L) = 2BL.$$

In practice, the bond length, hence the moment of inertia, increases slightly as $|L|$ increases, causing the spacing to decrease correspondingly. Bond lengths can be measured by comparing observed rotational spectra to theory with small corrections for bond stretching.

The sketch shows idealized energy levels (to scale). Both positive and negative values of L are possible, and both give energies ≥ 0 as shown.

The figure shows the observed rotational absorption spectrum of gas phase diatomic HCl molecules, with each line labeled by L. The lines in the experimental results fall into two branches. The branch at higher wave numbers $L \geq 0$ corresponds to transitions $\Delta L = +1$ and the branch at lower wave numbers $L \leq 0$ to transitions $\Delta L = -1$. The gap between the branches occurs because $L = -1$ and $L = 0$ levels have the same energy, hence no transition is possible.

L	E
-4, 3 ———	12B
-3, 2 ———	6B
-2, 1 ———	2B
-1, 0 ———	0

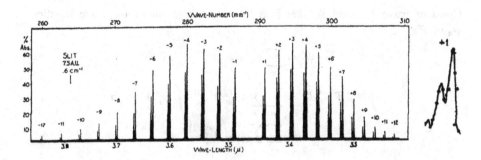

Reprinted figures from C. Meyer and A. Levin Phys. Rev. **34**, 44 (1929). Copyright 1929 by the American Physical Society.

A feature of the HCl spectrum is that each line is a doublet of two close lines, as seen in the high-resolution trace on the right labeled +1. Chlorine has two principal isotopes, one with mass number A = 35 (76 percent natural abundance) and the other with A = 37 (24 percent abundance). These HCl molecules have different moments of inertia, hence their spectral lines have slightly different wave numbers. The more intense absorption line in the doublet is due to molecules with the more abundant ^{35}Cl.

Rotational transitions are in the microwave region with wave numbers of a few cm^{-1} and were inaccessible to earlier experimenters who had only IR spectroscopy at their disposal. However, in combined vibration-rotation a large vibration energy can lift rotation energy levels into the IR for observation.

$$E = \underbrace{(n + \tfrac{1}{2})h\nu_0}_{large} + \underbrace{BJ(J + 1)}_{small}$$

IR absorption occurs between rotation states L grounded on vibration level n and states L' grounded on the next higher vibration level, as indicated in the sketch (not to scale). The solid arrow represents a $\Delta L = +1$ transition from $n = 0$, $L = 2$ to $n = 1$, $L' = 3$. The dashed arrow is a $\Delta L = -1$ transition from $n = 0$, $L = 2$ to $n = 1$, $L' = 1$.

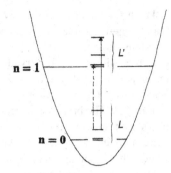

6.5   Matrix Diagonalization

Consider wave function ψ' that is a linear combination of the wave functions ψ_n of $\mathcal{F}$.

$$\psi_i' = \sum_n a_n^{(i)} \psi_n$$

$$\int \psi_j'^* \mathcal{F} \psi_i' \, d\tau = \sum_m \sum_n (a_m^{(j)})^* a_n^{(i)} \int \psi_m^* \mathcal{F} \psi_n \, d\tau$$

$$= \sum_m \sum_n (a_m^{(j)})^* a_n^{(i)} F_{mn}$$

In this formulation, ψ' is not an eigenfunction of $\mathcal{F}$, so the matrix may have nonzero off-diagonal elements. Equation (6.9) states that the eigenvalues F_n are the diagonal elements $\mathcal{F}_{nn}$ only if $\mathcal{F}$ is a diagonal matrix. How can we find the eigenvalues if $\mathcal{F}$ has off-diagonal elements?

The answer is to use a similarity transformation $S\mathcal{F}S^{-1}$ to convert $\mathcal{F}$ to a diagonal matrix $\mathcal{F}'$ so that the eigenvalues are $F_n = \mathcal{F}'_{nn}$. The process of diagonalizing a matrix is an established technique in classical mechanics known as the *principal axis transformation*. One example of its use is to diagonalize a moment of inertia ellipsoid, leaving only the nonzero moments of inertia along the principal symmetry axes.

Let the eigenfunctions of $\mathcal{F}'$ be ψ'. Now express the diagonal matrix of eigenvalues F in terms of the original matrix $\mathcal{F}$.

$$\mathcal{F}'\psi' = F\psi'$$

$$S\mathcal{F}S^{-1}\psi' = F\psi'$$

Multiply by S from the right.

$$S\mathcal{F}\psi' = FS\psi'$$

The relation between F and $\mathcal{F}$ is therefore

$$S\mathcal{F} = FS. \tag{6.22}$$

6.5.1 Example: Diagonalizing a Matrix

The method for solving Eq. (6.22) is more easily understood by a concrete numerical example. Consider the problem of diagonalizing the 2×2 Hermitian matrix $\mathscr{F}$ by means of a similarity transformation with matrix S.

$$\mathscr{F} = \begin{pmatrix} 4 & 6i \\ -6i & -1 \end{pmatrix}$$

$$S \mathscr{F} S^{-1} = E$$

$$S \mathscr{F} = E S$$

$$\begin{pmatrix} S_{11} & S_{12} \\ S_{21} & S_{22} \end{pmatrix} \begin{pmatrix} 4 & 6i \\ -6i & -1 \end{pmatrix} = \begin{pmatrix} E_1 & 0 \\ 0 & E_2 \end{pmatrix} \begin{pmatrix} S_{11} & S_{12} \\ S_{21} & S_{22} \end{pmatrix}$$

$$\begin{pmatrix} 4S_{11} - 6i\,S_{12} & 6i\,S_{11} - S_{12} \\ 4S_{21} - 6i\,S_{22} & 6i\,S_{21} - S_{22} \end{pmatrix} = \begin{pmatrix} E_1 S_{11} & E_1 S_{12} \\ E_2 S_{21} & E_2 S_{22} \end{pmatrix}$$

Equating corresponding matrix elements gives

$$(4 - E_1)S_{11} - 6i\,S_{12} = 0$$
$$6i\,S_{11} - (1 + E_1)S_{12} = 0 \qquad (6.23)$$
$$(4 - E_2)S_{21} - 6i\,S_{22} = 0$$
$$6i\,S_{21} - (1 + E_2)S_{22} = 0. \qquad (6.24)$$

Trivial solutions are $S_{ij} = 0$. Nontrivial solutions occur only for certain values of the eigenvalues E. The eigenvalues are found by setting the determinant of coefficients equal to zero so that Cramer's rule does not apply.

$$\begin{vmatrix} (4 - E) & -6i \\ 6i & -(1 + E) \end{vmatrix} = 0$$

This yields the secular equation

$$E^2 - 3E - 40 = 0.$$

Solutions for E are the two eigenvalues $E_1, E_2 = 8, -5$. Because $\mathscr{F}$ is Hermitian, the eigenvalues are real even though some elements of $\mathscr{F}$ are complex.

Now that the eigenvalues are known, Eqs. (6.23) and (6.24) can be solved for the elements of $\mathbf{S}$. The equations are homogeneous in $\mathbf{S}$, so only ratios of elements can be determined, but the values can be fixed by imposing a normalization condition. Details are left to the problems.

In some cases, a matrix to be diagonalized breaks up into finite blocks so that diagonalization can be accomplished by solving the secular equation for each block. If a secular equation is of high order, numerical methods are required.

6.6   Quantum Measurement

Despite the great success of quantum mechanics in explaining the microscopic world, its deeper meaning continues to be a subject of speculation, no more so than in the topics of measurement and the interpretation of the wave function.

6.6.1   Born's Rule

In 1926, Max Born gave an interpretation of the wave function known as *Born's Rule* now taken to be a key principle of quantum mechanics. According to Born, the square of a wave function $\psi^*\psi$ is a *probability density*. Born did not provide a proof but argued that his interpretation of the wave function was the only possible one.

Consider a wave function for an electron's position $\psi = \psi(x, y, z)$. Then

$$\iiint_V \psi^*\psi \, dx \, dy \, dz$$

is the probability that the electron is located within volume V. If $\psi^*\psi$ is to represent a probability density, it must be normalized to 1 when integrated over all space because probabilities are between 0 and 1.

In classical physics, the measuring equipment plays no essential role in a measurement. The instantaneous distance from the Earth to the Moon can be measured to an accuracy of a few cm or better by reflecting a laser beam from a mirror assembly left on the Moon in 1969 by the Apollo 11 astronauts. A vast number of photons transfer momentum to the Moon during the measurement, but the effect on the Moon's motion is unmeasurably small. In quantum physics, however, a single photon can change the state of an atom, so the effect of measurement on a quantum system cannot be neglected.

Suppose that $\psi(t)$ is the wave function of an idealized isolated system at time t. The wave function at some later time t' is given by the time-dependent Schrödinger equation Eq. (6.6),

$$\mathcal{H}\psi = i\hbar \frac{\partial \psi}{\partial t},$$

that has the solution

$$\psi(t') = \psi(t)e^{-\left(\frac{i}{\hbar}\right)(t'-t)\mathcal{H}}. \tag{6.25}$$

The initial condition $\psi_{t'} = \psi_t$ for $t' = t$ is satisfied. Equation (6.25) shows that the wave function of an isolated system develops continuously in time in a classical *causal* manner.

Now suppose that at some later time t'' the system interacts with a measuring instrument. Suppose further that the original wave function is an eigenfunction of operator $\mathcal{A}$ with eigenvalue A_1, called a *pure* state. The result of a measurement on a pure state of a system must be its eigenvalue A_1.

Suppose instead that the wave function ψ is a *superposition* of several wave functions, called a *mixed* state,

$$\psi = \sum_{n=1}^{N} c_n \psi_n,$$

where the c_n are numerical coefficients. Superposition is always possible because the wave functions of a system form a complete set so that any wave function can be expressed as their weighted sum, analogous to Fourier series. Conversely, the sum of several wave functions is also a wave function, a concept that has no meaning in classical physics.

Let each wave function in a mixed state be an eigenfunction ψ_n of operator $\mathcal{A}$ with eigenvalue A_n. It follows that

$$\mathcal{A}\psi = \sum_{n=1}^{N} c_n A_n \psi_n.$$

What will be the result if the mixed state is attached to a measuring apparatus? In a measurement, the system is no longer isolated but must interact with the measuring apparatus. Here is where probability enters quantum mechanics, much to the dislike of Einstein, who argued that indeterminacy has no place in a complete theory of nature. According to Born's Rule, the result of a measurement on a mixed state will be one of the eigenvalues A_k, with probability $|c_k|^2$. Quantum mechanics does not predict which eigenvalue results from the measurement.

Von Neumann provided a picture of quantum measurement by saying that the very act of measuring a mixed state *collapses* the mixed state to a pure state with a single eigenfunction. A corollary of von Neumann's picture is that if a wave function depends on two operators that do not commute, measuring one collapses the wave function and destroys all previous knowledge of the other so that they cannot be simultaneously measured.

The picture of state collapse (also called *quantum jumps*) is called the *Copenhagen interpretation* because Bohr, who lived and worked in Copenhagen, strongly supported this view. Quantum jumps were first observed in 1986 for single ions trapped in fields. The Copenhagen interpretation focuses on what can be measured and accepts indeterminacy as a fundamental property of nature.

Interaction between a quantum system and a measurement is a fundamental aspect of quantum mechanics and provides an answer to the perennial question: "Is the electron a wave or a particle?" According to quantum measurement theory, an electron is a wave if measured with equipment that measures waves, such as two-slit interference. It is a particle if measured with equipment that measures particles, such as the scattering of a photon from an electron (Compton effect). The Copenhagen interpretation implies that there is no objective reality until a measurement is performed.

6.7 The EPR Paradox and Entanglement

In 1935 Albert Einstein and his coworkers Boris Podolsky and Nathan Rosen published a paper that argued that quantum mechanics is an incomplete theory of physical reality. They supported their argument with a model based on quantum mechanics that in their opinion predicted an "unreasonable" result. The apparent contradiction, known as the *EPR Paradox*, has generated a vast literature, alternative theories, and many experiments.

Appendix D follows the EPR paper in detail. Here is a summary of the EPR argument. Consider a system I described by variables α_I with wave functions $u(\alpha_I)$ and a second system II described by variables β_{II} with wave functions $\psi(\beta_{II})$. The combined wave function is written

$$\Psi(\alpha_I, \beta_{II}) = \sum_{n=-\infty}^{\infty} \psi_n(\beta_{II}) u_n(\alpha_I).$$

The systems are assumed to interact for a time, after which they propagate according to Eq. (6.20) in different directions. Eventually they are so far apart that the time for light to travel between them is longer than the time required for a measurement (spacelike separation). System I is then measured, causing Ψ to collapse to a single term $n = k$. The collapse leaves system I in state u_k and system II in state ψ_k. EPR conclude that this must be a failure of quantum mechanics because it seems unreasonable that a far distant system II can be changed because of a measurement on system I.

Soon after the EPR paper, Schrödinger introduced a term to describe quantum systems that influence one another even though in spacelike separation. He took *entanglement* as his own translation of German *Verschränkung* that can mean "folded arms" (a clever human model of entanglement). According to quantum mechanics, entangled particles constitute a single quantum state, so measurement of one part collapses the entire state to the same measured value. Einstein in an oft-quoted remark called it "spooky action at a distance" (German *spukhafte Fernwirkung*) because it contradicted the accepted view that interactions between systems must be *local*, unaffected by actions occurring far away. Schrödinger wrote that entanglement is instead a real effect and *the* characteristic trait that distinguishes quantum physics from classical physics. According to EPR, a result depending on a distant measurement is not in accord with any reasonable definition of reality. Entanglement may not be "reasonable," but it is nevertheless an inherent feature of quantum physics that has been verified by experiment.

Ironically, EPR were the first to predict quantum entanglement, but they dismissed their result as unreasonable.

6.7.1   Entangled Wave Functions

To define the wave function of an entangled state, start by looking at a wave function that is not entangled. Consider a neutron n^0. It has intrinsic spin $\frac{1}{2}$, a magnetic dipole moment, and no measurable charge. Its spin quantum number S along an arbitrary axis is a dichotomic (two-valued) variable with two spin states S_z symbolized $|\uparrow\rangle$ and $|\downarrow\rangle$.

Suppose that Alice provides $n^o{}_A$ in a general spin state with probability $|a_1^2|$ of $|\uparrow\rangle$ and probability $|a_2^2|$ of $|\downarrow\rangle$.

$$|\psi_A\rangle = a_1\,|\uparrow\rangle + a_2\,|\downarrow\rangle$$
$$\equiv \begin{pmatrix} a_1 \\ 0 \end{pmatrix} + \begin{pmatrix} 0 \\ a_2 \end{pmatrix} = \begin{pmatrix} a_1 \\ a_2 \end{pmatrix},$$

where $a_1^2 + a_2^2 = 1$. Coworker Bob provides another neutron $n^o{}_B$ in its most general spin state

$$|\psi_B\rangle = b_1\,|\uparrow\rangle + b_2\,|\downarrow\rangle$$
$$\equiv \begin{pmatrix} b_1 \\ 0 \end{pmatrix} + \begin{pmatrix} 0 \\ b_2 \end{pmatrix} = \begin{pmatrix} b_1 \\ b_2 \end{pmatrix},$$

where $b_1^2 + b_2^2 = 1$. The combined wave function of the two neutrons is written in terms of a Kronecker product $\otimes$.

$$|\Psi_{AB}\rangle = |\psi_A\rangle \otimes |\psi_B\rangle$$
$$\begin{pmatrix} \Psi_1 \\ \Psi_2 \\ \Psi_3 \\ \Psi_4 \end{pmatrix} = \begin{pmatrix} a_1 \\ a_2 \end{pmatrix} \otimes \begin{pmatrix} b_1 \\ b_2 \end{pmatrix}$$
$$= \begin{pmatrix} a_1 b_1 \\ a_1 b_2 \\ a_2 b_1 \\ a_2 b_2 \end{pmatrix}$$

Note that

$$\Psi_1 \Psi_4 - \Psi_2 \Psi_3 = (a_1 b_1)(a_2 b_2) - (a_1 b_2)(a_2 b_1)$$
$$= 0.$$

The matrix elements a_i and b_j are numbers, so multiplication order is unimportant. If state $|\Psi_{AB}\rangle$ satisfies this criterion, it is said to be *separable* (unentangled): a measurement on state A is unaffected by state B.

If A and B are allowed to interact, an *entangled* state can be formed. Here is the wave function of an entangled state. The factor $\frac{1}{\sqrt{2}}$ is for normalization.

$$|\Psi_{AB}^{(ent)}\rangle = \frac{1}{\sqrt{2}}(|\uparrow\rangle_A \otimes |\uparrow\rangle_B + |\downarrow\rangle_A \otimes |\downarrow\rangle_B)$$
$$= \frac{1}{\sqrt{2}}\left(\begin{pmatrix} a \\ 0 \end{pmatrix} \otimes \begin{pmatrix} b \\ 0 \end{pmatrix} + \begin{pmatrix} 0 \\ a \end{pmatrix} \otimes \begin{pmatrix} 0 \\ b \end{pmatrix} \right)$$

$$\begin{pmatrix} \Psi_1 \\ \Psi_2 \\ \Psi_3 \\ \Psi_4 \end{pmatrix} = \begin{pmatrix} ab \\ 0 \\ 0 \\ 0 \end{pmatrix} + \begin{pmatrix} 0 \\ 0 \\ 0 \\ ab \end{pmatrix}$$

$$= \begin{pmatrix} ab \\ 0 \\ 0 \\ ab \end{pmatrix}$$

$$\Psi_1 \Psi_4 - \Psi_2 \Psi_3 = (ab)^2$$

$$\neq 0$$

This state is not separable; it is entangled. In an entangled state, a measurement of A affects the state B. The example is a state, called a *Bell state*, for two *qubits* that are entangled to the maximum.

6.7.2   Hidden Variables

In classical physics, the kinetic theory of gases is a *statistical* theory that predicts averages and statistical distributions. It is indeterminate and does not predict exact results. However, if the motions of all the gas molecules were accurately known, it would be determinate.

EPR argued that quantum physics seems statistical but might be causal and local if it included as yet unknown additional parameters called *hidden variables*. Quantum physics augmented by hidden variables would then be a determinate theory with no uncertainties. The hidden variables carried along with a photon would predetermine its actions; for example, whether a particular photon is reflected or transmitted by a half-silvered mirror.

6.7.3   Correlation of Entangled Photons: Hidden Variables

The possible existence of hidden variables was a question for theory until advances in instrumentation made it a subject of experiment. A typical experiment measures the polarization of a pair of entangled photons, one of which travels to station A (Alice) and the other an equal distance to station B (Bob), arriving simultaneously at each station. The two stations have spacelike separation guaranteeing that light could not send information between the stations in time to influence the measurements. Such an experiment follows the EPR model.

In bra-ket notation, the state of a vertically polarized photon is symbolized $|\updownarrow\rangle$ and the state of a horizontally polarized photon is $|\leftrightarrow\rangle$. Polarizer cubes have two orthogonal axes and can divert $|\updownarrow\rangle$ photons and $|\leftrightarrow\rangle$ photons individually, as indicated in the sketch, so that light detectors can register the counts separately.

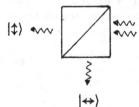

In hidden variable theory (HVT) the correlation of polarized photons is deterministic and can be treated by a vector model of polarization. The sketch shows the axes $\hat{\mathbf{a}}, \hat{\mathbf{a}}'$ of Alice's polarizer and the axes $\hat{\mathbf{b}}, \hat{\mathbf{b}}'$ of Bob's polarizer. Bob's polarizer is set to some angle θ. If $\theta = 0°$, the axes of the A and B polarizers are parallel so that both stations simultaneously detect two photons $|\updownarrow\rangle$ or two photons $|\leftrightarrow\rangle$, giving maximum correlation $C_{cor} = +1$. If $\theta = 90°$, a photon $|\updownarrow\rangle$ detected by A and a photon $|\leftrightarrow\rangle$ detected by B at the same

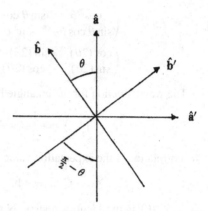

instant are anticorrelated so C_{cor} is the minimum -1. If $\theta = 45°$, there is probability $\frac{1}{2}$ of detecting $|\updownarrow\rangle$ and probability $\frac{1}{2}$ of detecting $|\leftrightarrow\rangle$. There is no correlation, and $C_{cor} = 0$.

For a general angle θ, the correlation for measuring $|\updownarrow\rangle$ on both detectors is $\hat{\mathbf{a}} \cdot \hat{\mathbf{b}} = \cos\theta$ from Malus' law for single-axis polarizers. For two-axis polarizers, it can be seen from the sketch that $|\leftrightarrow\rangle$ polarization along axis $\hat{\mathbf{b}}'$ contributes to $|\updownarrow\rangle$ along $\hat{\mathbf{a}}$ in the amount $\hat{\mathbf{a}} \cdot \hat{\mathbf{b}}' = \cos\left(\theta - \frac{\pi}{2}\right) = \sin\theta$, which must be subtracted to give the net correlation.

$$C_{cor} = \cos\theta - \sin\theta \quad \text{(HVT)} \tag{6.26}$$

6.7.4 Correlation of Entangled Photons: Quantum Mechanics

Quantum mechanics (QM) predicts a different result from HVT for the correlation of entangled photons.

The following matrices describe photon polarization along the $\hat{\mathbf{b}}$ and $\hat{\mathbf{b}}'$ axes with respect to the $|\updownarrow\rangle$ and $|\leftrightarrow\rangle$ axes.

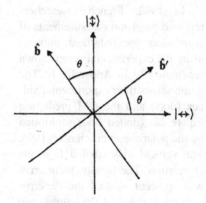

polarized along $\hat{\mathbf{b}}'$:

$$\begin{pmatrix} \cos\theta \\ \sin\theta \end{pmatrix} \begin{pmatrix} |\leftrightarrow\rangle \\ |\updownarrow\rangle \end{pmatrix}$$

polarized along $\hat{\mathbf{b}}$:

$$\begin{pmatrix} -\sin\theta \\ \cos\theta \end{pmatrix} \begin{pmatrix} |\leftrightarrow\rangle \\ |\updownarrow\rangle \end{pmatrix}$$

Form the Kronecker product of each matrix with itself, and subtract.

$$T(\theta) = \begin{pmatrix} \cos\theta \\ \sin\theta \end{pmatrix} \otimes \begin{pmatrix} \cos\theta \\ \sin\theta \end{pmatrix} - \begin{pmatrix} -\sin\theta \\ \cos\theta \end{pmatrix} \otimes \begin{pmatrix} -\sin\theta \\ \cos\theta \end{pmatrix}$$

$$= \begin{pmatrix} \cos^2\theta & \sin\theta\cos\theta \\ \sin\theta\cos\theta & \sin^2\theta \end{pmatrix} - \begin{pmatrix} \sin^2\theta & -\cos\theta\sin\theta \\ -\sin\theta\cos\theta & \cos^2\theta \end{pmatrix}$$

$$= \begin{pmatrix} \cos(2\theta) & \sin(2\theta) \\ \sin(2\theta) & -\cos(2\theta) \end{pmatrix} \tag{6.27}$$

The wave function for the entangled photons is

$$|\Phi^-\rangle = \frac{1}{\sqrt{2}}(|\leftrightarrow\rangle_A|\leftrightarrow\rangle_B - |\updownarrow\rangle_A|\updownarrow\rangle_B).$$

The correlation is the expectation value matrix element

$$C_{cor} = \langle\Phi^-| T_A(0)T_B(\theta) |\Phi^-\rangle ,$$

where $T_A(0)$ is the identity matrix. Note also that the matrix elements are just the diagonal elements of the matrix Eq. (6.27).

$$
\begin{aligned}
C_{cor} &= \frac{1}{2}\langle\leftrightarrow| T_B(\theta) |\leftrightarrow\rangle - \frac{1}{2}\langle\updownarrow| T_B(\theta) |\updownarrow\rangle \\
&= \frac{1}{2}\cos(2\theta) + \frac{1}{2}\cos(2\theta) \\
&= \cos(2\theta) \qquad \text{(QM)} \tag{6.28}
\end{aligned}
$$

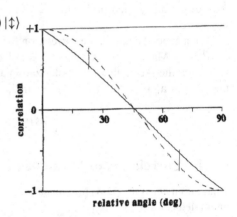

The graph is a plot of the photon correlation according to HVT (solid line, Eq. (6.26)) and QM (dashed line, Eq. (6.28)). The HVT plot is nearly a straight line. The two plots differ, especially near relative angles 22.5° and 67.5° as indicated by the short vertical lines.

In 1982, French researchers reported precision measurements of correlation for polarized photons using the experimental setup shown schematically in Appendix E. The graph shows their experimental values (dots) and the QM prediction Eq. (6.28) (dotted line) multiplied by the polarimeter efficiencies 0.98. The vertical axis labeled **E** is the correlation. The measurement error was 1 percent, so small that the error bars (two standard deviations) are submerged by the dots. The measurements clearly support QM over HVT.

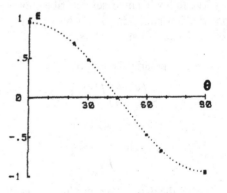

Reprinted figure with permission from A. Aspect *et al.* Phys. Rev. Lett. **49**, 91 (1982). Copyright 1982 by the American Physical Society.

None of the later increasingly sophisticated experiments has shown any discrepancy with the predictions of quantum mechanics. We are left with the successful but strange quantum theory that is not classical, not local, and not determinate.

6.8 Brief Bio

Erwin Schrödinger (1887–1961) was born in Vienna, Austria. Although a native speaker of German, he learned English so well at an early age that he was able to write in excellent English. In addition to research papers he wrote several books on philosophy and science intended for general readership. In World War I he served as an artillery officer in the Austrian army. During the 1930s his dislike for government policies led him to spend 15 years in Ireland at the Dublin Institute for Advanced Studies. He shared the 1933 Nobel Prize in Physics with Paul Dirac, cited "for the discovery of new productive forms of atomic theory." He retired to Austria after the war and was buried in Alpbach. The grave marker shows his time-dependent wave equation.
– photo courtesy of Karl Gruber/Wikimedia Commons/Licensed under the Creative Commons Attribution-Share Alike 3.0 Unported

Summary of Chapter 6

Chapter 6 discusses Schrödinger's wave mechanics, an alternative to matrix mechanics but fundamentally equal to it. Also discussed are measurement in quantum mechanics and the phenomenon of entanglement, a mysterious but characteristic feature of quantum mechanics.

a) De Broglie's idea that matter has a wave nature caused Schrödinger to think of particles as nothing but wave packets.

b) The wave nature of matter has been demonstrated experimentally by two-slit interference of electrons even when only one electron at a time reaches the slits.

c) Schrödinger's simplest derivation of his wave equation starts from the classical wave equation modified by the condition that the group velocity of the wave packet must be equal to the velocity of the particle, a condition achieved by dispersion of the waves.

d) In Schrödinger's wave mechanics, every system carries its own particular quantization condition that requires wave functions to satisfy boundary conditions and to be well behaved. These rules are illustrated by the wave mechanical solution for a dumbbell rotating about a fixed axis and for a diatomic molecule free to rotate about any axis.

e) If a Hamiltonian is purely diagonal, its diagonal elements represent the allowed energies of the system. If the Hamiltonian has off-diagonal matrix elements, it has to be diagonalized by a similarity transformation. The diagonal-element energies are the roots of a secular equation.

a) Born considered that the square of the wave function $\psi^*\psi$ is the probability density of a particle's location. It must be normalized to 1 because the particle is certainly located somewhere.

b) In quantum physics, the measuring equipment must be included as part of the system. When a mixed state sum of wave functions $\Sigma c_n A_n \psi_n$ is measured by operator $\mathcal{A}$, the wave function collapses, yielding one of the eigenvalues with probability $|c_k|^2$. This picture is called the Copenhagen interpretation.

c) Einstein and his collaborators (EPR) published a thought experiment with two interacting quantum systems I and II. They showed that even after the two systems are separated by a great distance, measurement of I collapses the mutual wave function, instantaneously putting II in a definite state related to the state measured for I. EPR concluded that quantum mechanics must be in error if it predicts that systems even very far apart can influence one another.

d) Schrödinger called the EPR phenomenon entanglement and said that it is a real effect. It has been demonstrated experimentally, for example by correlation of photon polarizations.

e) The combined wave function of two quantum systems is expressed by the direct (Kronecker) product. The Kronecker product of two dichotomic systems is a four-element column matrix. If $\Psi_1 \Psi_4 - \Psi_2 \Psi_3 \neq 0$, the systems are entangled.

f) In an effort to eliminate indeterminacy from quantum mechanics, EPR argued that yet unmeasured hidden variables (HVT) would make quantum mechanics (QM) determinate if they were known. The degree of correlation between the photons is different for HVT and for QM. Experiments on entangled photons rule out HVT and support QM.

Problems and Exercises

6.1   A particle of mass m is moving freely along the x-axis (no external forces). What is its wave function?

6.2 Consider a 1-dimensional harmonic oscillator with mass m moving along the x-axis. Its angular frequency is $\omega = \sqrt{\frac{k}{m}}$, where k is the spring constant. The wave function of its ground state $n = 0$ is ($\beta = \frac{\hbar}{m\omega}$).

$$\psi(x) = \frac{1}{(\pi\beta)^{\frac{1}{4}}} e^{-\frac{x^2}{2\beta}}$$

(a) Show that this wave function is normalized.
(b) What is the expectation value $\langle x \rangle$ of x?

6.3 Consider a 1-dimensional harmonic oscillator with mass m moving along the x-axis. Its angular frequency is $\omega = \sqrt{\frac{k}{m}}$, where k is the spring constant. The wave function of its first excited state $n = 1$ is ($\beta = \frac{\hbar}{m\omega}$).

$$\psi(x) = \frac{1}{(\pi\beta)^{\frac{1}{4}}} \sqrt{\frac{2}{\beta}} x\, e^{-\frac{x^2}{2\beta}}$$

Use Schrödinger's wave equation to calculate the energy of this state.

6.4 The wave function for the ground state of hydrogen $n = 1$ is

$$\psi(r) = \frac{1}{\sqrt{a_0^3 \pi}} e^{-\frac{r}{a_0}}.$$

Take the reduced mass to be $\approx m_e$.
(a) Show that this wave function is normalized.
(b) What is the expectation value $\langle r \rangle$ of r?

6.5 The wave function for the first excited state of hydrogen $n = 2$, $\ell = 0$ is

$$\psi(r) = \frac{1}{4\sqrt{2a_0^3 \pi}} \left(2 - \frac{r}{a_0} \right) e^{-\frac{r}{2a_0}}$$

Take the reduced mass to be $\approx m_e$.
(a) Show that this wave function is normalized.
(b) What is the expectation value $\langle r \rangle$ of r?

6.6 The wave function for the first excited state of hydrogen $n = 2$, $\ell = 0$ is

$$\psi(r) = \frac{1}{4\sqrt{2a_0^3 \pi}} \left(2 - \frac{r}{a_0} \right) e^{-\frac{r}{2a_0}}.$$

Take the reduced mass to be $\approx m_e$.
Use Schrödinger's wave equation to calculate the energy of this state.

6.7   Consider the row vector $\langle A|$ and column vector $|B\rangle$:

$$\langle A| = \begin{pmatrix} 3i & 4 & -2i \end{pmatrix} \qquad |B\rangle = \begin{pmatrix} 3 \\ -2i \\ 0 \end{pmatrix}$$

(a) Calculate the inner product $\langle A|B\rangle$.
(b) Calculate the outer product $|B\rangle\langle A|$.

6.8   Consider the row vector $\langle A|$ and column vector $|B\rangle$:

$$\langle A| = \begin{pmatrix} 2 & 2i & -3i \end{pmatrix} \qquad |B\rangle = \begin{pmatrix} 5 \\ 3 \\ -2i \end{pmatrix}$$

(a) Calculate the inner product $\langle A|B\rangle$.
(b) Calculate the outer product $|B\rangle\langle A|$.

6.9   For the rigid rotor dumbbell in Section 6.4.1 (arbitrary amplitude A), normalize the wave function.

6.10   The wave function for one of the excited states of a freely rotating rigid diatomic molecule (Section 6.4.2) is

$$= \sqrt{\frac{3}{8\pi}} e^{i\phi} \sin\theta$$

Use Schrödinger's wave equation to calculate the energy of the system.

6.11   For the diagonalization example in Section 1.5.1 solve for the elements of matrix $\mathbf{S}$ and normalize to obtain definite values. Several methods have been used to normalize a matrix. For this problem, normalize by making the determinant of the matrix equal to 1.

6.12   Diagonalize the reducible matrix for the operation $\mathbf{T}x = -x$ introduced in Section 1.2.

$$\begin{pmatrix} 0 & 1 \\ 1 & 0 \end{pmatrix}$$

What are the irreducible representations?

6.13   Consider the matrix.

$$\begin{pmatrix} 2 & -1 \\ 0 & 1 \end{pmatrix}$$

(a) Diagonalize the matrix.
(b) Show that diagonalization does not change the character (trace) of the matrix.

6.14 Consider the wave function.

$$|\Psi_{AB}^{(ent)}\rangle = \frac{1}{\sqrt{2}}(|\downarrow\rangle_A \otimes |\uparrow\rangle_B) - |\uparrow\rangle_A \otimes |\downarrow\rangle_B)$$

Show that this is an entangled wave function. (It is one of the maximally entangled Bell states.)

6.15 Consider two matrices S and T.

$$S = \begin{pmatrix} a & b \\ c & d \end{pmatrix} \qquad T = \begin{pmatrix} A & B \\ C & D \end{pmatrix}$$

Evaluate $S \otimes T$ and show that the result can be written in the form

$$\mathbf{S} \otimes \mathbf{T} = \begin{pmatrix} aT & bT \\ cT & dT \end{pmatrix}.$$

6.16 Consider two matrices A and B.

$$A = \begin{pmatrix} 3 & 2 \\ 2 & 1 \end{pmatrix} \qquad B = \begin{pmatrix} 4 & 2 \\ 3 & 1 \end{pmatrix}$$

Evaluate $A \otimes B$ and $B \otimes A$. Are they equal?

7

ROTATION

7.1 Introduction

This chapter treats rotation to lay the groundwork for Chapter 8 on angular momentum, the subject that historically led physicists to recognize the power of group theory in quantum mechanics.

7.2 Two Ways of Looking at Rotation

The next two sections show that rotating a vector counterclockwise (positive angle) in a fixed coordinate system gives the same result as keeping the vector fixed and rotating the coordinate system clockwise (negative angle).

7.2.1 Rotated Vector, Fixed Axes

In a 2-dimensional Cartesian coordinate system, vector **A** from the origin is written $A_x\,\hat{\mathbf{i}} + A_y\,\hat{\mathbf{j}}$ with respect to the Cartesian unit vectors. ψ is the angle between **A** and $\hat{\mathbf{i}}$ as shown, and let A be the magnitude of **A**.

Suppose that **A** is rotated counterclockwise about the origin around the z-axis by an additional angle θ, resulting in a new vector **A′** that points to a new point (x', y'). This picture of rotation is called the *alibi* picture because vector **A** can argue that it was somewhere else at the time:

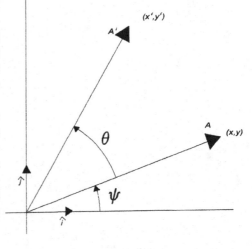

$$\mathbf{A} = A_x \hat{\mathbf{i}} + A_y \hat{\mathbf{j}}$$

$$= A \cos \psi \, \hat{\mathbf{i}} + A \sin \psi \, \hat{\mathbf{j}}$$

$$\mathbf{A}' = A_x' \hat{\mathbf{i}} + A_y' \hat{\mathbf{j}}$$

$$= A \cos (\theta + \psi) \hat{\mathbf{i}} + A \sin (\theta + \psi) \hat{\mathbf{j}}$$

$$= A(\cos \theta \cos \psi - \sin \theta \sin \psi)\hat{\mathbf{i}} + A(\sin \theta \cos \psi + \cos \theta \sin \psi)\hat{\mathbf{j}}$$

$$= A \cos \psi \left(\cos \theta \, \hat{\mathbf{i}} + \sin \theta \, \hat{\mathbf{j}} \right) + A \sin \psi \left(-\sin \theta \, \hat{\mathbf{i}} + \cos \theta \, \hat{\mathbf{j}} \right)$$

$$= A_x \left(\cos \theta \, \hat{\mathbf{i}} + \sin \theta \, \hat{\mathbf{j}} \right) + A_y \left(-\sin \theta \, \hat{\mathbf{i}} + \cos \theta \, \hat{\mathbf{j}} \right)$$

$$= (A_x \cos \theta - A_y \sin \theta)\hat{\mathbf{i}} + (A_x \sin \theta + A_y \cos \theta)\hat{\mathbf{j}}$$

or in matrix form with $A_z' = A_z$,

$$\begin{pmatrix} A_x' \\ A_y' \\ A_z' \end{pmatrix} = \begin{pmatrix} \cos \theta & -\sin \theta & 0 \\ \sin \theta & \cos \theta & 0 \\ 0 & 0 & 1 \end{pmatrix} \begin{pmatrix} A_x \\ A_y \\ A_z \end{pmatrix}. \tag{7.1}$$

A and **A**′ both define the same length from the origin, so a vector transformation is identified as a rotation if it causes no change in the vector's magnitude. Another identifier of rotation is that the matrix in Eq. (7.1) has determinant $\cos^2 \theta + \sin^2 \theta = 1$.

Rotation Notation

A convenient notation for counterclockwise rotation by angle α about axis n is $R(\alpha, n)$. The notation specifies both the angle and the axis of rotation. For example, rotation by θ about the z-axis as in Eq. (7.1) is $R(\theta, z)$. More generally, n can be unit vector $\hat{\mathbf{n}}$ to specify an arbitrary axis.

7.2.2 Rotated Axes, Fixed Vector

In a second way of looking at rotation, the object remains fixed and the axes are rotated clockwise by θ about the z-axis in this example. As the figure illustrates, the vector **A** has the same relation to the axes as in Section 7.2.1, but now **A** is described by its components with respect to the unit vectors $\hat{\mathbf{i}}', \hat{\mathbf{j}}', \hat{\mathbf{k}}'$ of the new coordinate system. This picture is called the *alias* picture because vector **A** can argue it doesn't have the same name as the vector being sought.

Both pictures are valid, but one or the other may prove more convenient in a particular problem.

With the help of the sketch, the unit vectors $\hat{\mathbf{i}}', \hat{\mathbf{j}}'$ in the (x', y') frame can be expressed in terms of the unit vectors $\hat{\mathbf{i}}, \hat{\mathbf{j}}$ in the (x, y) frame:

$$\xi + \psi = \frac{\pi}{2} - \theta$$

$$\cos\left(\xi + \psi\right) = \sin\theta$$

$$\hat{\mathbf{i}}' = \cos\theta\,\hat{\mathbf{i}} - \cos\left(\xi + \psi\right)\hat{\mathbf{j}}$$

$$= \cos\theta\,\hat{\mathbf{i}} - \sin\theta\,\hat{\mathbf{j}}$$

$$\hat{\mathbf{j}}' = \cos\left(\xi + \psi\right)\hat{\mathbf{i}}$$

$$+ \cos\theta\,\hat{\mathbf{j}}$$

$$= \sin\theta\,\hat{\mathbf{i}} + \cos\theta\,\hat{\mathbf{j}}$$

or in matrix form, with $\hat{\mathbf{k}}' = \hat{\mathbf{k}}$,

$$\begin{pmatrix} \hat{\mathbf{i}}' \\ \hat{\mathbf{j}}' \\ \hat{\mathbf{k}}' \end{pmatrix} = \begin{pmatrix} \cos\theta & -\sin\theta & 0 \\ \sin\theta & \cos\theta & 0 \\ 0 & 0 & 1 \end{pmatrix} \begin{pmatrix} \hat{\mathbf{i}} \\ \hat{\mathbf{j}} \\ \hat{\mathbf{k}} \end{pmatrix}.$$

The inverse is found by setting $\theta = -\theta$.

$$\begin{pmatrix} \hat{\mathbf{i}} \\ \hat{\mathbf{j}} \\ \hat{\mathbf{k}} \end{pmatrix} = \begin{pmatrix} \cos\theta & \sin\theta & 0 \\ -\sin\theta & \cos\theta & 0 \\ 0 & 0 & 1 \end{pmatrix} \begin{pmatrix} \hat{\mathbf{i}}' \\ \hat{\mathbf{j}}' \\ \hat{\mathbf{k}}' \end{pmatrix} \tag{7.2}$$

Note that the product of these two matrices is the identity, proving the inverse.

Vector **A** is stationary in this picture with components in the two frames.

$$\mathbf{A} = A_x\,\hat{\mathbf{i}} + A_y\,\hat{\mathbf{j}}$$

$$= A_x'\,\hat{\mathbf{i}}' + A_y'\,\hat{\mathbf{j}}'$$

Use Eq. (7.2) to express the components of **A′** in terms of the components of **A**.

$$A_x'\,\hat{\mathbf{i}}' + A_y'\,\hat{\mathbf{j}}' = A_x\,\hat{\mathbf{i}} + A_y\,\hat{\mathbf{j}}$$

$$= A_x\left(\cos\theta\,\hat{\mathbf{i}}' + \sin\theta\,\hat{\mathbf{j}}'\right) + A_y\left(-\sin\theta\,\hat{\mathbf{i}}' + \cos\theta\,\hat{\mathbf{j}}'\right)$$

$$= \left(A_x\cos\theta - A_y\sin\theta\right)\hat{\mathbf{i}}' + \left(A_x\sin\theta + A_y\cos\theta\right)\hat{\mathbf{j}}'$$

$$A_x' = A_x\cos\theta - A_y\sin\theta$$

$$A_y' = A_x\sin\theta + A_y\cos\theta$$

In matrix form

$$\begin{pmatrix} A_x' \\ A_y' \\ A_z' \end{pmatrix} = \begin{pmatrix} \cos\theta & -\sin\theta & 0 \\ \sin\theta & \cos\theta & 0 \\ 0 & 0 & 1 \end{pmatrix} \begin{pmatrix} A_x \\ A_y \\ A_z \end{pmatrix}, \tag{7.3}$$

Note that here the rotation angle is clockwise positive.

Setting $A_x = x, A_y = y, A_z = z, R(\theta, z)$ transforms the point $\{x, y, z\}$ to the point $\{x', y'z'\}$.

$$\begin{pmatrix} x' \\ y' \\ z' \end{pmatrix} = \begin{pmatrix} \cos \theta & -\sin \theta & 0 \\ \sin \theta & \cos \theta & 0 \\ 0 & 0 & 1 \end{pmatrix} \begin{pmatrix} x \\ y \\ z \end{pmatrix} \tag{7.4}$$

7.2.3 Inverse of a Rotation

The inverse of $R(\theta, z)$ is $R^{-1}(\theta, z) = R(-\theta, z)$ because rotating a vector counterclockwise by θ about a given axis and then clockwise by θ about the same axis causes no net change. With $\cos(-\theta) = \cos\theta$ and $\sin(-\theta) = -\sin\theta$, the rotation matrices are

$$R^{-1}(\theta, z)R(\theta, z) = R(-\theta, z)R(\theta, z)$$

$$= \begin{pmatrix} \cos \theta & \sin \theta & 0 \\ -\sin \theta & \cos \theta & 0 \\ 0 & 0 & 1 \end{pmatrix} \begin{pmatrix} \cos \theta & -\sin \theta & 0 \\ \sin \theta & \cos \theta & 0 \\ 0 & 0 & 1 \end{pmatrix}$$

$$= \begin{pmatrix} \cos^2 \theta + \sin^2 \theta & -\cos\theta\sin\theta + \sin\theta\cos\theta & 0 \\ -\sin\theta\cos\theta + \cos\theta\sin\theta & \sin^2\theta + \cos^2\theta & 0 \\ 0 & 0 & 1 \end{pmatrix}$$

$$= \begin{pmatrix} 1 & 0 & 0 \\ 0 & 1 & 0 \\ 0 & 0 & 1 \end{pmatrix}.$$

Here is a more general way to express the inverse of a rotation matrix. Consider the rotation of vector $\mathbf{A}$ by rotation matrix $R(\alpha, n)$ to give vector $\mathbf{A}' = R(\alpha, n)\mathbf{A}$.

$$\mathbf{A} = A_x \hat{\mathbf{i}} + A_y \hat{\mathbf{j}} + A_z \hat{\mathbf{k}}$$
$$\mathbf{A}' = A_x' \hat{\mathbf{i}} + A_y' \hat{\mathbf{j}} + A_z' \hat{\mathbf{k}}$$

The scalar product of $\mathbf{A}$ with itself is written in matrix form as the product of the row vector of its coefficients times the column vector:

$$\begin{pmatrix} A_x & A_y & A_z \end{pmatrix} \begin{pmatrix} A_x \\ A_y \\ A_z \end{pmatrix} = \tilde{\mathbf{A}} \cdot \mathbf{A},$$

where $\tilde{\mathbf{A}}$ is the transpose of $\mathbf{A}$.

Pure rotation does not change the length of $\mathbf{A}$, so that

$$\tilde{\mathbf{A}} \cdot \mathbf{A} = \tilde{\mathbf{A}}' \cdot \mathbf{A}'$$

$$= \widetilde{(R(\alpha, n)\mathbf{A})} \cdot \mathbf{A}'$$

$$= \widetilde{(R(\alpha, n)\mathbf{A})} \cdot (R(\alpha, n)\mathbf{A}$$

$$= (\tilde{\mathbf{A}}\tilde{R}(\alpha, n)) \cdot (R(\alpha, n)\mathbf{A}),$$

using the general relation $\widetilde{BC} = \tilde{C}\tilde{B}$. Hence

$$\tilde{R}(\alpha, n) \cdot R(\alpha, n) = 1$$
$$R^{-1}(\alpha, n) = \tilde{R}(\alpha, n).$$

The proof shows that the inverse of any rotation matrix is its transpose, a result based on the fundamental property that rotation does not change the magnitude of a vector. As an example, the explicit matrix form shows that $R^{-1}(\theta, z)$ is indeed $\tilde{R}(\theta, z)$.

$$R(\theta, z) = \begin{pmatrix} \cos\theta & -\sin\theta & 0 \\ \sin\theta & \cos\theta & 0 \\ 0 & 0 & 1 \end{pmatrix} \quad R^{-1}(\theta, z) = \begin{pmatrix} \cos\theta & \sin\theta & 0 \\ -\sin\theta & \cos\theta & 0 \\ 0 & 0 & 1 \end{pmatrix} = \tilde{R}(\theta, z)$$

7.2.4 Rotation about an Arbitrary Axis

Take the arbitrary rotation axis to be specified by a unit vector $\hat{\mathbf{n}}$ in a system with fixed axes. $\hat{\mathbf{n}}$ can be expressed in spherical coordinates by polar angle θ (measured from the z-axis) and azimuthal angle ϕ (in the x-y plane).

$$\hat{\mathbf{n}} = \sin\theta\cos\phi\,\hat{\mathbf{i}} + \sin\theta\sin\phi\,\hat{\mathbf{j}} + \cos\theta\,\hat{\mathbf{k}}$$

$\hat{\mathbf{n}}$ can be written in matrix form by making the z-axis unit vector $\hat{\mathbf{k}}$ coincident with $\hat{\mathbf{n}}$. First rotate $\hat{\mathbf{k}}$ by θ about the y-axis and then rotate the result by ϕ about the z-axis.

$$\hat{\mathbf{n}} = R(\phi, z)R(\theta, y)\,\hat{\mathbf{k}}$$
$$\begin{pmatrix} n_x \\ n_y \\ n_z \end{pmatrix} = \begin{pmatrix} \cos\phi & -\sin\phi & 0 \\ \sin\phi & \cos\phi & 0 \\ 0 & 0 & 1 \end{pmatrix} \begin{pmatrix} \cos\theta & 0 & \sin\theta \\ 0 & 1 & 0 \\ -\sin\theta & 0 & \cos\theta \end{pmatrix} \begin{pmatrix} 0 \\ 0 \\ 1 \end{pmatrix} \tag{7.5}$$

We wish to find an expression for $R(\alpha, \hat{\mathbf{n}})$, the rotation operator for rotation by angle α about the $\hat{\mathbf{n}}$ axis. The solution can be summarized in three main steps using five rotation operators.

$$R(\alpha, \hat{\mathbf{n}}) = R(\phi, z)\,R(\theta, y)\,[R(\alpha, z)]\,\tilde{R}(\theta, y)\,\tilde{R}(\phi, z) \tag{7.6}$$

(1) The two operators on the right $\tilde{R}(\theta, y)\tilde{R}(\phi, z)$ undo Eq. (7.5) and move $\hat{\mathbf{n}}$ back to the z-axis.
(2) The operator $R(\alpha, z)$ in square brackets executes rotation by α about $\hat{\mathbf{n}}$, which is now coincident with the z-axis.
(3) The last step is to use the two operators on the left, $R(\phi, z)R(\theta, y)$, to restore $\hat{\mathbf{n}}$ to its original position.

7.3   Rotation of a Function

This section shows how to evaluate a rotated algebraic function. Take as an example rotation about the z-axis.

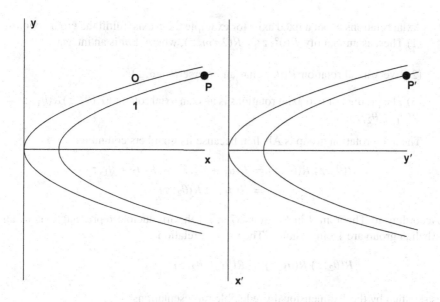

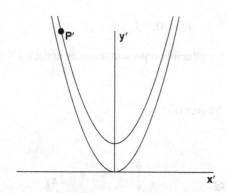

Consider the function $f(x, y) = x - y^2$. The top sketch shows the contours $f = 0$ and $f = 1$. Point $P(x, y, z)$ on the $f = 1$ contour has coordinates $P(5, 2, 0)$.

Now consider counterclockwise rotation of the function by 90°. This is the same as clockwise rotation of the axes, as shown in the middle sketch. From Eq. (7.4) $\mathbf{r} = \tilde{R}(90°, z)\,\mathbf{r}'$, so

$$x = y' \quad y = -x' \quad z = z'.$$

Hence the rotated function is $f' = y' - (-x')^2$. The bottom sketch is the same as the middle sketch but redrawn to make the y'-axis vertical. Comparison with the top sketch shows that the function has indeed been rotated counterclockwise by 90°.

Solving algebraically or directly from Eq. (7.4),

$$x' = -y \quad y' = x \quad z' = z,$$

so the point $P(x, y, z) = P(5, 2, 0)$ becomes $P'(x', y', z') = P'(-2, 5, 0)$. It is on the contour $f' = y' - (-x')^2 = 5 - (-2)^2 = 1$ as expected.

7.4 The Axial Rotation Group

The discussion of the rotation group begins with the simple case of *axial rotation*, which is rotation about a single axis fixed in space.

Axial rotations about a fixed axis, for example the z-axis, fulfill the group axioms:
(1) There is an identity $R(0°, z) = R(2\pi m, z)$, where $\pm m$ is an integer.

(2) Every axial rotation $R(\theta, z)$ has an inverse $R(-\theta, z)$.

(3) The product of two axial rotations is also an axial rotation $R(\theta_2, z) R(\theta_1, z) = R(\theta_1 + \theta_2, z)$.

The axial rotation group is Abelian because its members commute:

$$R(\theta_2, z)\, R(\theta_1, z) = R(\theta_1 + \theta_2, z) = R(\theta_2 + \theta_1, z)$$
$$= R(\theta_1, z)\, R(\theta_2, z).$$

According to Theorem 9 in Section 2.7.4, all the irreducible representations of an Abelian group are 1-dimensional. The product relation

$$R(\theta_2, z)\, R(\theta_1, z) = R(\theta_1 + \theta_2, z)$$

is satisfied by the 1-dimensional irreducible representations

$$D(R(\theta, z)) = e^{im\theta} \qquad m = 0, \pm 1, \pm 2, \ldots.$$

The axial group has an infinite number of irreducible representations as expected for a continuous group.

7.4.1   Vibration of a Hydrogen Molecule

As an application of the axial rotation group, consider the vibrations of a hydrogen molecule H_2, using the method developed in Chapter 3. Take the hydrogen atoms to have the same mass M (same isotope); it is therefore classed as *homonuclear diatomic*. With only two atoms, it is necessarily linear.

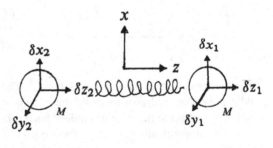

There are six virtual displacements $\delta x_1, \delta y_1, \ldots \delta z_2$ as shown. Basis functions for the molecule's symmetry group are linear combinations of these displacements. Here are the three members of the symmetry group operations where mirrors or inversions are not included.

(1) The identity **1** makes no change: $\delta x_1 \mapsto \delta x_1$, etc. The character $\chi(\mathbf{1}) = 6$.
(2) A 2-fold (180°) rotation $\mathbf{2_x}$ about the x-axis. The diagonal elements in the representation matrix are all 0 because neither atom stays at home: $\delta x_1 \mapsto \delta x_2$ etc. Hence character $\chi(\mathbf{2_x}) = 0$.

Table 7.1 Character table for $\infty 2$

basis functions		**1**	$\mathbf{2_x}$	$\mathbf{R}(\theta, z)$
$z, z^2, x^2 + y^2$	A_1	1	1	1
L_z	A_2	1	-1	1
x, y, xz, yz, L_x, L_y	E_1	2	0	$2\cos\theta$
$x^2 - y^2, xy$	E_2	2	0	$2\cos 2\theta$
$\vdots$	E_m	2	0	$2\cos m\theta$
	$\vdots$	$\vdots$	$\vdots$	$\vdots$

(3) An axial rotation $R(\theta, z)$ by θ about the z-axis. For $m = 1$, each atom contributes $2\cos\theta + 1$ to the character (Theorem 2, Section 2.6) for a total of $4\cos\theta + 2$. There are an infinite number of such representations. Table 7.1 gives representative characters of the group's irreducible representations.

This group is named $\infty 2$ (Hermann–Mauguin). There are two 1-dimensional irreducible representations A_1 and A_2, and 2-dimensional irreducible representations E_m.

The reducible representation matrices use all six basis functions, so the reducible matrices must be equivalent to the 6×6 regular representation. Recall that characters are invariant under similarity transformations and that the regular representation can be reduced to block diagonal form with every irreducible representation present a number of times equal to its dimension. It follows that the $\infty 2$ group can be reduced to $A_1 + A_2 + 2E_1$.

For the trivial $\omega = 0$ translation modes, Table 7.1 shows that x, y, z basis functions transform like $A_1 + E_1$. Subtracting the translations leaves $A_2 + E_1$. There are only two trivial rotation modes for a linear molecule, namely rotations about the x-axis and about the y-axis, both of which displace the atoms significantly. These rotations transform like E_1 according to Table 7.1. Rotation about the z axis (L_z) is not included among the rotation modes because it does not displace the atoms.

A_2 remains as the nontrivial vibration mode. There is only one normal mode for H_2 as expected from Newtonian mechanics. By conservation of momentum, the mode has the atoms simultaneously moving toward and away from one another. The corresponding basis function is therefore $\phi^{(A_2)} = \delta z_1 - \delta z_2$. Group theory can generate the basis function by using Eq. (3.13) from Section 3.5.1.

$$\phi^{(\alpha)} = \sum_T \chi^{(\alpha)*}(\mathbf{T})\mathbf{T}\phi$$

Let $\phi = \delta z_1$ and $\phi = \delta z_2$ be trial functions. The effects of the group operations are

$$\mathbf{1}: \delta z_1 \mapsto \delta z_1 \quad \mathbf{2_x}: \delta z_1 \mapsto -\delta z_2 \quad \mathbf{R}(\theta, \mathbf{z}): \delta z_1 \mapsto \delta z_1$$
$$\mathbf{1}: \delta z_2 \mapsto \delta z_2 \quad \mathbf{2_x}: \delta z_2 \mapsto -\delta z_1 \quad \mathbf{R}(\theta, \mathbf{z}): \delta z_2 \mapsto \delta z_2$$

so that

$$\phi^{(A_2)} = \sum_T \chi^{(A_2)*}(\mathbf{T})\mathbf{T}\delta z_1$$

$$= (1)(\delta z_1) + (-1)(-\delta z_2) + (1)(\delta z_1)$$

$$= 2\,\delta z_1 + \delta z_2$$

$$\phi^{(A_2)} = \sum_T \chi^{(A_2)*}(\mathbf{T})\mathbf{T}\delta z_2$$

$$= (1)(\delta z_2) + (-1)(-\delta z_1) + (1)(\delta z_2)$$

$$= 2\,\delta z_2 + \delta z_1.$$

Subtract to give equal coefficients to ensure conservation of linear momentum,

$$\phi^{(A_2)} = \delta z_1 - \delta z_2,$$

as expected: the atoms move 180° out of phase toward and away from each other.

7.5   The U(1) and SU(2) Groups

U(1) and SU(2) are *Lie groups* (pronounced "Lee") whose elements are unitary matrices, 1-dimensional in the case of U(1) and 2-dimensional for SU(2).

7.5.1   U(1)

The matrix elements of U(1) are 1×1 unitary matrices. They are complex numbers $e^{i\xi}$ of magnitude 1: $|e^{i\xi}| = 1$. The elements of U(1) commute, so U(1) is Abelian, from which it follows that its irreducible representations are all 1-dimensional.

$$e^{i\xi_1}e^{i\xi_2} = e^{i(\xi_1+\xi_2)}$$

$$= e^{i(\xi_2+\xi_1)}$$

$$= e^{i\xi_2}e^{i\xi_1}$$

U(1) is called the *circle group* because its elements represent points on a unit circle centered at the origin of the complex plane. Replacing ξ by θ makes this clear:

$$e^{i\theta} = \cos\theta + i\sin\theta.$$

7.5.2   SU(2)

The members of the SU(2) *Special Unitary* group are 2×2 unitary matrices, *Special* because they must have determinant +1. This section develops the SU(2) matrices and shows how they are related to physical rotations in 3-dimensional space.

The condition that the determinant should be +1 is important because it allows only proper (physical) rotation and excludes nonphysical improper rotation (proper rotation plus inversion) that has determinant -1.

If matrix A is unitary, its adjoint (transpose complex conjugate) equals its inverse $A^\dagger = \tilde{A}^* = A^{-1}$. The imposed conditions substantially restrict the possible form of an SU(2) matrix. Let A be a general 2×2 matrix with complex elements.

$$A = \begin{pmatrix} a & b \\ c & d \end{pmatrix} \qquad A^\dagger = \begin{pmatrix} a^* & c^* \\ b^* & d^* \end{pmatrix} \qquad AA^\dagger = \begin{pmatrix} aa^* + bb^* & ac^* + bd^* \\ ca^* + db^* & cc^* + dd^* \end{pmatrix}$$

From the unitarity condition $AA^\dagger = AA^{-1} = E$ so that

$$aa^* + bb^* = 1$$
$$ca^* + db^* = 0.$$

The determinant condition is

$$ad - bc = +1.$$

Solving,

$$ca^* = -db^*$$
$$c = -\frac{db^*}{a^*}$$
$$+1 = ad - bc = ad + \frac{bb^*d}{a^*} = (aa^* + bb^*)\frac{d}{a^*}$$
$$= \frac{d}{a^*}$$
$$d = a^* \qquad c = -b^*.$$

Every unitary 2×2 matrix with determinant $+1$ therefore has the form

$$A = \begin{pmatrix} a & b \\ -b^* & a^* \end{pmatrix}. \tag{7.7}$$

The elements are any complex numbers that satisfy the condition $aa^* + bb^* = +1$.

The matrices of SU(2) are members of a group.

(1) There is an identity group member: $a = a^* = 1$, $b = b^* = 0$.

(2) Every group member A has an inverse $A^{-1} = A^\dagger$.

(3) The product of two members of SU(2) is also a member of SU(2).

$$\begin{pmatrix} a_1 & b_1 \\ -b_1^* & a_1^* \end{pmatrix}\begin{pmatrix} a_2 & b_2 \\ -b_2^* & a_2^* \end{pmatrix} = \begin{pmatrix} a_1a_2 - b_1b_2^* & a_1b_2 + b_1a_2^* \\ -b_1^*a_2 - a_1^*b_2^* & -b_1^*b_2 + a_1^*a_2^* \end{pmatrix}$$
$$= \begin{pmatrix} c_1 & d_1 \\ -d_1^* & c_1^* \end{pmatrix}$$

The product matrix has determinant $+1$ because the determinant of products is the product of the determinants $(+1)(+1) = +1$.

If a and b are both real, the group is called the *S*pecial *O*rthogonal group SO(2). Physical rotations in three dimensions are represented by the group of 3×3 matrices SO(3). The rotation operator $R(\theta, z)$ Eq. (7.1) is a member of the SO(3) group.

All $n \times n$ unitary and orthogonal matrices have the property that if their rows or columns are considered to be the components of n-dimensional vectors with respect to a set of unit basis vectors $\hat{e}_1, \hat{e}_2, \ldots, \hat{e}_n$, the row and column vectors each form orthonormal sets.

Define a generalized scalar product to take complex components into account. Let u and v be vectors whose components are complex numbers. Their complex scalar product is defined as $\sum_i u_i v_i^*$. For a unitary matrix, the complex scalar product of a row (or column) with itself = 1, and the complex scalar product with a different row (or column) = 0. Take, for example, the first row of matrix A in Eq. (7.5). Its complex scalar product with itself is $(a\hat{e}_1 + b\hat{e}_2) \cdot (a^*\hat{e}_1 + b^*\hat{e}_2) = (aa^* + bb^*) = 1$. The complex scalar product of the first and second column vectors is $(a\hat{e}_1 - b^*\hat{e}_2) \cdot (b^*\hat{e}_1 + a\hat{e}_2) = (ab^* - b^*a) = 0$.

The irreducible matrix representations of an SU(2) group of dimension n are labeled $\frac{1}{2}(n-1)$. The representation for $n = 1$ is the identity representation:

$$D^{(0)} = (1).$$

The SU(2) matrices are not only group elements, but they are also an irreducible representation $n = 2$. This representation has the smallest dimension and is called the *fundamental* representation.

$$D^{\left(\frac{1}{2}\right)} = \begin{pmatrix} a & b \\ -b^* & a^* \end{pmatrix}$$

There is an irreducible representation for any n. Here is the matrix representation for $n = 3$:

$$D^{(1)} = \begin{pmatrix} a^2 & \sqrt{2}\,ab & b^2 \\ -\sqrt{2}\,ab^* & aa^* - bb^* & \sqrt{2}\,a^*b \\ b^{*2} & -\sqrt{2}\,a^*b^* & a^{*2} \end{pmatrix}. \tag{7.8}$$

Multiplying $D^{(1)}$ by its adjoint gives $D^{(1)} D^{(1)\dagger} = E$, so that $D^{(1)}$ is indeed unitary. If a and b satisfy the determinant condition $aa^* + bb^* = +1$, lengthy algebra shows that the determinant of $D^{(1)} = +1$, completing the proof that $D^{(1)}$ is a valid representation of SU(2). Chapter 8 uses angular momentum operators to derive $D^{(1)}$ in terms of physical angles to show that $D^{(1)}$ is a representation of SO(3) for physical rotations in 3-dimensional space.

The Importance of U(1) and SU(2)

If U(1) and SU(2) were important only for generating the circle and physical rotations, it would be easy to dismiss them as only an interesting sidelight on the mathematical theory of groups. In the twentieth century, however, theorists showed that they are some of the most powerful group-theoretic tools for understanding the fundamental structure of matter. These theories guided experimenters to use their accelerators to discover a multitude of new particles, culminating in the development of the Standard Model. The Standard Model organized the discoveries using group theory, and although incomplete, it is the best we have today toward a "Theory of Everything," Chapter 10 is an introduction to particle physics.

7.6 Pauli Matrices, SU(2), and Rotation

Consider the following set of three 2×2 matrices labeled by σ_μ, where the subscript μ can be either numeric $1, 2, 3$ or coordinate x, y, z.

$$\sigma_1 = \sigma_x = \begin{pmatrix} 0 & 1 \\ 1 & 0 \end{pmatrix} \qquad \sigma_2 = \sigma_y = \begin{pmatrix} 0 & -i \\ i & 0 \end{pmatrix} \qquad \sigma_3 = \sigma_z = \begin{pmatrix} 1 & 0 \\ 0 & -1 \end{pmatrix}$$

They are called the *Pauli (spin) matrices*, after German theorist Wolfgang Pauli (1900–58). They have zero trace, and any 2×2 matrix with zero trace can be expressed as a linear combination of the Pauli matrices.

The matrices $\frac{i}{2}\sigma$ plus the 2×2 identity are members of a group su(2), an example of a Lie algebra. Furthermore, the matrices are "infinitesimal" generators for SU(2), so their exponentials $e^{\frac{i}{2}\sigma}$ generate the elements of SU(2).

$$i\sigma_x = \begin{pmatrix} 0 & i \\ i & 0 \end{pmatrix} \qquad i\sigma_y = \begin{pmatrix} 0 & 1 \\ -1 & 0 \end{pmatrix} \qquad i\sigma_z = \begin{pmatrix} i & 0 \\ 0 & -i \end{pmatrix}$$

It is easy to check that their determinants are $\det(\sigma_i) = +1$. The three σ_i are also traceless $\mathrm{tr}(\sigma_i) = 0$, a property related to their determinants because for an operator **A**,

$$\det(e^{\mathbf{A}}) = e^{\mathrm{tr}\mathbf{A}},$$

so $\det(e^{\mathbf{A}}) = +1$ if $\mathrm{tr}\mathbf{A} = 0$.

Pauli matrices are called spin matrices because, when multiplied by $\frac{\hbar}{2}$, they are the components of the quantum spin operator **S** for spin $\frac{1}{2}$ particles such as electrons. For illustration, consider the eigenvalues s of $\frac{\hbar}{2}\sigma_z$. Diagonalization is not a problem here because σ_z is already diagonal. The secular equation is

$$\begin{vmatrix} \frac{\hbar}{2} - s & 0 \\ 0 & -\frac{\hbar}{2} - s \end{vmatrix} = 0,$$

which has solutions $s = \pm\frac{\hbar}{2}$ for the spin of an electron.

The eigenfunctions cannot be continuous functions, because the eigenvalues are discrete. A matrix representation of the spin eigenfunctions is clearly called for. The spin-up u_+ and spin-down u_- eigenfunctions can be represented by column matrices:

$$u_+ = \begin{pmatrix} 1 \\ 0 \end{pmatrix} \qquad u_- = \begin{pmatrix} 0 \\ 1 \end{pmatrix}.$$

Taking σ_z as an example,

$$\sigma_z u_+ = \begin{pmatrix} \frac{\hbar}{2} & 0 \\ 0 & -\frac{\hbar}{2} \end{pmatrix} \begin{pmatrix} 1 \\ 0 \end{pmatrix} = \frac{\hbar}{2} \begin{pmatrix} 1 \\ 0 \end{pmatrix}.$$

This result has the same mathematical structure as the generalized eigenvalue equation $\mathcal{F}\psi = F\psi$. Further discussion is left to Chapter 8 where a more complete theory of quantum angular momentum is developed.

Here is a clue that SO(3) could be related to SU(2). An irreducible representation of SU(2) for any dimension depends on two complex numbers a and b. Hence there are four independent numbers for a given representation matrix. Satisfying the determinant condition leaves three independent numbers, just the number needed to represent the three angles that describe a physical rotation. An equally valid accounting requires two parameters to specify the direction of a rotation axis and a third to specify the angle of rotation. Further, there is an infinite set of such independent parameters, hence an infinite number of representation matrices for any dimension, as expected for a continuous group.

We now establish the relation between SU(2) and the 3-dimensional rotation group SO(3). Let matrix V be the scalar product of $\mathbf{r}$ and the Pauli matrices.

$$V = \mathbf{r} \cdot \boldsymbol{\sigma} = x\,\sigma_x + y\,\sigma_y + z\,\sigma_z = \begin{pmatrix} z & x - iy \\ x + iy & -z \end{pmatrix}$$

The matrix V is not unitary and is therefore not a member of SU(2). As a hint of things to come, Chapter 8 will show that $(x + iy, x - iy, z)$ are basis functions for SO(3).

The matrix $U_z(\alpha)$

$$U_z(\alpha) = \begin{pmatrix} e^{i\frac{\alpha}{2}} & 0 \\ 0 & e^{-i\frac{\alpha}{2}} \end{pmatrix}$$

is unitary with determinant $+1$ and is therefore a member of SU(2); the factor 2 in the denominator of the argument is essential, as we shall show in Section 7.6.1.

Perform a similarity transformation of V with U_z to give matrix V'.

$$\begin{aligned}
V' &= \begin{pmatrix} z' & x' - iy' \\ x' + iy' & -z' \end{pmatrix} = U_z(\alpha) V U_z^{-1}(\alpha) = U_z(\alpha) V \tilde{U}_z^*(\alpha) \\
&= \begin{pmatrix} e^{-i\frac{\alpha}{2}} & 0 \\ 0 & e^{i\frac{\alpha}{2}} \end{pmatrix} \begin{pmatrix} z & x - iy \\ x + iy & -z \end{pmatrix} \begin{pmatrix} e^{i\frac{\alpha}{2}} & 0 \\ 0 & e^{-i\frac{\alpha}{2}} \end{pmatrix} \\
&= \begin{pmatrix} z & e^{-i\alpha}(x - iy) \\ e^{i\alpha}(x + iy) & -z \end{pmatrix}
\end{aligned} \tag{7.9}$$

The similarity transformation has, in fact, generated a physical rotation. As a proof, we first demonstrate that Eq. (7.9) is a rotation and then we find the explicit matrix for the rotation.

Calculate the determinants of V and V'.

$$|V| = \begin{vmatrix} z & x - iy \\ x + iy & -z \end{vmatrix} = -(x^2 + y^2 + z^2)$$

$$|V'| = \begin{vmatrix} z' & x' - iy' \\ x' + iy' & -z' \end{vmatrix} = -(x'^2 + y'^2 + z'^2) \tag{7.10}$$

The determinant of a square matrix is invariant under a similarity transformation (Section 2.6), so $|V| = |V'|$. The distance from the origin to point (x', y', z') is therefore equal to the distance from the origin to point (x, y, z), proving that the coordinate transformation is a rotation.

To find the rotation matrix, equate the elements of V' and $U_z V U_z^{-1}$ in Eq. (7.9).

$$x' - iy' = e^{-i\alpha}(x - iy)$$
$$x' + iy' = e^{i\alpha}(x + iy)$$
$$z' = z$$

The solution for (x', y', z') is

$$\begin{pmatrix} x' \\ y' \\ z' \end{pmatrix} = \begin{pmatrix} \cos(\alpha) & -\sin(\alpha) & 0 \\ \sin(\alpha) & \cos(\alpha) & 0 \\ 0 & 0 & 1 \end{pmatrix} \begin{pmatrix} x \\ y \\ z \end{pmatrix}. \tag{7.11}$$

It represents a rotation of a vector by $\theta = \alpha$ about the z-axis, justifying the subscript z on $U_z(\alpha)$ with α as its argument. (Compare Eq. (7.1).)

Similarly, the SU(2) group member $U_y(\beta)$ induces a physical rotation $R(\beta, y)$ by angle β about the y-axis.

$$U_y(\beta) = \begin{pmatrix} \cos\left(\frac{\beta}{2}\right) & -\sin\left(\frac{\beta}{2}\right) \\ \sin\left(\frac{\beta}{2}\right) & \cos\left(\frac{\beta}{2}\right) \end{pmatrix}$$

$$R(\beta, y) = \begin{pmatrix} \cos\beta & 0 & \sin\beta \\ 0 & 1 & 0 \\ -\sin\beta & 0 & \cos\beta \end{pmatrix} \tag{7.12}$$

Note that the half-angles in the unitary matrix generate full angles in the physical rotation matrix so that the physical rotation is invariant under a change in angle by 2π as it must be.

A sequence of such similarity transformations generates a corresponding sequence of rotations. Suppose that U_1 induces rotation R_1, and U_2 induces R_2. Applying similarity transformations with U_1 and U_2 sequentially gives

$$U_2(U_1 V U_1^{-1})U_2^{-1} = (U_2 U_1)V(U_2 U_1)^{-1}.$$

Hence $U_2 U_1$ induces the sequential rotations $R_2 R_1$.

7.6.1 Spinors

Section 7.5.1 shows that a similarity transformation with a member of SU(2) induces physical rotation in 3-dimensional space. One of the first things learned in the study of trigonometric functions is that they are unchanged by an additional rotation of 2π, as, for example, $\sin(\theta + 2\pi) = \sin\theta$. Not so for the elements of SU(2). In SU(2), additional rotation by 2π changes the sign of the elements: $U(\alpha + 2\pi) = -U(\alpha)$. The signs remain unchanged by rotation through 4π: $U(\alpha + 4\pi) = U(\alpha)$.

$$R(\alpha + 2\pi, z) = R(\alpha, z)$$

$$U_z(\alpha + 2\pi) = \begin{pmatrix} e^{i(\frac{\alpha}{2}+\pi)} & 0 \\ 0 & e^{-i(\frac{\alpha}{2}+\pi)} \end{pmatrix} = \begin{pmatrix} -e^{i\frac{\alpha}{2}} & 0 \\ 0 & -e^{-i\frac{\alpha}{2}} \end{pmatrix}$$

$$= -U_z(\alpha)$$

Physical rotation $R(\alpha, z)$ is unchanged by a 2π rotation, but the sign of U changes. Changing the sign of U does not change the result of a similarity transformation.

$$UVU^{-1} = (-U)V(-U^{-1})$$

For example, $U_z(\alpha)$ and $U_z(\alpha + \pi) = -U_z(\alpha)$ both induce $R(\alpha, z)$. The relation between U and the induced rotation R is therefore a homomorphism, with two values of U for each rotation operator.

Mathematical objects that change sign under 2π rotation are called *spinors*. Spinors do not live in ordinary physical space but in a complex space.

7.7 Euler Angles

Axial rotations describe rotation about a given fixed axis, but more general problems require rotations in 3-dimensional physical space about an arbitrary axis.

A 3×3 matrix has nine elements, not all independent. Three independent numbers (for example, three direction cosines) are needed to specify an arbitrary rotation axis. Recall that the matrices for SU(2) depend on three independent parameters, emphasizing their relation to the three angles of a physical rotation. The *Euler angles*, developed by Swiss-born mathematician Leonhard Euler in 1765, are a widely used and easily visualized method for describing 3-dimensional rotations. His system is a particular sequence of three rotations about various axes. There are several different conventions in use, so care must be taken when consulting the literature.

7.7.1 Euler Angles and Stationary Axes

In many calculations it is convenient to describe the rotation of an object with respect to stationary lab axes. Here is the definition of Euler angles using stationary axes according to the alibi picture, called the *zyz* definition, following the designation of the axes of rotation. The sketches show how the Euler angles with respect to fixed axes map an initial point **P** to an arbitrary final point **P'''** using three sequential rotations.

(1) Rotate **P** about the z-axis by γ to point **P'** located on the arc that intersects the x axis. This is counterclockwise rotation $R(\gamma, z)$ by γ about the z-axis.

(2) Rotate **P'** about the y axis by β to point **P''** at the same latitude as the final point **P'''**. This is counterclockwise rotation $R(\beta, y)$ by β about the y-axis.

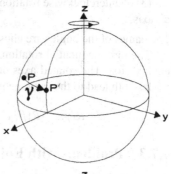

(3) Rotate **P''** about the z-axis by α to reach the final point **P'''**. This is counterclockwise rotation $R(\alpha, z)$ by α about the z-axis.

This definition of Euler angles is with respect to stationary axes, so its matrix for the rotation of a vector **A** can be written as the product $\mathbf{A'''} = R(\alpha, z) R(\beta, y) R(\gamma, z) \mathbf{A}$ because all three matrices refer to the same set of coordinate axes. It is important to maintain the order. Axial rotations are about a fixed axis and commute, but Section 7.8 presents a proof that finite rotations about different axes do not commute.

$$R(\gamma, z) = \begin{pmatrix} \cos \gamma & -\sin \gamma & 0 \\ \sin \gamma & \cos \gamma & 0 \\ 0 & 0 & 1 \end{pmatrix} \quad (7.13)$$

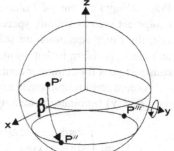

$$R(\beta, y) = \begin{pmatrix} \cos \beta & 0 & \sin \beta \\ 0 & 1 & 0 \\ -\sin \beta & 0 & \cos \beta \end{pmatrix} \quad (7.14)$$

$$R(\alpha, z) = \begin{pmatrix} \cos \alpha & -\sin \alpha & 0 \\ \sin \alpha & \cos \alpha & 0 \\ 0 & 0 & 1 \end{pmatrix} \quad (7.15)$$

$$\mathbf{A'''} = \begin{pmatrix} \cos \alpha & -\sin \alpha & 0 \\ \sin \alpha & \cos \alpha & 0 \\ 0 & 0 & 1 \end{pmatrix} \begin{pmatrix} \cos \beta & 0 & \sin \beta \\ 0 & 1 & 0 \\ -\sin \beta & 0 & \cos \beta \end{pmatrix} \begin{pmatrix} \cos \gamma & -\sin \gamma & 0 \\ \sin \gamma & \cos \gamma & 0 \\ 0 & 0 & 1 \end{pmatrix} \mathbf{A} \quad (7.16)$$

7.7.2 Euler Angles and Rotated Axes

The definition of Euler angles with respect to rotated axes consists of three sequential rotations according to the alias picture where a vector **A** is stationary and the coordinate axes are rotated. The three rotations are

(1) counterclockwise rotation $R(\alpha, z)$ of the coordinate axes by α about the z-axis

(2) counterclockwise rotation $R(\beta, y')$ of the coordinate axes by β about the new y'-axis

(3) counterclockwise rotation $R(\gamma, z'')$ of the coordinate axes by γ about the new z''-axis.

The names of the angles are chosen to be in agreement with the stationary axes case.

The three sequential rotations are written $R(\gamma, z'')R(\beta, y')R(\alpha, z)$, with $R(\alpha, z)$ applied first. The rotated axes definition and the stationary axes definition of Euler angles both lead to the same geometry, so

$$R(\gamma, z'')R(\beta, y')R(\alpha, z) = R(\alpha, z)R(\beta, y)R(\gamma, z). \qquad (7.17)$$

7.7.3   Problems with Euler Angle Applications

Practical applications of Euler angles involve setting the orientation (*attitude*) of equipment such as aircraft, spacecraft, and cameras for computer animation. In practice, the Euler equations are solved to find the Euler angles that give a specified attitude. A problem is that normally two sets of Euler angles give the same attitude, causing issues for computer control.

A more serious problem is a condition called "gimbal lock." If $\beta = 0$ in Eq. (7.16), then $R(\beta, y)$ becomes the identity and

$$\mathbf{A}''' = \begin{pmatrix} \cos\alpha & -\sin\alpha & 0 \\ \sin\alpha & \cos\alpha & 0 \\ 0 & 0 & 1 \end{pmatrix} \begin{pmatrix} \cos\gamma & -\sin\gamma & 0 \\ \sin\gamma & \cos\gamma & 0 \\ 0 & 0 & 1 \end{pmatrix} \mathbf{A}$$

$$= \begin{pmatrix} \cos\alpha\cos\gamma - \sin\alpha\sin\gamma & -\cos\alpha\sin\gamma - \sin\alpha\sin\gamma & 0 \\ \sin\alpha\cos\gamma + \cos\alpha\sin\gamma & -\sin\alpha\sin\gamma + \cos\alpha\cos\gamma & 0 \\ 0 & 0 & 1 \end{pmatrix} \mathbf{A}$$

$$= \begin{pmatrix} \cos(\alpha+\gamma) & -\sin(\alpha+\gamma) & 0 \\ \sin(\alpha+\gamma) & \cos(\alpha+\gamma) & 0 \\ 0 & 0 & 1 \end{pmatrix} \mathbf{A}.$$

The result shows that if $\beta = 0$, then α and γ are no longer independent angles – only their sum contributes, and there is no way to choose the proper α and γ to set the attitude. Attempts to recover from gimbal lock are dangerous because efforts of the control system to reset the attitude can cause violent and possibly damaging maneuvers.

7.8   Finite Rotations Don't Commute

Introductory texts in mechanics make it plausible that finite rotations don't commute by presenting a sketch of an object having an asymmetrical form that clearly distinguishes rotations. The sketch typically shows the object undergoing two successive rotations about different axes compared with the same rotations in reverse order. The outcome is seen to be different, proving the theorem.

The proof by sketch is not needed here because rotation matrices are available. Use Eqs. (7.11) and (7.12) to compare the products $R(\psi, z)R(\theta, y)$ and $R(\theta, y)R(\psi, z)$.

$$
R(\psi, z)R(\theta, y) = \begin{pmatrix} \cos\psi & -\sin\psi & 0 \\ \sin\psi & \cos\psi & 0 \\ 0 & 0 & 1 \end{pmatrix} \begin{pmatrix} \cos\theta & 0 & \sin\theta \\ 0 & 1 & 0 \\ -\sin\theta & 0 & \cos\theta \end{pmatrix}
$$

$$
= \begin{pmatrix} \cos\psi\cos\theta & -\sin\psi & \cos\psi\sin\theta \\ \sin\psi\cos\theta & \cos\psi & \sin\psi\sin\theta \\ \sin\theta & 0 & \cos\theta \end{pmatrix} \tag{7.16}
$$

$$
R(\theta, y)R(\psi, z) = \begin{pmatrix} \cos\theta & 0 & \sin\theta \\ 0 & 1 & 0 \\ -\sin\theta & 0 & \cos\theta \end{pmatrix} \begin{pmatrix} \cos\psi & -\sin\psi & 0 \\ \sin\psi & \cos\psi & 0 \\ 0 & 0 & 1 \end{pmatrix}
$$

$$
= \begin{pmatrix} \cos\psi\cos\theta & -\sin\psi\cos\theta & \sin\theta \\ \sin\psi & \cos\psi & 0 \\ -\cos\psi\sin\theta & \sin\psi\sin\theta & \cos\theta \end{pmatrix} \tag{7.17}
$$

The product matrices are not equal for arbitrary angles, proving that finite rotations about different axes do not commute.

7.9 …But Rotations Do Commute to First Order

If an angle ξ is small, symbolized by $\delta\xi$, *first-order approximations* are $\sin(\delta\xi) \approx \delta\xi$ and $\cos(\delta\xi) \approx 1$. Suppose that the angles θ and ψ in Eqs. (7.16) and (7.17) are small. To first order, noting that the product of two small angles is zero to first order,

$$
R(\delta\psi, z)R(\delta\theta, y) \approx \begin{pmatrix} 1 & -\delta\psi & \delta\theta \\ \delta\psi & 1 & 0 \\ -\delta\theta & 0 & 1 \end{pmatrix}
$$

$$
R(\delta\theta, y)R(\delta\psi, z) \approx \begin{pmatrix} 1 & -\delta\psi & \delta\theta \\ \delta\psi & 1 & 0 \\ -\delta\theta & 0 & 1 \end{pmatrix}.
$$

The products are equal, showing that rotations commute to first order.

Expressing quantities to first order is a useful mathematical tool because, in the limit $\to 0$, all higher-order terms vanish, so it is unnecessary to consider them. Some texts use the term *infinitesimal* for the limit of small-angle rotations, but *first order* is more descriptive of the mathematical process.

7.10 Brief Bios

Leonhard Euler (pronounced "Oiler") (1707–83) was a mathematician of exceptional ability and productivity. In his enormous volume of mathematical results, he introduced notation that has become standard: examples are $f(x)$, e, i, π. He worked for many years at the St. Petersburg Academy of Sciences, the first national science Academy to be directed by a woman (tsarina Catherine II, the Great). He was blind the last years of his life but continued to work at a great pace with the help of assistants and his ability to carry out extensive calculations in his head.

Physicist Wolfgang Pauli (1900–58) was born in Vienna. He was awarded the 1945 Nobel Prize in Physics, cited "for the discovery of the Exclusion Principle, also called the Pauli Principle." In the 1920s, Pauli worked with Max Born on the development of quantum mechanics. Pauli was initially not accepting of matrix mechanics, but he changed his mind when he was able to derive the Balmer formula using matrix mechanics. Pauli's work embraced many fields, including relativity and atomic structure, and he predicted the existence of the neutrino more than 25 years before it was detected experimentally. He left Europe in 1940 for the Institute for Advanced Study in Princeton and became an American citizen. After World War II he returned to Zurich to resume his professorship at the same federal university where Einstein had been a student.

Summary of Chapter 7

Chapter 7 describes rotations according to group theory and matrices.

a) There are two equivalent ways of describing rotation about a fixed axis. In the alibi picture, the coordinate axes are fixed in space and an object (for example, a position vector) rotates with respect to the axes. In the alias picture, the vector is fixed in space and the axes rotate.

b) Rotation of the coordinate axes clockwise is physically the same as rotation of the vector counterclockwise.

c) Rotation about a fixed axis can be represented by a 3×3 matrix. An element 1 on the main diagonal corresponds to the axis of the rotation, because a vector parallel to the axis of rotation is unchanged by the rotation.

d) The inverse matrix of a rotation about a fixed axis is represented by reversing the sign of the rotation angle. Rotation does not change the magnitude of a vector, and because of this special property the inverse matrix of a rotation is the transpose of the rotation matrix.

e) If a function $f(x, y, z)$ is rotated, its value can be found using the rotation matrix on the original coordinates.

f) Rotations about an axis fixed in space form a group called the axial rotation group. The axial rotation group is Abelian, hence its irreducible representations are 1-dimensional and equal to $e^{im\theta}$, where $\pm m$ is an integer.

g) The symmetry group of a vibrating linear molecule includes axial rotation about the linear axis. The nontrivial vibration mode can be found using the character table from references.

h) The SU(2) group is the continuous group of 2×2 unitary matrices having determinant +1. The general form of a member of SU(2) is

$$\begin{pmatrix} a & b \\ -b^* & a^* \end{pmatrix},$$

where the complex numbers a and b satisfy the determinant condition $aa^* + bb^* = +1$.

i) The SU(2) group has irreducible matrix representations of any dimension. The matrix elements of any of its representations depend upon two complex numbers a and b that satisfy the condition $aa^* + bb^* = +1$.

j) All 3-dimensional physical rotations can be generated by similarity transformations of a coordinate matrix with a member of the SU(2) group. The elements of the coordinate matrix come from the scalar product of the position vector **R** and the traceless 2×2 Pauli matrices.

k) Any rotation in three dimensions can be represented by the set of three Euler angles. The Euler angles can be defined using an alibi picture, where coordinate axes are fixed and the vector is rotated, or equivalently by an alias picture, where a vector is fixed in space and the axes are rotated. Equation (7.14) relates the two pictures.

l) Rotations by finite angles do not commute unless they have the same axis of rotation. Rotations do commute to first order in the angles.

Problems and Exercises

7.1 Consider the vector **A**.

$$\mathbf{A} = 2\hat{\mathbf{i}} - 4\hat{\mathbf{j}} + 3\hat{\mathbf{k}}$$

Calculate

$$\mathbf{A}' = R(60°, z)\,\mathbf{A}.$$

7.2 Consider the vector **A**.

$$\mathbf{A} = 2\hat{\mathbf{i}} - 4\hat{\mathbf{j}} + 3\hat{\mathbf{k}}$$

Calculate

$$\mathbf{A}' = R(-30°, z)\,\mathbf{A}.$$

7.3 Calculate the matrix for $R(\theta, x)$ for clockwise rotation of the axes and also for counterclockwise rotation of a vector.

7.4 Calculate the matrix for $R(\theta, y)$ for clockwise rotation of the axes and also for counterclockwise rotation of a vector.

7.5 The line $y = x$ makes an angle of 45° with both the x- and y-axes. It is evident that rotating the line by $R(45°, z)$ makes it coincide with the y-axis. Prove by calculation that this conclusion is correct.

7.6   Consider the hyperbola.

$$\frac{x^2}{4} - y^2 = 1$$

Calculate its equation after rotation by $R(90°, z)$.

7.7   Give an argument to show that the three Pauli spin matrices do not form a group.

7.8   (a) Show that the three Pauli spin matrices are Hermitian.
        (b) Show that the square of each Pauli matrix is the identity.

7.9   For the Pauli matrices, show that

$$\sigma_i \sigma_j = i P \sigma_k,$$

where the subscripts i, j, k are all different $P = +1$ if the subscripts are an even permutation and -1 if the permutation is odd, as defined in Section (1.4.3).

7.10 Consider the spin vector $\boldsymbol{\sigma}_S$.

$$\boldsymbol{\sigma}_S = \frac{\hbar}{2}(\sigma_x \hat{\mathbf{i}} + \sigma_y \hat{\mathbf{j}} + \sigma_z \hat{\mathbf{k}})$$

Calculate the matrix σ_S^2 and determine its eigenvalues.
Note that the numerical factor of the eigenvalue has the form $m(m+1)$ with $m = \frac{1}{2}$. We shall show in Chapter 8 that σ_S^2 is the square of an electron's spin angular momentum.

7.11 With reference to Eq. (7.12), show by calculation that the similarity transformation of matrix V by $U_y(\beta)$ generates the rotation matrix $R(\beta, y)$.

7.12 Diagonalize this Pauli spin matrix,

$$\frac{\hbar}{2}\begin{pmatrix} 0 & -i \\ i & 0 \end{pmatrix},$$

and find its eigenvalues.

8

QUANTUM ANGULAR MOMENTUM

8.1 Introduction

Chapter 8 discusses quantum-mechanical angular momentum, angular momentum operators, irreducible representations of the rotation group, combining quantum-mechanical angular momenta, and selection rules for electron transitions. This chapter is especially long because the material is central to the application of group representation theory to quantum mechanics. It is all of a piece and not easily divided into two chapters.

The chapter begins by describing an experiment in 1922, several years before the advent of modern quantum theory, that demonstrated directional quantization of an atom's angular momentum in a magnetic field.

8.2 Stern and Gerlach: An Important Experiment (1922)

Beginning in 1920 the experimentalist Walther Gerlach (1889–1979) and theorist Otto Stern (1888–1969) carried out experiments on the magnetic properties of neutral atoms that were eventually correctly interpreted as powerful support for the later quantum mechanics. Stern and Gerlach performed their work, now known as the *Stern–Gerlach experiment*, at the University of Frankfurt. Their humble nondescript workroom is classed as a historic science site.

Before modern quantum mechanics, the classical model of the atom due to Ernest Rutherford (1871–1937, Nobel laureate in chemistry 1908) portrayed electrons circulating about a central massive nucleus, and it was natural to think that an electrically neutral atom might have magnetic properties due to moving charges. Stern and

Gerlach investigated these properties using the tool of atomic/molecular beams they initiated and that has been responsible ever since for important discoveries including several Nobel Prizes.

Disclosure: The author carried out his doctoral research at Harvard in the molecular beams laboratory of Norman Ramsey (1915–2011, Nobel laureate in physics 1989).

8.2.1 The Apparatus

As shown in the schematic diagram, the apparatus of the Stern–Gerlach experiment consisted of an electrically heated oven that vaporized the material to be studied. The atoms passed out of

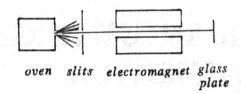

oven   slits   electromagnet   glass plate

the oven through a small orifice and then through a series of slits to define a narrow beam. After passing between the poles of an electromagnet, the atoms deposited on a glass plate. The apparatus was enclosed in an evacuated chamber at a pressure so low (10^{-5} mm Hg) that atoms in the beam had little probability of colliding with residual gas molecules in their passage.

The 1922 paper shows in their Fig. 2 a deposit of silver atoms with the electromagnet turned off to demonstrate the slit geometry. The deposit in the magnified photo was only 1.1 mm high and less than 0.1 mm wide. A run time of 5 hours was required because of the beam's low intensity. Story has it that sulfur in the smoke from Stern's cigar converted the metallic silver to more visible black silver sulfide, but their 1924 paper said a chemical solution was used to develop the trace.

8.2.2 Magnetic Moment in a Magnetic Field

A finite region of circulating currents gives rise to a *magnetic moment vector* μ that characterizes the magnetic field generated by the currents. As an example, a flat circular loop of radius r carrying current $\mathcal{I}$ produces a magnetic moment $\mu = \mathcal{I} \times \pi r^2$ with μ normal to the plane of the loop.

If the current is due to an electron e circulating with speed v, then $\mathcal{I} = \frac{ev}{2\pi r}$. The electron has angular momentum $L = mvr$ along the line of μ.

$$\frac{\mu}{L} = \left(\frac{ev}{2\pi r}\right)(\pi r^2)\left(\frac{1}{mvr}\right) = \frac{e}{2m}$$

The ratio of magnetic moment to angular momentum is called the *gyromagnetic ratio*. $\mathbf{L}/L$ is a unit vector along $\mathbf{L}$, so in vector form the magnetic moment is

$$\mu = \mu \left(\frac{\mathbf{L}}{L} \right).$$

A magnetic field **B** transverse to the beam produces a torque $\mu \times \mathbf{B}$ that causes μ to precess (the *Larmor precession*) about the field direction. As the sketch suggests, the resultant moment along the line of **B** is

$\mu_{res} = \mu \cos \theta$, where θ is the cone half-angle. A magnetic moment in a field gradient experiences a force proportional to the moment and along the direction of the gradient.

$$F_x = \mu_{res} \cdot \frac{\partial \mathbf{B}_x}{\partial x}$$

Stern and Gerlach used shaped pole pieces to produce a gradient so that an atom in the beam would be deflected transversely, thus making the trajectory a "spectrometer" for an atom's magnetic moment. The larger the value of μ_{res}, the more an atom is deflected transverse to the main beam direction.

Fig. 1 from the 1922 paper shows the cross section of the poles. The gap was very small; the distance from the tip of the angled pole to the face of the neighboring pole was only 1 mm, requiring great care in alignment. The rectangle B between the poles is the cross section of the beam.

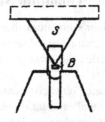

8.2.3   The Result

Classically, silver atoms in the beam would be expected to have their resultant moments pointing in all possible directions transverse to the beam, causing the deposit on the glass plate to be a bounded continuous smear, $-1 \leq \cos \theta \leq 1$. Fig. 3 is their photo of the result from an 8-hour run with the magnet energized.

Instead of the continuous smear expected classically, the silver trace was split into two parts with no visible deposit between them. Stern and Gerlach interpreted their result as showing that the atom's angular momentum (hence its magnetic moment) was *quantized* – pointing only with or against the field direction. The title of their 1922 paper can be translated "Experimental Proof of Directional Quantization in a Magnetic Field." They clearly understood that quantum physics was involved.

The magnetic moment of an orbiting electron is

$$\frac{\mu}{L} = g_L \frac{e\,\hbar}{2m_e} = (1)\mu_B$$

$$\mu_B = \frac{e\,\hbar}{2m_e}, \tag{8.1}$$

where μ_B is called the *Bohr magneton*, and $g_L = 1$ is the *orbital angular momentum g-factor*. After painstaking calibration, Stern and Gerlach showed that the measured splitting of the silver trace was consistent with one Bohr magneton to ± 10 percent accuracy. From modern experiments, the Bohr magneton is

$$\mu_B \approx 9.274 \times 10^{-24} \text{ J} \cdot \text{T}^{-1},$$

where J (joule) is the unit of energy, and T (tesla) is the unit of magnetic field strength. 1 T = 10,000 gauss = 10 kG.

Credits: The illustrations labeled Fig. 1, Fig. 2, and Fig. 3 are reprinted from W. Gerlach and O. Stern, *Z. Phys.* **9**, 349 (1922). Figs. 2 and 3 have been contrast enhanced for clarity. With permission of the Licensed Content Publisher Springer Nature.

8.2.4   Quantum Numbers

In Bohr's quantum theory (Section 5.2), the quantum number n = integer originally characterized an orbiting electron's quantized angular momentum but came to be called the *principal quantum number* that specifies the size and energy of the orbit.

Later physicists extended Bohr's model to elliptic orbits and introduced the *azimuthal* or *orbital quantum number* L = integer $0, 1, 2, \ldots$ for quantized orbital angular momentum in units of $\hbar$. Zero angular momentum was excluded from the old theory because it implies an orbit with the electron heading straight for the nucleus. The *magnetic quantum number* m is a projection of the angular momentum along an axis and takes integer values $-L, -L + 1, \ldots, 0, \ldots, L - 1, L$.

In 1925 Dutch-born Samuel Goudsmit (1902–78) and George Uhlenbeck (1900–88), young physics graduate students at University of Leiden, took up earlier ideas and postulated a fourth quantum number, *spin angular momentum S*, to explain detailed features of atomic spectra. Unlike orbital angular momentum, which refers to the physical motion of a particle, spin angular momentum is an intrinsic property of a particle independent of its motion and cannot be altered. The spin angular momentum S of an electron has only two possible values $\pm\frac{1}{2}\hbar$. According to modern scattering experiments, the electron is a dimensionless point particle yet has mass and angular momentum contrary to classical mechanics.

Relativistic theory of the electron predicts its gyromagnetic ratio to be twice what would be expected classically for a spinning ball of charge. This nonclassical behavior is accounted for by the dimensionless *spin g-factor* g_S so that $\mu_e = g_S \frac{eS}{2m_e}$ where $g_S \approx 2.002 \ldots$ The difference $(g_S - 2)$ has been measured to an accuracy of a few parts in 10^{13}, and in 1948 Harvard University theorist Julian Schwinger (1918–94) used quantum electrodynamics (QED) to calculate

$$\mu_e = g_S \frac{e\,S}{2m_e}$$

$$\approx 2\left(1 + \frac{\alpha}{2\pi} - \ldots\right)\frac{e\,(\hbar/2)}{2m_e}$$

$$\approx (2.002)\frac{e\,\hbar}{4m_e}$$

$$\text{where} \quad \alpha = \frac{\mu_0 c e^2}{2h} \approx 7.297 \times 10^{-3} \approx \frac{1}{137}.$$

α is the dimensionless *fine structure constant*. Here μ_0 is the *permeability of vacuum* defined as $4\pi \times 10^{-7}$ SI units, and c is the speed of light in vacuum $\approx 3 \times 10^8$ m · s^{-1}.

The total angular momentum **J** of an atom arises from two sources: *orbital angular momentum* **L**, due to physical motion of electrons, and intrinsic *spin angular momentum* **S**, so that $\mathbf{J} = \mathbf{L} + \mathbf{S}$.

μ_e is approximately one Bohr magneton because the spin quantum number $S = \frac{1}{2}$ so that $(2.002)\left(\frac{1}{2}\right)\mu_B \approx \mu_B$. The Stern–Gerlach experiment had to be reinterpreted. According to later research, a silver atom in its ground state has a single $L = 0$ electron in the outermost $n = 5$ shell. The other 46 electrons are in closed shells and contribute zero angular momentum. The angular momentum of a silver atom is due to the intrinsic spin angular momentum of the outermost electron. The two separated silver traces in the Stern–Gerlach experiment actually correspond to electron spin angular momentum, with spin quantum numbers $+\frac{1}{2}$ (μ_e parallel to **B**) and $-\frac{1}{2}$ (μ_e antiparallel to **B**). The Stern–Gerlach experiment measured a magnetic moment of approximately one Bohr magneton Eq. (8.1). It was numerically accurate but physically incorrect.

8.3 Rotation and Angular Momentum Operators

The founders of quantum mechanics were steeped in classical mechanics and used it as a guide to the quantum world. A deep reason for the appearance of angular momentum in quantum mechanics is that rotational symmetry is such a fundamental property that it must be involved in a wide range of physical theories.

In this chapter, unspecified **I** symbolizes a general quantum angular momentum, either orbital or spin. This not a restriction because in group theory every angular momentum obeys the same relations.

Let a vector $\mathbf{A} = x\hat{\mathbf{i}} + y\hat{\mathbf{j}} + z\hat{\mathbf{k}}$ be rotated counterclockwise about the z-axis by angle θ. Let the rotated vector be $\mathbf{A}' = x'\hat{\mathbf{i}} + y'\hat{\mathbf{j}} + z'\hat{\mathbf{k}}$. From Eq. (7.1)

$$\begin{pmatrix} x' \\ y' \\ z' \end{pmatrix} = \begin{pmatrix} \cos\theta & -\sin\theta & 0 \\ \sin\theta & \cos\theta & 0 \\ 0 & 0 & 1 \end{pmatrix} \begin{pmatrix} x \\ y \\ z \end{pmatrix}.$$

Assume that θ is a small angle $\delta\theta$ so to first order $\cos(\delta\theta) \approx 1$ and $\sin(\delta\theta) \approx \delta\theta$. Thus

$$\begin{pmatrix} x' \\ y' \\ z' \end{pmatrix} = \begin{pmatrix} \cos(\delta\theta) & -\sin(\delta\theta) & 0 \\ \sin(\delta\theta) & \cos(\delta\theta) & 0 \\ 0 & 0 & 1 \end{pmatrix} \begin{pmatrix} x \\ y \\ z \end{pmatrix} \approx \begin{pmatrix} 1 & -\delta\theta & 0 \\ \delta\theta & 1 & 0 \\ 0 & 0 & 1 \end{pmatrix} \begin{pmatrix} x \\ y \\ z \end{pmatrix}$$

so that

$$\begin{aligned} x' &= x - y\delta\theta & \implies x' - x &= -y\delta\theta \\ y' &= x\delta\theta + y & \implies y' - y &= x\delta\theta \\ z' &= z & \implies z' - z &= 0. \end{aligned}$$

Consider the rotation of a function $f(x, y, z)$ by θ and expand in Taylor's series.

$$R(\theta, z)f(x, y, z) = f(x', y', z')$$

$$R(\delta\theta, z) \approx f(x, y, z) + \left.\frac{\partial f}{\partial x}\right|_x (x' - x) + \left.\frac{\partial f}{\partial y}\right|_y (y' - y) + \left.\frac{\partial f}{\partial z}\right|_z (z' - z)$$

$$\approx f(x, y, z) - y\delta\theta\frac{\partial f}{\partial x} + x\delta\theta\frac{\partial f}{\partial y} \tag{8.2}$$

$$R(\delta\theta, z)f(x, y, z) \approx f(x, y, z) - \delta\theta\left(y\frac{\partial}{\partial x} - x\frac{\partial}{\partial y}\right)f(x, y, z)$$

$$R(\delta\theta, z) = 1 - \delta\theta\left(y\frac{\partial}{\partial x} - x\frac{\partial}{\partial y}\right) \tag{8.3}$$

The operator in parentheses in Eq. (8.3) is the operator for quantum orbital angular momentum. If a particle has position vector $\mathbf{r}$ and quantum-mechanical momentum $\mathbf{p} = -i\hbar\nabla = -i\hbar(\hat{\mathbf{i}}\frac{\partial}{\partial x} + \hat{\mathbf{j}}\frac{\partial}{\partial y} + \hat{\mathbf{k}}\frac{\partial}{\partial z})$, its orbital angular momentum $\mathbf{I} = \mathbf{r} \times \mathbf{p}$ in determinant form is

$$\mathbf{I} = -i\hbar \begin{vmatrix} \hat{\mathbf{i}} & \hat{\mathbf{j}} & \hat{\mathbf{k}} \\ x & y & z \\ \frac{\partial}{\partial x} & \frac{\partial}{\partial y} & \frac{\partial}{\partial z} \end{vmatrix}.$$

The Cartesian components of the quantum *orbital angular momentum operator* are thus

$$I_x = -i\hbar \left(y\frac{\partial}{\partial z} - z\frac{\partial}{\partial y} \right)$$

$$I_y = -i\hbar \left(z\frac{\partial}{\partial x} - x\frac{\partial}{\partial z} \right)$$

$$I_z = -i\hbar \left(x\frac{\partial}{\partial y} - y\frac{\partial}{\partial x} \right). \tag{8.4}$$

In Eq. (8.4) the operators in parentheses are dimensionless. The dimension of angular momentum is contained entirely in the factor $\hbar$ because quantized angular momentum comes in integer or half odd integer multiples of $\hbar$.

Comparing Eqs. (8.3) and (8.4),

$$R(\delta\theta, z) = 1 - i\delta\theta\, \frac{I_z}{\hbar}, \tag{8.5}$$

and similarly for I_x and I_y. Equation (8.5) is correct to first order in $\delta\theta$. It is an operator equation with 1 standing for the identity.

To find the angular momentum operator for a finite angle, note that angles add for successive rotations about a given fixed axis so that a finite angle can be expressed as the sum of angles $\sum_n \left(\frac{\theta}{n}\right)$ in the limit $n \to \infty$. From Eq. (8.5), $R(\theta, z)$ for a finite angle θ is

$$R(\theta, z) = \lim_{n\to\infty} \left(1 - i\frac{\theta}{n}\frac{I_z}{\hbar} \right)^n.$$

The result is an exponential function of negative argument,

$$R(\theta, z) = e^{-i\theta\frac{I_z}{\hbar}}, \tag{8.6}$$

and similarly for any axis of rotation. Note that the argument of the exponential is dimensionless, as it must be. Following the model of Eq. (8.6), the fixed axis definition of the Euler angles can be written as

$$R(\alpha, z)R(\beta, y)R(\gamma, z) = e^{-i\alpha\frac{I_z}{\hbar}} e^{-i\beta\frac{I_y}{\hbar}} e^{-i\gamma\frac{I_z}{\hbar}}.$$

As Eq. (8.6) demonstrates, exponential angular momentum operators generate rotations. The operator in the exponent is called the *generator* of the rotation. In Eq. (8.6) the generator is $\frac{L_z}{\hbar}$ and generates rotation about z. Such generators are sometimes called "infinitesimal" generators because of the way finite results are developed by infinite sums.

Exponential operators occur frequently in quantum mechanics, but they do not usually obey the common algebra of exponents. To evaluate an exponential operator, it is necessary to expand it in series. For an operator $\mathbf{A}$,

$$e^{-i\mathbf{A}} = 1 - i\mathbf{A} + \frac{(-i\mathbf{A})^2}{2!} + \cdots.$$

As we have seen in this section, the relations obeyed by exponential operators can be derived by expanding to the first few orders because the low-order terms are also operators with the same symmetry. As an example of this method, we now show that the generator $\mathbf{G}$ of a unitary operator $\mathbf{U}$ is Hermitian $\mathbf{G}^\dagger = \mathbf{G}$.

$$\mathbf{U} = e^{i\theta\mathbf{G}}$$
$$\approx 1 + i\theta\mathbf{G}$$
$$\mathbf{U}^\dagger\mathbf{U} \approx (1 - i\theta\mathbf{G}^\dagger)(1 + i\theta\mathbf{G})$$
$$= 1 - i\theta(\mathbf{G}^\dagger - \mathbf{G})$$

Because $\mathbf{U}$ is unitary, it obeys $\mathbf{U}^\dagger\mathbf{U} = 1$. Combining results,

$$1 = \mathbf{U}^\dagger\mathbf{U} = 1 - i\theta(\mathbf{G}^\dagger - \mathbf{G})$$
$$\mathbf{G}^\dagger = \mathbf{G}.$$

Because the generator is Hermitian, it is physically observable. Furthermore, if $\mathbf{U}$ commutes with the Hamiltonian $\mathcal{H}$ and does not depend explicitly on time, its time derivative is zero according to Eq. (5.18). If $\mathbf{U}$ is constant in time, so is the generator $\mathbf{G}$. It follows that $\mathbf{G}$, a Hermitian observable, thus obeys a conservation law.

8.4 Commutation Relations

Finite rotations about different axes do not commute, and the same is true of the rotation operators for finite angles. Consider rotations $R(\xi, x) R(\psi, z)$ followed by the reverse to give a total sequence $R(-\xi, x) R(-\psi, z) R(\xi, x) R(\psi, z)$. If rotations about different axes commuted, the terms in ψ and in ξ would each reduce to the identity.

The rotation sequence in this example can be written in terms of angular momentum operators by expanding the definition Eq. (8.6) for each axis to second order, $e^x \approx 1 + x + \frac{1}{2}x^2$. Products of more than two small angles can be neglected to second order.

$$e^{i\xi\frac{I_x}{\hbar}} e^{i\psi\frac{I_z}{\hbar}} e^{-i\xi\frac{I_x}{\hbar}} e^{-i\psi\frac{I_z}{\hbar}}$$

$$\approx \left(1 + i\xi\frac{I_x}{\hbar} + \frac{1}{2}(i\xi\frac{I_x}{\hbar})^2\right)\left(1 + i\psi\frac{I_z}{\hbar} + \frac{1}{2}(i\psi\frac{I_z}{\hbar})^2\right)$$

$$\times \left(1 - i\xi\frac{I_x}{\hbar} + \frac{1}{2}(-i\xi\frac{I_x}{\hbar})^2\right)\left(1 - i\psi\frac{I_z}{\hbar} + \frac{1}{2}(-i\psi\frac{I_z}{\hbar})^2\right)$$

Be sure to retain the order of the angular momentum operators, for reasons that will become obvious. Many terms cancel, and the final result is

$$1 - \frac{\xi\psi}{\hbar^2}(I_x I_z - I_z I_x),$$

correct to second order.

The term in parentheses would be zero if rotations commuted. The discrepancy $I_x I_z - I_z I_x$ is called the *commutator* of I_x and I_z. Commutators are symbolized using square brackets and a comma, taking care to maintain the order of the operators.

$$[I_1, I_2] \equiv I_1 I_2 - I_2 I_1$$

It follows from the group axioms that the commutator of angular momentum operators must be a member of the group. To evaluate the commutator of angular momentum operators, start from the first-order matrices for the operators and then use matrix multiplication.

Consider, for example, the matrix for $R(\theta, z)$ to first order.

$$R(\theta, z) = \begin{pmatrix} \cos\theta & -\sin\theta & 0 \\ \sin\theta & \cos\theta & 0 \\ 0 & 0 & 1 \end{pmatrix}$$

$$R(\delta\theta, z) = \begin{pmatrix} 1 & -\delta\theta & 0 \\ \delta\theta & 1 & 0 \\ 0 & 0 & 1 \end{pmatrix}$$

$$= \begin{pmatrix} 1 & 0 & 0 \\ 0 & 1 & 0 \\ 0 & 0 & 1 \end{pmatrix} + \begin{pmatrix} 0 & -\delta\theta & 0 \\ \delta\theta & 0 & 0 \\ 0 & 0 & 0 \end{pmatrix} \qquad (8.7)$$

Comparing Eqs. (8.7) and (8.5),

$$R(\delta\theta, z) = 1 - i\delta\theta \frac{I_z}{\hbar}$$

$$i\frac{I_z}{\hbar} = \begin{pmatrix} 0 & 1 & 0 \\ -1 & 0 & 0 \\ 0 & 0 & 0 \end{pmatrix}$$

$$I_z = \hbar \begin{pmatrix} 0 & -i & 0 \\ i & 0 & 0 \\ 0 & 0 & 0 \end{pmatrix}.$$

Using the same approach,

$$I_x = \hbar \begin{pmatrix} 0 & 0 & 0 \\ 0 & 0 & -i \\ 0 & i & 0 \end{pmatrix} \qquad\qquad I_y = \hbar \begin{pmatrix} 0 & 0 & i \\ 0 & 0 & 0 \\ -i & 0 & 0 \end{pmatrix}.$$

It is easy to show from these matrices that the angular momentum operators I_x, I_y, I_z are Hermitian $I_\alpha^\dagger = I_\alpha$; proof is left to the problems. They therefore have real eigenvalues.

Use the first-order matrices to evaluate the commutators. Take, for example, $[I_x, I_y]$.

$$[I_x, I_y] = I_x I_y - I_y I_x$$

$$= \hbar^2 \begin{pmatrix} 0 & 0 & 0 \\ 0 & 0 & -i \\ 0 & i & 0 \end{pmatrix} \begin{pmatrix} 0 & 0 & i \\ 0 & 0 & 0 \\ -i & 0 & 0 \end{pmatrix} - \hbar^2 \begin{pmatrix} 0 & 0 & i \\ 0 & 0 & 0 \\ -i & 0 & 0 \end{pmatrix} \begin{pmatrix} 0 & 0 & 0 \\ 0 & 0 & -i \\ 0 & i & 0 \end{pmatrix}$$

$$= \hbar^2 \begin{pmatrix} 0 & 1 & 0 \\ -1 & 0 & 0 \\ 0 & 0 & 0 \end{pmatrix}$$

$$= i\hbar I_z$$

The remaining commutators are readily found from $[I_x, I_y] = i\hbar I_z$ by cycling the subscripts $x \to y \to z \to x$.

$$[I_x, I_y] = i\hbar I_z \qquad (8.8)$$

$$[I_y, I_z] = i\hbar I_x \qquad (8.9)$$

$$[I_z, I_x] = i\hbar I_y \qquad (8.10)$$

According to quantum mechanics, two quantities can be measured simultaneously only if the operators commute so that their commutator vanishes. None of the commutators is zero, showing that two different components of angular momentum cannot be measured simultaneously.

Using the commutation relations proves that the squared magnitude I^2 of total angular momentum $\mathbf{I}$ commutes with any of the angular momentum components. Take I_z as an example.

$$I^2 = I_x^2 + I_y^2 + I_z^2$$

$$[I^2, I_z] = I_x[I_x, I_z] + I_y[I_y, I_z] - [I_z, I_x]I_x - [I_z, I_y]I_y$$

$$= -iI_x I_y + iI_y I_x - iI_y I_x + iI_x I_y$$

$$= 0,$$

using $[I_z^2, I_z] = 0$ and the commutation relation $i\hbar I_z = [I_x, I_y]$ to eliminate I_z.

I^2 commutes with any component of the angular momentum operators. In the language of Lie groups I^2 is called a *Casimir operator*, an operator that commutes with all the components. Lie group theory shows that the 3-dimensional rotation group has only the one Casimir operator I^2. I^2 commutes with any component, generating *good* quantum numbers that can be measured simultaneously so that wave functions can be labeled with both quantum numbers. Only one angular momentum component can be measured simultaneously with I^2 because the angular momentum operators do not commute with one another.

8.5   The Axial Rotation Group Again

The formal development of the irreducible representations of the rotation group begins with the simple case of rotation about a fixed axis $R(\theta, z)$. The axial rotation group is Abelian because rotations about a single axis commute. Its irreducible representations are therefore all 1-dimensional and can be written

$$D(R(\theta, z)) = e^{\pm im\theta} \qquad m = 0, 1, 2, \ldots.$$

Let u_m be the basis function corresponding to m. To first order,

$$\begin{aligned} R(\delta\theta, z) u_m &= e^{-im\,\delta\theta} u_m \\ &\approx (1 - im\,\delta\theta) u_m. \end{aligned} \tag{8.11}$$

From the relation between $R(\theta, z)$ and I_z,

$$\begin{aligned} R(\delta\theta, z) u_m &= e^{-i\delta\theta \frac{I_z}{\hbar}} u_m \\ &\approx (1 - i\delta\theta \frac{I_z}{\hbar}) u_m. \end{aligned} \tag{8.12}$$

From Eqs. (8.11) and (8.12),

$$I_z u_m = (m\hbar) u_m.$$

Thus, $m\hbar$ is the eigenvalue of I_z, and u_m is the corresponding eigenfunction.

8.6 Raising and Lowering (Ladder) Operators

Spherical coordinates with unit radius $r = 1$ are

$$x = \sin\theta \cos\phi \qquad y = \sin\theta \sin\phi \qquad z = \cos\theta,$$

where θ is the polar angle measured from the z-axis, and ϕ is the azimuthal angle in the x-y plane. The variables $x \pm iy$ are

$$x \pm iy = e^{\pm i\phi} \sin\theta.$$

In a similar way, introduce the *raising operator* I_+ and *lowering operator* I_-. The operators I_+ and I_- are called *ladder operators*.

$$I_\pm = I_x \pm i I_y$$

From Eq. (8.4),

$$I_\pm = \mp\hbar(x \pm iy)\frac{\partial}{\partial z} \pm \hbar z\left(\frac{\partial}{\partial x} \pm i\frac{\partial}{\partial y}\right). \tag{8.13}$$

In terms of spherical polar coordinates θ, ϕ, the $I_\pm$ are

$$I_\pm = \hbar e^{\pm i\phi}\left(\pm\frac{\partial}{\partial\theta} + \frac{i}{\tan\theta}\frac{\partial}{\partial\phi}\right). \tag{8.14}$$

The $I_\pm$ are not Hermitian.

With straightforward algebra using the commutators for I_x, I_y, I_z from Eqs. (8.8), (8.9), (8.10), the commutators involving $I_\pm$ are

$$[I_+, I_-] = 2\hbar I_z \tag{8.15}$$

$$[I_z, I_+] = \hbar I_+ \tag{8.16}$$

$$[I_z, I_-] = -\hbar I_-. \tag{8.17}$$

Leaving details to the problems, it can be shown that

$$[I^2, I_\pm] = 0. \tag{8.18}$$

We proved in Section 8.4 that I_z commutes with I^2. Equation (8.18) shows that I_+ and I_- also commute with I^2. However, I_+ and I_- do not commute with each other, Eq. (8.15).

Note: Carrying $\hbar$ through equations decreases readability and obscures the sense of relations. $\hbar$ will be dropped in most cases from here on. It can be reinserted by checking the dimensions of angular momentum on both sides of an equation.

Using $I_z u_m = m\, u_m$ and Eq. (8.17),

$$(I_z I_- - I_- I_z)\, u_m = -I_-\, u_m$$
$$I_z(I_- u_m) = (m-1)(I_- u_m) \tag{8.19}$$
$$I_z u_{m-1} = (m-1)u_{m-1}.$$

Equation (8.19) shows that $I_- u_m$ is an eigenfunction of I_z with eigenvalue $(m-1)$. This is why I_- is called the lowering operator – it changes eigenfunction u_m to u_{m-1}.

$$I_- u_m = u_{m-1} \tag{8.20}$$

Note that the label m is a variable that can be set to any allowed value throughout an expression. Setting $m \to m-1$ in Eq. (8.20),

$$I_- u_{m-1} = u_{m-2}.$$

Continuing this process generates a chain of basis functions with their m labels sequentially lowered by one unit at a time. Assume that the chain is finite and comes to an end at $m = -j$, so that I_- cannot lower m further. Consequently $u_{-j} = 0$.

Use the same trick with I_+, which raises m to $m+1$. From Eq. (8.18),

$$I_z I_+ u_m - I_+ I_z u_m = I_+ u_m$$
$$I_z(I_+ u_m) = (m+1)(I_+ u_m) \tag{8.21}$$
$$I_+ u_m \propto u_{m+1}$$
$$= \rho_m u_{m+1}, \tag{8.22}$$

where ρ_m is a proportionality factor depending on m. The reason for writing $I_+ u_m$ with a proportionality factor rather than as an equality is to avoid inconsistencies with $I_- u_m = u_{m-1}$ in Eq. (8.20).

From Eq. (8.22),

$$\rho_m u_{m+1} = I_+ u_m$$
$$\rho_{m-1} u_m = I_+ u_{m-1} \tag{8.23}$$
$$= I_+ I_- u_m.$$

Using Eqs. (8.15) and (8.21),

$$\rho_{m-1} u_m = (I_- I_+ + 2I_z) u_m$$
$$= \rho_m I_- u_{m+1} + 2m u_m$$
$$= \rho_m u_m + 2m u_m$$
$$\rho_{m-1} = \rho_m + 2m. \tag{8.24}$$

To solve the difference equation Eq. (8.24) for ρ_m in terms of integer m, try a series solution with a few terms.

$$\rho_m = a_0 + a_1 m + a_2 m^2$$

Inserting the assumed solution into Eq. (8.24) gives

$$-a_1 - 2a_2 m + a_2 = 2m.$$

For the coefficient of m to be the same on both sides, $a_2 = -1$, from which it follows that $a_1 = -1$.

$$\rho_m = a_0 - m^2 - m$$

Let $m = j$ be the highest value of m so that ρ_j must satisfy the boundary condition $\rho_j = 0$ because it cannot be raised further. Then

$$a_0 = j^2 + j$$
$$\rho_m = j(j + 1) - m(m + 1).$$

To satisfy $\rho_m = 0$, the upper bound of m is $+j$. This result and the earlier ones show that m has both upper and lower bounds, $-j \le m \le +j$. The number of basis functions is therefore $(2j + 1)$, so $(2j + 1)$ is also the dimension of the irreducible representation $D^{(j)}$ for the group of physical rotations SO(3). Furthermore, $2j + 1$ is an integer, so j can only be an integer $j = 0, 1, 2, \ldots$ or half an odd integer $j = \frac{1}{2}, \frac{3}{2}, \ldots$.

Here half odd integer quantum numbers automatically enter quantum mechanics and are the only possible fractional quantum numbers.

Because the quantum-mechanical operators I^2 and I_z commute, u_m must be a basis function for both I^2 and I_z.

$$I_z u_m = m u_m$$
$$I^2 = I_x^2 + I_y^2 + I_z^2$$
$$= \frac{1}{2}(I_+ I_- + I_- I_+) + I_z^2$$

From the commutation relation Eq. (8.8),

$$I_z = \tfrac{1}{2}(I_+ I_- - I_- I_+)$$
$$\tfrac{1}{2} I_+ I_- = I_z + \tfrac{1}{2} I_- I_+$$
$$I^2 = I_z^2 + I_z + I_- I_+ \tag{8.25}$$
$$(I_z^2 + I_z + I_- I_+) u_m = (m^2 + m + j(j+1) - m(m+1)) u_m$$
$$I^2 u_m = j(j+1) u_m. \tag{8.26}$$

The basis functions can therefore be labeled u_{jm} with both j and m. The label j is identified as the angular momentum quantum number, and m is the magnetic quantum number.

8.7 Angular Momentum Operators and Representations of the Rotation Group

The goal of this section and the next is to use angular momentum operators to develop explicit irreducible matrix representations of the group SO(3) of physical rotations.

Angular momentum operators cannot change the quantum number j of the u_{jm}. Physical rotations can only transform u_{jm} into basis functions $u_{jm'}$ with the same j but possibly different m. For a fixed j, the u_{jm} transform only into themselves under rotation. They therefore span an invariant subspace and hence constitute an irreducible block labeled by j on the diagonal of a rotation matrix.

The action of angular momentum operators on the basis functions must be reevaluated because now the eigenfunctions u_{jm} depend on both j and m. Consider

$$I_+ u_{jm} = \kappa_m u_{j\,m+1},$$

where κ_m is to be determined.

We now show that all the u_{jm} can be normalized by a suitable choice of κ_m.

$$I^2 u_{jm} = j(j+1)\, u_{jm}$$
$$I_z\, u_{jm} = m u_{jm}$$
$$(I^2 - I_z^2 - I_z)\, u_{jm} = (j(j+1) - m(m+1))\, u_{jm}$$
$$I^2 - I_z^2 - I_z = I_- I_+$$
$$(j(j+1) - m(m+1))\, u_{jm} = I_- I_+\, u_{jm}$$
$$= \kappa_m I_-\, u_{j\,m+1}$$

Take the complex conjugate, multiply by u_{jm}, and integrate for normalization. Use Theorem 10 from Eq. (6.12), Section 6.3.2 and use $I_-^* = I_+$.

$$(j(j+1) - m(m+1)) \int u_{jm}^* u_{jm} \, d\tau = \kappa_m \int (I_- u_{j\,m+1})^* u_{jm} \, d\tau$$

$$= \kappa_m \int u_{j\,m+1}^* I_+ u_{jm} \, d\tau$$

$$= \kappa_m^2 \int u_{j\,m+1}^* u_{j\,m+1} \, d\tau$$

All the u_{jm} are therefore normalized if

$$\kappa_m^2 = j(j+1) - m(m+1)$$
$$\kappa_m = \sqrt{j(j+1) - m(m+1)}.$$

In Dirac bracket notation,

$$I_+ |j\,m\rangle = \sqrt{j(j+1) - m(m+1)} \,|j\,m+1\rangle \qquad (8.27)$$

$$I_- |j\,m\rangle = \sqrt{j(j+1) - m(m-1)} \,|j\,m-1\rangle \qquad (8.28)$$

$$I_z |j\,m\rangle = m \,|j\,m\rangle. \qquad (8.29)$$

If $m = +j$, then $I_+ |j\,j\rangle = 0$, and if $m = -j$, then $I_- |j\,-j\rangle = 0$ in agreement with the bounds on m derived earlier. Equations (8.27), (8.28), and (8.29) hold for j an integer or half odd integer.

8.7.1 Matrix Representation of $D^{(1)}$

This section uses angular momentum operators to express a group representation matrix for SO(3) in terms of physical angles using operators $R(\alpha, z) R(\beta, y) R(\gamma, z)$ corresponding to the Euler angle description of physical rotations with fixed coordinate axes.

Take as an example $j = 1$ with representation matrices $D^{(1)}$. For $j = 1$ there are $(2j + 1) = 3$ basis functions $u_m = u_1, u_0, u_{-1}$ corresponding to $m = 1, 0, -1$. Matrices for $D^{(j)}$ have dimensions $(2j + 1) \times (2j + 1)$, so the $D^{(1)}$ matrices are 3×3. The irreducible matrix representations for physical rotations SO(3) have only odd-numbered dimensions.

If $(2j + 1)$ is even, then $2j$ is odd and j is half odd integer. Representations with j equal to half odd integers correspond to spinors and do not represent physical rotations.

From Eqs. (8.27), (8.28), and (8.29) the angular momentum operators act on the basis functions to give

$$I_+ u_{11} = 0 \qquad\qquad I_- u_{11} = \sqrt{2}\,u_{10} \qquad\qquad I_z u_{11} = u_{11}$$

$$I_+ u_{10} = \sqrt{2}\,u_{11} \qquad\qquad I_- u_{10} = \sqrt{2}\,u_{1-1} \qquad\qquad I_z u_{10} = 0$$

$$I_+ u_{1-1} = \sqrt{2}\,u_{10} \qquad\qquad I_- u_{1-1} = 0 \qquad\qquad I_z u_{1-1} = -u_{1-1}. \;.$$

Evaluating the matrix representation in the Euler sequence needs the matrices for I_y and for I_z. Using $I_y = \frac{i}{2}(I_- - I_+)$ and $I_z u_m = m u_m$, the above results give (with $\hbar$ temporarily reinserted)

$$\frac{I_y}{\hbar} = \frac{i}{\sqrt{2}}\begin{pmatrix} 0 & 1 & 0 \\ -1 & 0 & 1 \\ 0 & -1 & 0 \end{pmatrix} \qquad \frac{I_z}{\hbar} = \begin{pmatrix} 1 & 0 & 0 \\ 0 & 0 & 0 \\ 0 & 0 & -1 \end{pmatrix}.$$

The matrix representation of a rotation is developed by evaluating the exponential function for a rotation operator Eq. (8.6). For example,

$$D^{(1)}(\gamma, z) = e^{-i\gamma \frac{I_z}{\hbar}}$$

$$= 1 + \sum_{n=1}^{\infty}(-1)^n \frac{(i\gamma)^n}{n!}\left(\frac{I_z}{\hbar}\right)^n,$$

where 1 stands for the 3×3 identity matrix. I_z^n is needed to evaluate this expression. Note that $(-1)^n = 1$ if n is even and $= -1$ if n is odd.

$$\left(\frac{I_z}{\hbar}\right)^2 = \begin{pmatrix} 1 & 0 & 0 \\ 0 & 0 & 0 \\ 0 & 0 & 1 \end{pmatrix} \quad \left(\frac{I_z}{\hbar}\right)^3 = \begin{pmatrix} 1 & 0 & 0 \\ 0 & 0 & 0 \\ 0 & 0 & -1 \end{pmatrix} \quad \cdots$$

$$D^{(1)}(\gamma, z) = \begin{pmatrix} 1 & 0 & 0 \\ 0 & 1 & 0 \\ 0 & 0 & 1 \end{pmatrix} + \begin{pmatrix} \sum_{n=1}^{\infty}(-1)^n \frac{(i\gamma)^n}{n!} & 0 & 0 \\ 0 & 0 & 0 \\ 0 & 0 & \sum_{n=1}^{\infty}\frac{(i\gamma)^n}{n!} \end{pmatrix}$$

$$= \begin{pmatrix} e^{-i\gamma} & 0 & 0 \\ 0 & 1 & 0 \\ 0 & 0 & e^{i\gamma} \end{pmatrix} \tag{8.30}$$

To find $D^{(1)}(\alpha, z)$, replace γ by α.

$$D^{(1)}(\alpha, z) = \begin{pmatrix} e^{-i\alpha} & 0 & 0 \\ 0 & 1 & 0 \\ 0 & 0 & e^{i\alpha} \end{pmatrix} \tag{8.31}$$

Finding $D^{(1)}(\beta, y)$ is more complicated because I_y^n takes different forms, depending on whether n is even or odd. The even factors give rise to terms involving $\cos\beta$, and the odd factors yield terms in $\sin\beta$.

$$\frac{I_y}{\hbar} = \frac{i}{\sqrt{2}}\begin{pmatrix} 0 & 1 & 0 \\ -1 & 0 & 1 \\ 0 & -1 & 0 \end{pmatrix} \left(\frac{I_y}{\hbar}\right)^2 = -\frac{1}{2}\begin{pmatrix} 1 & 0 & -1 \\ 0 & 2 & 0 \\ -1 & 0 & 1 \end{pmatrix}$$

$$\left(\frac{I_y}{\hbar}\right)^3 = \frac{i}{\sqrt{2}}\begin{pmatrix} 0 & 1 & 0 \\ -1 & 0 & 1 \\ 0 & -1 & 0 \end{pmatrix}$$

$$\left(\frac{I_y}{\hbar}\right)^{(n\,odd)} = \frac{i}{\sqrt{2}}\begin{pmatrix} 0 & 1 & 0 \\ -1 & 0 & 1 \\ 0 & -1 & 0 \end{pmatrix} \qquad \left(\frac{I_y}{\hbar}\right)^{(n\,even)}_y = \frac{1}{2}\begin{pmatrix} 1 & 0 & -1 \\ 0 & 2 & 0 \\ -1 & 0 & 1 \end{pmatrix}$$

$$D^{(1)}(\beta, y) = \begin{pmatrix} \frac{1+\cos\beta}{2} & -\frac{\sin\beta}{\sqrt{2}} & \frac{1-\cos\beta}{2} \\ \frac{\sin\beta}{\sqrt{2}} & \cos\beta & -\frac{\sin\beta}{\sqrt{2}} \\ \frac{1-\cos\beta}{2} & \frac{\sin\beta}{\sqrt{2}} & \frac{1+\cos\beta}{2} \end{pmatrix} \tag{8.32}$$

Multiplying the matrices in Eqs. (8.30), (8.31), and (8.32) gives

$$D^{(1)}(\alpha, \beta, \gamma) = \begin{pmatrix} e^{-i\alpha}\left(\frac{1+\cos\beta}{2}\right)e^{-i\gamma} & -e^{-i\alpha}\left(\frac{\sin\beta}{\sqrt{2}}\right) & e^{-i\alpha}\left(\frac{1-\cos\beta}{2}\right)e^{i\gamma} \\ \left(\frac{\sin\beta}{\sqrt{2}}\right)e^{-i\gamma} & \cos\beta & -\left(\frac{\sin\beta}{\sqrt{2}}\right)e^{i\gamma} \\ e^{i\alpha}\left(\frac{1-\cos\beta}{2}\right)e^{-i\gamma} & e^{i\alpha}\left(\frac{\sin\beta}{\sqrt{2}}\right) & e^{i\alpha}\left(\frac{1+\cos\beta}{2}\right)e^{i\gamma} \end{pmatrix}.$$
$$\tag{8.33}$$

The matrix elements $D^{(j)}_{m'm}$ are labeled so that the upper left-hand element has $m' = m = j$ and the lower right-hand element has $m' = m = -j$.

$$D^{(1)} = \begin{pmatrix} D^{(1)}_{11} & D^{(1)}_{10} & D^{(1)}_{1-1} \\ D^{(1)}_{01} & D^{(1)}_{00} & D^{(1)}_{0-1} \\ D^{(1)}_{-11} & D^{(1)}_{-10} & D^{(1)}_{-1-1} \end{pmatrix}$$

An application of the $D^{(1)}$ matrix is to rotate a basis function u_{1k}, using the fundamental functional relation Eq. (2.1).

$$R(\alpha, \beta, \gamma) u_{1k} = \sum_i D^{(1)}_{ik}(\alpha, \beta, \gamma) u_{1i}$$

8.7.2 Rotation Matrices and SU(2)

The discussion of the SU(2) group in Section 7.6 displayed $D^{(1)}$ in Eq. (7.8) as a function of two complex numbers a and b that satisfy the determinant condition $aa^* + bb^* = 1$.

$$D^{(1)}(a, b) = \begin{pmatrix} a^2 & \sqrt{2}\,ab & b^2 \\ -\sqrt{2}\,ab^* & aa^* - bb^* & \sqrt{2}\,a^*b \\ b^{*2} & -\sqrt{2}\,a^*b^* & a^{*2} \end{pmatrix}$$

Take

$$a = e^{-i\frac{(\alpha+\gamma)}{2}}\cos\frac{\beta}{2}$$
$$b = -e^{-i\frac{(\alpha-\gamma)}{2}}\sin\frac{\beta}{2}.$$

With some trigonometry, $D^{(1)}(a, b)$ reduces to Eq. (8.33).

The same identification expresses the representation $D^{\left(\frac{1}{2}\right)}(a,b)$ in terms of Euler angles.

$$D^{\left(\frac{1}{2}\right)}(a,b) = \begin{pmatrix} a & b \\ -b^* & a^* \end{pmatrix}$$

$$D^{\left(\frac{1}{2}\right)}(\alpha,\beta,\gamma) = \begin{pmatrix} e^{-i\frac{(\alpha+\gamma)}{2}}\cos\frac{\beta}{2} & -e^{-i\frac{(\alpha-\gamma)}{2}}\sin\frac{\beta}{2} \\ e^{i\frac{(\alpha-\gamma)}{2}}\sin\frac{\beta}{2} & e^{i\frac{(\alpha+\gamma)}{2}}\cos\frac{\beta}{2} \end{pmatrix}$$

The half angles reflect the double-valued nature of spinors.

8.8   The u_{jm} Are Spherical Harmonics

What are the basis functions expressed as explicit algebraic functions? Strictly speaking, this question never needs to be asked in group theory. As shown throughout this text, basis functions in group theory are defined by how they transform under operations as in the examples

$$R u_{jm} = \sum_{m'} D^{(j)}_{m'm} u_{jm'} \quad \text{and} \quad I^2 u_{jm} = j(j+1) u_{jm}.$$

Algebraic functions act as basis functions if they "transform like" basis functions under group operations. Nevertheless, it can be convenient in some calculations to have explicit algebraic forms for the u_{jm}. Explicit algebraic forms also establish a relation between group theory and Schrödinger's wave mechanics.

8.8.1   Spherical Harmonics and Group Theory

In courses that do not employ group theory, basis functions for the rotation group are derived from partial differential equations and are shown to be spherical harmonics $Y_\ell^m(\theta,\phi)$. This section derives spherical harmonics from group theory using angular momentum operators. Take as an example $\ell = 1$ so there are three basis functions $m = +1, 0, -1$.

Section 7.6 showed that the matrix

$$V = \begin{pmatrix} z & x-iy \\ x+iy & -z \end{pmatrix}$$

generates a 3-dimensional physical rotation under a similarity transformation with a member of SU(2).

Apply the lowering operator I_- to $(x-iy) = \sin\theta e^{-i\phi}$ using Eq. (8.14).

$$I_-(\sin\theta e^{-i\phi}) = \hbar e^{-i\phi}\left(-\frac{\partial}{\partial\theta} + \frac{i}{\tan\theta}\frac{\partial}{\partial\phi}\right)(\sin\theta e^{-i\phi})$$

$$= e^{-2i\phi}\hbar(-\cos\theta + \cos\theta)$$

$$= 0$$

This result shows that the lowering operator cannot lower $\sin \theta e^{-i\phi}$ any further. Hence the basis function u_{1-1} must be proportional to $\sin \theta e^{-i\phi} \propto Y_1^{-1}$.

Apply the raising operator I_+ to u_{1-1}, transforming u_{1-1} to u_{10}.

$$u_{10} = I_+ u_{1-1}$$
$$\propto I_+ (\sin \theta e^{-i\phi})$$

From Eq. (8.14),

$$I_+ (\sin \theta e^{-i\phi}) = \hbar e^{i\phi} \left(\frac{\partial}{\partial \theta} + \frac{i}{\tan \theta} \frac{\partial}{\partial \phi} \right) (\sin \theta e^{-i\phi})$$
$$= \hbar (\cos \theta + \cos \theta)$$
$$\propto \cos \theta$$
$$\propto Y_1^0.$$

Leaving details to the problems, applying I_+ to $u_{10} \propto \cos \theta$ gives $u_{11} \propto e^{i\phi} \sin \theta \propto Y_1^1$.

As the example illustrates, the spherical harmonic basis functions Y_ℓ^m for any ℓ and m can be generated (apart from normalization constants and algebraic signs) by starting from the basis function $u_{\ell-\ell} = (x - iy)^\ell = (\sin \theta e^{-i\phi})^\ell$ at the bottom of the ladder and then sequentially applying the raising operator to generate the remaining 2ℓ basis functions. For Cartesian coordinates, use Eq. (8.13), and for spherical coordinates, use Eq. (8.14).

$$Y_\ell^m(x, y) = \underbrace{I_+ \dots I_+}_{2\ell \text{ times}} (x - iy)^\ell \tag{8.34}$$

$$Y_\ell^m(\theta, \phi) = \underbrace{I_+ \dots I_+}_{2\ell \text{ times}} (\sin \theta e^{-i\phi})^\ell \tag{8.35}$$

8.8.2 Spherical Harmonics and Differential Equations

Spherical harmonics are the angular part of solutions of Laplace's equation. Laplace's equation is spherically symmetric, so it is not surprising that it is related to rotations.

As shown in Section 6.4.2, spherical harmonics can be expressed as the product of two functions $\Theta(\theta)$ and $\Phi(\phi)$, where θ is the polar angle measured from the z-axis and ϕ is the azimuthal angle in the x-y plane measured from the x-axis.

Φ is the solution of the differential equation

$$\frac{d^2\Phi}{d\phi^2} + m^2\Phi = 0, \tag{8.36}$$

where m must be an integer to satisfy $e^{\pm im(\phi+2\pi)} = e^{\pm im\phi}$.

Θ is the solution of

$$\frac{1}{\sin \theta} \frac{d}{d\theta} \left(\sin \theta \frac{d\Theta}{d\theta} \right) \left(\ell(\ell+1) - \frac{m^2}{\sin^2 \theta} \right) \Theta = 0. \tag{8.37}$$

Solutions of the Θ equation are well-behaved only if ℓ is an integer that satisfies the condition $-m \leq \ell \leq m$. The Θ are associated Legendre functions $P_\ell^m(\cos\theta)$.

The normalized spherical harmonics for $\ell = 0, 1,$ and 2 are

$$Y_0^0 = \sqrt{\frac{1}{4\pi}} \qquad Y_1^1 = -\sqrt{\frac{3}{8\pi}}e^{i\phi}\sin\theta \qquad Y_2^2 = \sqrt{\frac{15}{32\pi}}e^{2i\phi}\sin^2\theta$$

$$Y_1^0 = \sqrt{\frac{3}{4\pi}}\cos\theta \qquad Y_2^1 = -\sqrt{\frac{15}{8\pi}}e^{i\phi}\sin\theta\cos\theta$$

$$Y_1^{-1} = \sqrt{\frac{3}{8\pi}}e^{-i\phi}\sin\theta \qquad Y_2^0 = \sqrt{\frac{5}{16\pi}}\left(3\cos^2\theta - 1\right)$$

$$Y_2^{-1} = \sqrt{\frac{15}{8\pi}}e^{-i\phi}\sin\theta\cos\theta$$

$$Y_2^{-2} = \sqrt{\frac{15}{32\pi}}e^{-2i\phi}\sin^2\theta.$$

The algebraic signs are arbitrary. The choice here is the *Condon–Shortley phase* defined as $(-1)^m$ for $m > 0$ and $+1$ for $m \leq 0$. The literature has extensive listings of spherical harmonics and a discussion of their many properties.

Spherical harmonics are basis functions for the rotation group SO(3) and appear explicitly in the irreducible representation matrices. In the complex conjugate of the rotation matrix, the middle column $m\,0$ is proportional to the spherical harmonics Y_ℓ^m. To demonstrate, take the complex conjugate of $D^{(1)}$ in Eq. (8.33).

$$\left(D_{m0}^{(1)}(\phi,\theta,0)\right)^* = \begin{pmatrix} \cdots & -\left(\dfrac{\sin\theta}{\sqrt{2}}\right)e^{i\phi} & \cdots \\ \cdots & \cos\theta & \cdots \\ \cdots & \left(\dfrac{\sin\theta}{\sqrt{2}}\right)e^{-i\phi} & \cdots \end{pmatrix}$$

Only matrix elements in the middle column are shown, to focus attention on the terms proportional to Y_1^1, Y_1^0, Y_1^{-1}. Basis functions for any ℓ always occupy the middle column of $D^{(\ell)*}$.

8.9   Spin Basis Functions and Pauli Matrices

The total angular momentum operator $\mathbf{J}$ can have contributions from both orbital angular momentum $\mathbf{L}$ and spin angular momentum $\mathbf{S}$ so that $\mathbf{J} = \mathbf{L} + \mathbf{S}$.

Spherical harmonics are basis functions for SO(3), where the orbital angular momentum quantum number ℓ is an integer. The spherical harmonics are continuous functions of the angles θ, ϕ. In contrast, the Stern–Gerlach experiment is consistent with discrete spin angular momentum that has only two values $\pm\frac{\hbar}{2}$. What are the basis functions for angular momentum of half an odd integer?

Consider the case of spin $\frac{1}{2}$ and let the basis functions be $u_{1/2}$ and $u_{-1/2}$. Spin is angular momentum, so the angular momentum operators apply.

$$I_+ u_{1/2} = 0 \qquad\qquad I_+ u_{-1/2} = u_{1/2}$$
$$I_- u_{1/2} = u_{-1/2} \qquad\qquad I_- u_{-1/2} = 0$$
$$I_z u_{1/2} = \tfrac{1}{2} u_{1/2} \qquad\qquad I_z u_{-1/2} = -\tfrac{1}{2} u_{1/2}$$

Continuous functions cannot represent discrete values, so it is natural to look to matrices to express the spin functions explicitly. They are two-component spinors.

$$u_{1/2} = \begin{pmatrix} 1 \\ 0 \end{pmatrix} \qquad\qquad u_{-1/2} = \begin{pmatrix} 0 \\ 1 \end{pmatrix}$$

$$I_x u_{1/2} = \tfrac{1}{2}(I_+ + I_-) u_{1/2} = \tfrac{1}{2} u_{-1/2}$$
$$I_x u_{-1/2} = \tfrac{1}{2}(I_+ + I_-) u_{-1/2} = \tfrac{1}{2} u_{1/2}$$

$$I_y u_{1/2} = \tfrac{i}{2}(I_- - I_+) u_{1/2} = \tfrac{i}{2} u_{-1/2}$$
$$I_y u_{-1/2} = \tfrac{i}{2}(I_- - I_+) u_{-1/2} = -\tfrac{i}{2} u_{1/2}$$

$$I_z u_{1/2} = \tfrac{1}{2} u_{1/2}$$
$$I_z u_{-1/2} = -\tfrac{1}{2} u_{-1/2}$$

The corresponding matrices are

$$\left(\frac{I_x}{\hbar}\right) = \tfrac{1}{2}\begin{pmatrix} 0 & 1 \\ 1 & 0 \end{pmatrix} \qquad \left(\frac{I_y}{\hbar}\right) = \tfrac{1}{2}\begin{pmatrix} 0 & -i \\ i & 0 \end{pmatrix} \qquad \left(\frac{I_z}{\hbar}\right) = \tfrac{1}{2}\begin{pmatrix} 1 & 0 \\ 0 & -1 \end{pmatrix}.$$

These matrices are the components of spin $\frac{1}{2}$ angular momentum. Without the factor of $\frac{1}{2}$ they are the zero trace Pauli spin matrices introduced in Section 7.5.1.

8.10 Coupling (Adding) Angular Momenta

In Newtonian mechanics, adding two angular momentum vectors is a straightforward application of classical vector addition. Not so in quantum mechanics – the result of coupling angular momenta must be consistent with the quantum-mechanical angular momentum operators.

8.10.1 Example: Positronium

Short-lived *positronium* is formed when a negatively charged electron e^- binds with its positively charged *positron* antiparticle e^+. Both particles have spin $\frac{1}{2}$, so the basis functions of each particle belong to the irreducible representation $D^{(\frac{1}{2})}$ of the rotation group. This section derives the basis functions W_M^S that characterize positronium's total spin angular momentum $\mathbf{S} = \mathbf{S}^{(e-)} + \mathbf{S}^{(e+)}$ in a state with zero orbital angular momentum.

Let $u_m^{(\frac{1}{2})}$ ($m = \pm\frac{1}{2}$) be the basis functions for $e-$ and let $v_{m'}^{(\frac{1}{2})}$ ($m' = \pm\frac{1}{2}$) be the basis functions for $e+$. The basis functions $W_M^{(S)}$ belong to the direct product $D^{(\frac{1}{2})} \otimes D^{(\frac{1}{2})}$. Section 2.8 discusses the direct product, where it is proved that the direct product of two groups is a group, and where it is stated without proof that if the u and v are each basis functions for their corresponding irreducible representations, then the W are basis functions for the irreducible representations of the direct product group.

The basis functions u and the basis functions v are assumed to be orthonormal. u and v refer to different spaces, so the electron spin operator $\mathbf{S}^{(e-)}$ operates only on $u_m^{(s)}$ and the positron spin operator $\mathbf{S}^{(e+)}$ operates only on $v_{m'}^{(s')}$. u and v therefore commute.

The elements of the direct product group are

$$u_{1/2}^{(\frac{1}{2})} v_{1/2}^{(\frac{1}{2})} \qquad\qquad u_{1/2}^{(\frac{1}{2})} v_{-1/2}^{(\frac{1}{2})}$$

$$u_{-1/2}^{(\frac{1}{2})} v_{1/2}^{(\frac{1}{2})} \qquad\qquad u_{-1/2}^{(\frac{1}{2})} v_{-1/2}^{(\frac{1}{2})}.$$

$u_m^{(s)}$ and $v_{m'}^{(s')}$ refer to different spaces, so

$$S_z(u_m^{(s)} v_{m'}^{(s')}) = (S_z^{(e-)} u_m^{(s)}) v_{m'}^{(s')} + u_m^{(s)} (S_z^{(e+)} v_{m'}^{(s')})$$

$$= m u_m^{(s)} v_{m'}^{(s')} + m' u_m^{(s)} v_{m'}^{(s')}$$

$$= (m + m') u_m^{(s)} v_{m'}^{(s')}.$$

Hence $u_m^{(s)} v_{m'}^{(s')}$ belongs to $W_M^{(S)}$ where $M = m + m'$.

$$u_{1/2}^{(\frac{1}{2})} v_{1/2}^{(\frac{1}{2})} \to M = 1 \qquad\qquad u_{1/2}^{(\frac{1}{2})} v_{-1/2}^{(\frac{1}{2})} \to M = 0$$

$$u_{-1/2}^{(\frac{1}{2})} v_{1/2}^{(\frac{1}{2})} \to M = 0 \qquad\qquad u_{-1/2}^{(\frac{1}{2})} v_{-1/2}^{(\frac{1}{2})} \to M = -1$$

All four products are accounted for by the set of magnetic quantum numbers $M = 1$, 0, -1 for $W_M^{(1)}$ and $M = 0$ for $W_M^{(0)}$. Thus, the net angular momentum can only be 1 or 0 from two $\frac{1}{2}$ spins.

To find the combinations of $u_m^{(s)}$ and $v_{m'}^{(s')}$ that make up $W_M^{(1)}$, use raising and lowering operators and Eqs. (8.27) and (8.28). The state with $S = 1$ and $M = 1$ must be

$$W_1^{(1)} = u_{1/2}^{(\frac{1}{2})} v_{1/2}^{(\frac{1}{2})}. \tag{8.38}$$

Applying the lowering operator $\mathbf{S}_-$ Eq. (8.28) to the left-hand side of Eq. (8.38) gives

$$\mathbf{S}_- W_1^{(1)} = \sqrt{(1)(2) - (1)(0)}\ W_0^{(1)}$$
$$= \sqrt{2}\ W_0^{(1)}.$$

Applying the lowering operator $\mathbf{S}_- = \mathbf{S}_-^{(e^-)} + \mathbf{S}_-^{(e^+)}$ to the right-hand side of Eq. (8.38) gives

$$\left(\mathbf{S}_-^{(e^-)} u_{1/2}^{(\frac{1}{2})}\right) v_{1/2}^{(\frac{1}{2})} + u_{1/2}^{(\frac{1}{2})}\left(\mathbf{S}_-^{(e^+)} v_{1/2}^{(\frac{1}{2})}\right) = u_{-1/2}^{(\frac{1}{2})} v_{1/2}^{(\frac{1}{2})} + u_1^{(\frac{1}{2})} v_{-1/2}^{(\frac{1}{2})}$$

$$\text{so}\quad W_0^{(1)} = \sqrt{\tfrac{1}{2}}\, u_{-1/2}^{(\frac{1}{2})} v_{1/2}^{(\frac{1}{2})} + \sqrt{\tfrac{1}{2}}\, u_{1/2}^{(\frac{1}{2})} v_{-1/2}^{(\frac{1}{2})}.$$

The sum of the squares of the coefficients is $\frac{1}{2} + \frac{1}{2} = 1$, showing that W is normalized if the u and v are normalized. The only possibility for $W_{-1}^{(1)}$ is

$$W_{-1}^{(1)} = u_{-1/2}^{(\frac{1}{2})} v_{-1/2}^{(\frac{1}{2})}. \tag{8.39}$$

Finding an expression for $W_0^{(0)}$ in terms of u and v is more complicated because a state with $m + m' = 0$ occurs both in $W^{(1)}$ and in $W^{(0)}$. One approach is to use orthogonality. Because of the condition $M = m + m' = 0$, the only possibilities are $m = 1/2, m' = -1/2$ and $m = -1/2, m' = 1/2$.

$$W_0^{(0)} = u_{1/2}^{(\frac{1}{2})} v_{-1/2}^{(\frac{1}{2})} + C u_{-1/2}^{(\frac{1}{2})} v_{1/2}^{(\frac{1}{2})},$$

where C is a constant to be determined. By orthogonality,

$$0 = \int W_0^{(1)} W_0^{(0)}\, d\tau$$
$$= \int \left(\sqrt{\tfrac{1}{2}}\, u_{-1/2}^{(\frac{1}{2})} v_{1/2}^{(\frac{1}{2})} + \sqrt{\tfrac{1}{2}}\, u_{1/2}^{(\frac{1}{2})} v_{-1/2}^{(\frac{1}{2})}\right)\left(u_{1/2}^{(\frac{1}{2})} v_{-1/2}^{(\frac{1}{2})} + C u_{-1/2}^{(\frac{1}{2})} v_{1/2}^{(\frac{1}{2})}\right) d\tau$$
$$= \sqrt{\tfrac{1}{2}} + C \sqrt{\tfrac{1}{2}}$$
$$C = -1$$
$$W_0^{(0)} \propto u_{1/2}^{(\frac{1}{2})} v_{-1/2}^{(\frac{1}{2})} - u_{-1/2}^{(\frac{1}{2})} v_{1/2}^{(\frac{1}{2})}.$$

Normalizing,

$$W_0^{(0)} = \sqrt{\tfrac{1}{2}}\, u_{1/2}^{(\frac{1}{2})} v_{-1/2}^{(\frac{1}{2})} - \sqrt{\tfrac{1}{2}}\, u_{-1/2}^{(\frac{1}{2})} v_{1/2}^{(\frac{1}{2})}.$$

In all these expressions, the coefficients of the basis functions u and v are called the *Wigner 3-j coefficients*. In matrix form, W is expressed in terms of u and v as

$$
\begin{pmatrix} W_1^{(1)} \\ W_0^{(1)} \\ W_0^{(0)} \\ W_{-1}^{(1)} \end{pmatrix} = \begin{pmatrix} 1 & 0 & 0 & 0 \\ 0 & \sqrt{\frac{1}{2}} & \sqrt{\frac{1}{2}} & 0 \\ 0 & \sqrt{\frac{1}{2}} & -\sqrt{\frac{1}{2}} & 0 \\ 0 & 0 & 0 & 1 \end{pmatrix} \begin{pmatrix} u_{1/2}^{(\frac{1}{2})} v_{1/2}^{(\frac{1}{2})} \\ u_{1/2}^{(\frac{1}{2})} v_{-1/2}^{(\frac{1}{2})} \\ u_{-1/2}^{(\frac{1}{2})} v_{1/2}^{(\frac{1}{2})} \\ u_{-1/2}^{(\frac{1}{2})} v_{-1/2}^{(\frac{1}{2})} \end{pmatrix}, \tag{8.40}
$$

where the elements of the 4×4 matrix are the *3-j* coefficients.

The *3-j* coefficient matrix is real orthogonal (real unitary), so to solve for u and v in terms of W, multiply both sides by the transpose (here equal to the inverse) of the coefficient matrix. This is a general approach, but in the special case $s = s' = \frac{1}{2}$ the coefficient matrix and its transpose happen to be equal.

$$
\begin{pmatrix} 1 & 0 & 0 & 0 \\ 0 & \sqrt{\frac{1}{2}} & \sqrt{\frac{1}{2}} & 0 \\ 0 & \sqrt{\frac{1}{2}} & -\sqrt{\frac{1}{2}} & 0 \\ 0 & 0 & 0 & 1 \end{pmatrix} \begin{pmatrix} W_1^{(1)} \\ W_0^{(1)} \\ W_0^{(0)} \\ W_{-1}^{(1)} \end{pmatrix} = \begin{pmatrix} 1 & 0 & 0 & 0 \\ 0 & \sqrt{\frac{1}{2}} & \sqrt{\frac{1}{2}} & 0 \\ 0 & \sqrt{\frac{1}{2}} & -\sqrt{\frac{1}{2}} & 0 \\ 0 & 0 & 0 & 1 \end{pmatrix} \begin{pmatrix} 1 & 0 & 0 & 0 \\ 0 & \sqrt{\frac{1}{2}} & \sqrt{\frac{1}{2}} & 0 \\ 0 & \sqrt{\frac{1}{2}} & -\sqrt{\frac{1}{2}} & 0 \\ 0 & 0 & 0 & 1 \end{pmatrix} \begin{pmatrix} u_{1/2}^{(\frac{1}{2})} v_{1/2}^{(\frac{1}{2})} \\ u_{1/2}^{(\frac{1}{2})} v_{-1/2}^{(\frac{1}{2})} \\ u_{-1/2}^{(\frac{1}{2})} v_{1/2}^{(\frac{1}{2})} \\ u_{-1/2}^{(\frac{1}{2})} v_{-1/2}^{(\frac{1}{2})} \end{pmatrix}
$$

$$
\begin{pmatrix} 1 & 0 & 0 & 0 \\ 0 & \sqrt{\frac{1}{2}} & \sqrt{\frac{1}{2}} & 0 \\ 0 & \sqrt{\frac{1}{2}} & -\sqrt{\frac{1}{2}} & 0 \\ 0 & 0 & 0 & 1 \end{pmatrix} \begin{pmatrix} W_1^{(1)} \\ W_0^{(1)} \\ W_0^{(0)} \\ W_{-1}^{(1)} \end{pmatrix} = \begin{pmatrix} 1 & 0 & 0 & 0 \\ 0 & 1 & 0 & 0 \\ 0 & 0 & 1 & 0 \\ 0 & 0 & 0 & 1 \end{pmatrix} \begin{pmatrix} u_{1/2}^{(\frac{1}{2})} v_{1/2}^{(\frac{1}{2})} \\ u_{1/2}^{(\frac{1}{2})} v_{-1/2}^{(\frac{1}{2})} \\ u_{-1/2}^{(\frac{1}{2})} v_{1/2}^{(\frac{1}{2})} \\ u_{-1/2}^{(\frac{1}{2})} v_{-1/2}^{(\frac{1}{2})} \end{pmatrix}
$$

The expression for uv in terms of W is thus

$$
\begin{pmatrix} u_{1/2}^{(\frac{1}{2})} v_{1/2}^{(\frac{1}{2})} \\ u_{1/2}^{(\frac{1}{2})} v_{-1/2}^{(\frac{1}{2})} \\ u_{-1/2}^{(\frac{1}{2})} v_{1/2}^{(\frac{1}{2})} \\ u_{-1/2}^{(\frac{1}{2})} v_{-1/2}^{(\frac{1}{2})} \end{pmatrix} = \begin{pmatrix} 1 & 0 & 0 & 0 \\ 0 & \sqrt{\frac{1}{2}} & \sqrt{\frac{1}{2}} & 0 \\ 0 & \sqrt{\frac{1}{2}} & -\sqrt{\frac{1}{2}} & 0 \\ 0 & 0 & 0 & 1 \end{pmatrix} \begin{pmatrix} W_1^{(1)} \\ W_0^{(1)} \\ W_0^{(0)} \\ W_{-1}^{(1)} \end{pmatrix}. \tag{8.41}
$$

Note that all the results in this section depend only on angular momentum operators with no reference to the details of the physical system being considered.

The results show that positronium has two possible spin states: $S = 0$ (*para* positronium, spins opposed) and $S = 1$ (*ortho* positronium, spins parallel). The two states exhibit different physical properties when the electron and positron annihilate.

Para positronium decays principally into two gamma rays in $\approx 1.2 \times 10^{-10}$ s. By conservation of mass-energy, each gamma ray has an energy of 0.51 MeV closely equal to the rest mass of an electron or positron, less a few eV for the binding energy of positronium. A photon has intrinsic spin 1, so by conservation of linear momentum and angular momentum the gamma ray photons are emitted back-to-back with opposed spins to satisfy $S = 0$.

Ortho positronium is longer lived, decaying in $\approx 1.4 \times 10^{-7}$ s. Decay into two gamma rays cannot conserve angular momentum when $S = 1$ and the principal decay mode is three gamma rays, a slower process.

8.10.2   Wigner *3-j* Coefficients

As the positronium example illustrates, the purpose of Wigner *3-j* coefficients is to couple quantum-mechanical angular momentum of two different systems.

$$W_M^{(J)} = \sum_m (jj'mm'|JM)\, u_m^{(j)} v_{m'}^{(j')}, \tag{8.42}$$

where the *3-j* coefficients are symbolized

$$(j\, j'\, m\, m'|JM) \quad \text{or as} \quad \begin{pmatrix} j & j' & J \\ m & m' & M \end{pmatrix}.$$

Using $m' = M - m$, Eq. (8.42) can be written

$$W_M^{(J)} = \sum_m (jj'm\, M - m|JM)\, u_m^{(j)} v_{M-m}^{(j')}$$

The name *3-j* arises from the symbol, which shows dependence on three angular momentum j, j', J.

In Eq. (8.42) the parameters j, j', J, M are fixed values chosen beforehand. *3-j* coefficients are zero unless their parameters satisfy certain conditions. In a *3-j* coefficient the magnetic quantum number m is restricted to the values $-j \le m \le j$, and similarly for m'. A *3-j* coefficient is 0 unless $M = m + m'$ (equivalently $m' = M - m$) consistent with $J_z = j_z + j_z'$. It is also 0 unless j and j' can combine to give J according to the *triangle condition* $j + j' \ge J \ge |j - j'|$. The *vector model* in the sketch illustrates the triangle condition with $j = j' = 1$, where the only possible values of J are 0, 1, 2.

Because the $W_M^{(J)}$, $u_m^{(j)}$, and $v_{m'}^{(j')}$ are all orthonormal sets, the *3-j* symbols obey orthogonality relations such as

$$\sum_m (j\, j'\, m\, M - m|J\, M)(j\, j'\, m\, M' - m|J'\, M') = \delta_{JJ'}\delta_{MM'}. \tag{8.43}$$

To prove Eq. (8.43), start from the orthogonality of the $W_M^{(J)}$ and then use Eq. (8.42) twice to express the W in terms of the uv. The $W, u,$ and v basis functions and the *3-j* coefficients are all real, so taking the complex conjugate is unnecessary.

$$\delta_{JJ'}\delta_{MM'} = \int W_M^{(J)} W_{M'}^{(J')}\, d\tau$$

$$= \int \sum_m \sum_{m'} (jj'm\, M-m|JM)\, u_m^{(j)} v_{M-m}^{(j')}$$

$$(jj'm'\, M'-m'|J'M')\, u_{m'}^{(j)} v_{M'-m'}^{(j')}\, d\tau$$

u and v are both orthonormal sets (in different spaces), so

$$= \sum_m \sum_{m'} (jj'm\, M-m|JM)(jj'm'\, M'-m'|J'M')\, \delta_{mm'}\, \delta_{M-m\, M'-m'}.$$

Summing over m' and using $\delta_{mm} = 1$,

$$= \sum_m (jj'm\, M-m|JM)(jj'm\, M'-m|J'M')\, \delta_{M-m\, M'-m}$$

$$= \sum_m (jj'm\, M-m|JM)(jj'm\, M'-m|J'M')\, \delta_{MM'}.$$

$\delta_{MM'}$ on the right can be dropped because it performs the same function as $\delta_{MM'}$ on the left, completing the proof of Eq. (8.43).

Another orthogonality relation, stated without proof, is

$$\sum_{J,M} (jj'm_1 m_2|JM)(jj'm_1'm_2'|JM) = \delta_{m_1 m_1'}\, \delta_{m_2 m_2'}. \tag{8.44}$$

Equation (8.44) can be used to express u and v in terms of the product basis functions W. Multiply Eq. (8.42) by $(jj'm\, M-m|JM')$ and sum over J, M.

$$\sum_{J,M} (jj'm\, M-m|JM)\, W_M^{(J)} = \sum_{m'} \sum_{J,M} (jj'm\, M-m|JM)$$

$$(jj'm'\, M-m'|JM)\, u_{m'}^{(j)} v_{M-m'}^{(j')}$$

Using Eq. (8.44), the sum over J, M on the right gives $\delta_{mm'}$.

$$\sum_{J,M} (jj'm\, M-m|JM)\, W_M^{(J)} = \sum_{m'} \delta_{mm'} u_{m'}^{(j)} v_{M-m'}^{(j')}$$

The sum over m' has only one term $m' = m$. Hence

$$u_m^{(j)} v_{M-m}^{(j')} = \sum_{J,M} (jj'm\, M-m|JM)\, W_M^{(J)}. \tag{8.45}$$

The example of positronium in Section 8.10.1 demonstrated how Wigner *3-j* coefficients for $D^{(\frac{1}{2})} \otimes D^{(\frac{1}{2})}$ can be derived. The sets of all important *3-j* coefficients can be found in the literature. Appendix F lists *3-j* coefficients for $D^{(\frac{1}{2})} \otimes D^{(\frac{1}{2})}$, $D^{(1)} \otimes D^{(1/2)}$, and $D^{(1)} \otimes D^{(1)}$ in tabular form. Some references use different normalizations or phase factors. The equations in this text agree with the tables in Appendix F.

8.10.3 Clebsch–Gordan Coefficients

In addition to Wigner *3-j* coefficients, Clebsch–Gordan coefficients and other sets are also widely used for coupling quantum-mechanical angular momentum. Such coefficients are in general called *vector-coupling* coefficients. They all have much in common because they all accomplish the same task, although they are specifically designed to simplify certain calculations. They differ mainly in symmetry of the notation, normalizations, and phase factors. All the coefficients are real numbers.

Different sources may use different definitions and symbols for vector-coupling coefficients. In some texts, *3-j* coefficients and Clebsch–Gordan coefficients are taken to be identical, while other sources may introduce a multiplicative constant and a phase factor to enhance symmetry properties. Any coefficient tabulation should not be used indiscriminately without checking the assumed definitions.

The literature also lists coefficients for coupling more than two angular momenta; for instance, the Wigner *6-j* and *9-j* coefficients.

8.11 Wigner–Eckart Theorem

Matrix elements are important in quantum mechanics. One reason is that the squared magnitude of a matrix element gives the probability that a system makes a transition from one quantum state to another.

A common task in quantum mechanics is evaluation of matrix elements such as

$$\int \psi^*(\alpha, j, m) \, \mathbf{T}_{J,M} \, \psi(\alpha', j', m') \, d\tau$$

for an operator $\mathbf{T}$. Here the wave functions $\psi(\alpha, j, m)$ and $\psi(\alpha', j'm')$ depend on angular momentum quantum numbers j, j', m, m' and possibly on other quantum numbers α, α' that are independent of angular momentum.

The Wigner–Eckart theorem simplifies the evaluation of matrix elements by expressing the dependence on m, m' entirely in a *3-j* coefficient. A proof is presented in Appendix G.

8.11.1 Theorem and Notation

Several notations are in use for the Wigner–Eckart theorem. This text uses *3-j* coefficients. Many sources use the older and less symmetric Clebsch–Gordan coefficients, sometimes with notation that looks like a *3-j* coefficient.

Consider a vector with Cartesian components.

$$T_1^{(1)} = x + iy \quad T_0^{(1)} = z \quad T_{-1}^{(1)} = x - iy$$

$T_q^{(1)}$ is a *spherical tensor* written in general notation $T_q^{(k)}$, where k is called the *rank*. $T_0^{(0)}$ is a scalar with only one member $q = 0$ in the set. $T_q^{(1)}$ is a vector with $2k + 1 = 3$ members in the set $q = +1, 0, -1$.

$T_q^{(k)}$ is defined as any operator that satisfies the commutation relations

$$[I_z, T_q^{(k)}] = q T_q^{(k)} \tag{8.46}$$

$$[I_\pm, T_q^{(k)}] = \sqrt{k(k+1) - q(q \pm 1)}\, T_{q\pm1}^{(k)}. \tag{8.47}$$

These expressions have the same form as Eqs. (8.27), (8.28), and (8.29) for $I_\pm, I_z$ acting on $|j\,m\rangle$. k plays the role of an angular momentum quantum number, and q acts like a magnetic quantum number. It is therefore not surprising that $T_q^{(k)}$ is related to *3-j* coefficients.

The Wigner–Eckart theorem can be written

$$\langle \alpha, j, m | \mathbf{T}_q^{(k)} | \alpha', j', m' \rangle \equiv (j\,k\,m\,q | j'\,m') \, \langle \alpha\,j || T^{(k)} || \alpha'\,j' \rangle. \tag{8.48}$$

The Wigner–Eckart theorem is a powerful tool because it expresses a matrix element's dependence on m, m', q by known *3-j* symbols. The expression on the left in Eq. (8.48) is a matrix element to be evaluated. The operator in the matrix element must be a spherical tensor or a linear combination of spherical tensors, as, for example, $x = \frac{1}{2}(T_1^{(1)} + T_{-1}^{(1)})$ in a matrix element for x.

The factor $(j\,k\,m\,q | j'\,m')$ on the right is a *3-j* coefficient and is zero unless $m' = m + q$. The *3-j* coefficient is also zero unless j, k, j' satisfy the triangle condition. The *double bar* quantity $\langle \alpha\,j || T^{(k)} || \alpha'\,j' \rangle$ is called the *reduced matrix element*, an algebraic factor independent of m, m', q and not actually a conventional matrix element. The value of q is fixed by the nature of the operator on the left.

8.11.2   How It Is Used

In one application, the matrix element on the left of Eq. (8.48) is evaluated once for any j, j', m, m', and fixed q, a difficult calculation in practice because accurate wave functions are needed.

$$\langle \alpha, j || T^{(k)} || \alpha', j' \rangle = \frac{\langle \alpha, j, m | \mathbf{T}_q^{(k)} | \alpha', j', m' \rangle}{(j\,k\,m\,q | j'\,m')}$$

The right-hand side is then known, so this one-time calculation delivers the value of the reduced matrix element. Matrix elements for all other m, m' can then be expressed, given the known *3-j* coefficients.

A simpler and more widely used application is the determination of *selection rules* for electric dipole and other multipole transitions. In this application, the reduced matrix element is a fixed (unknown) number that depends on j, j' independent of m, m'. Because multipole transitions do not depend on j, j', the selection rules to be discussed in Section 8.12.2 can be found entirely from known *3-j* coefficients.

The essence of the Wigner–Eckart theorem is that once q is chosen, the reduced matrix element has no dependence on m, m'; all the dependence on m, m' is contained in the *3-j* coefficient.

8.12 Selection Rules

A given quantum number may or may not change in a transition from one quantum state to another. A *selection rule* for the transition specifies what change in the quantum numbers is permissible and what the result could be. If an assumed transition is in accord with the selection rule, the transition is said to be *allowed*. A selection rule says nothing about the probability of the transition, so an allowed transition may or may not be observable with current technology. *Allowed* should be broadly interpreted as *not forbidden*. If the change of a quantum number in an assumed transition is not in accord with the transition's selection rule, the transition is *forbidden* or equivalently, *not allowed*.

A photon has intrinsic spin 1, so emission of spectral lines from an atom involves electric fields and angular momentum. Conservation of angular momentum places constraints, expressed by selection rules, on whether a transition can occur. This section shows how selection rules arise, first by consideration of spherical harmonics, then by using the Wigner–Eckart theorem, and finally by parity-inversion.

8.12.1 From Spherical Harmonics

Selection rules for multipole transitions can be derived from matrix elements and the properties of spherical harmonics. Section 8.8 shows that spherical harmonics $Y_\ell^m(\theta, \phi)$ are basis functions for the rotation group. The spherical harmonics are an orthonormal set with regard to both indices:

$$\int_0^{2\pi} \int_0^{\pi} \left(Y_\ell^m\right)^*(\theta, \phi) Y_{\ell'}^{m'}(\theta, \phi) \sin\theta \, d\theta \, d\phi = \delta_{\ell\ell'}\delta_{mm'}.$$

In addition, the spherical harmonics are a *complete set*, meaning that any well-behaved function of θ, ϕ can be represented by a linear combination of spherical harmonics.

For a transition from a state characterized by ℓ, m to a state ℓ', m', the *matrix element* of an operation $\mathcal{T}$ is

$$\int_0^{2\pi} \int_0^{\pi} \left(Y_\ell^m\right)^*(\theta, \phi) \, \mathcal{T} \, Y_{\ell'}^{m'}(\theta, \phi) \sin\theta \, d\theta \, d\phi.$$

If the matrix element equals 0 for certain indices, that transition is forbidden and has zero probability of happening. It is allowed if the matrix element $\neq 0$.

The most intense spectral lines typically occur because of time-varying electric dipole moments. Electric dipole moment $\boldsymbol{p}$ is a vector that transforms like (x, y, z).

$$\boldsymbol{p} = q\left(x\hat{\mathbf{i}} + y\hat{\mathbf{j}} + z\hat{\mathbf{k}}\right)$$

The matrix element

$$\iint \left(Y_\ell^m\right)^*(\theta, \phi)(x, y, z) Y_{\ell'}^{m'}(\theta, \phi) \, d\tau$$

expresses an electric dipole transition between states ℓ, m and ℓ', m'. Omitting the factor r that plays no role in rotation,

$$x \propto \cos \phi \sin \theta \qquad y \propto \sin \phi \sin \theta \qquad z \propto \cos \theta \qquad x \pm iy \propto e^{\pm i\phi} \sin \theta.$$

Because only the dependence of spherical harmonics on θ, ϕ is needed here, the following listing omits normalization constants and algebraic signs.

$$Y_0^0 \propto 1 \qquad Y_1^1 \propto e^{i\phi} \sin \theta \qquad Y_2^2 \propto e^{2i\phi} \sin^2 \theta$$
$$Y_1^0 \propto \cos \theta \qquad Y_2^1 \propto e^{i\phi} \sin \theta \cos \theta$$
$$Y_1^{-1} \propto e^{-i\phi} \sin \theta \qquad Y_2^0 \propto 3\cos^2 \theta - 1$$
$$Y_2^{-1} \propto e^{-i\phi} \sin \theta \cos \theta$$
$$Y_2^{-2} \propto e^{-2i\phi} \sin^2 \theta$$

Consider the matrix element for $z \propto \cos \theta$ in a transition from state $\ell=1$, $m=0$ to the state $\ell = 0$, $m = 0$. Using $\cos \theta \, Y_0^0 \propto Y_1^0$,

$$\iint \left(Y_1^0\right)^* \cos \theta \, Y_0^0 \, d\tau \propto \iint \left(Y_1^0\right)^* Y_1^0 \, d\tau \neq 0.$$

This is therefore an allowed transition, with selection rules $\Delta \ell = -1$, $\Delta m = 0$.

Next, consider the matrix element for $x + iy = e^{i\phi} \sin \theta$ in a transition from state $\ell = 2$, $m = 2$ to the state $\ell = 1$, $m = 1$. Using $e^{i\phi} \sin \theta \, Y_1^1 \propto Y_2^2$,

$$\iint \left(Y_2^2\right)^* e^{i\phi} \sin \theta \, Y_1^1 \, d\tau \propto \iint \left(Y_2^2\right)^* Y_2^2 \, d\tau \neq 0,$$

which gives the selection rules $\Delta \ell = -1$, $\Delta m = -1$. Similar examples show that a transition $\Delta \ell = 1$, $\Delta m = -1$ is allowed.

Finally, consider the matrix element for z in a transition from state $\ell = 1$, $m = 0$ to state $\ell = 1$, $m = 1$. Using $\cos \theta \, Y_1^1 \propto Y_2^1$,

$$\iint \left(Y_1^0\right)^* \cos \theta \, Y_1^1 \, d\tau \propto \iint \left(Y_1^0\right)^* Y_2^1 \, d\tau = 0.$$

This is a forbidden transition. If electric dipole radiation is forbidden, radiation from higher multipoles might be allowed, but they have smaller probability, hence give lower intensity. The next most probable electric multipole transition after electric dipole is electric quadrupole radiation. Magnetic dipole radiation has roughly the same transition probability as electric quadrupole radiation and can dominate if electric dipole and electric quadrupole radiation are forbidden.

In summary, the selection rules for electric dipole transitions are

$$\Delta j = 0, \pm 1 \quad \text{but the transition } j = 0 \text{ to } j' = 0 \text{ is forbidden}$$
$$\Delta m = 0, \pm 1 \tag{8.49}$$
$$\Delta s = 0.$$

Electric dipole radiation is forbidden in a transition between state $j = 0$ and state $j' = 0$. The reason is that a photon has intrinsic spin 1, so it is impossible to satisfy conservation of angular momentum in transitions between two states both having 0 angular momentum.

The physical reason for the selection rule $\Delta s = 0$ is that an electric field $\mathcal{E}$ cannot exert a torque $\boldsymbol{\tau} = \boldsymbol{\wp} \times \mathcal{E}$ to flip an electron's spin because the electron is a point particle with no measurable electric dipole moment. An interaction that contains spin operators can flip electron spin, in which case spherical harmonics are no longer exact basis functions.

Magnetic dipole transition is an example of a spin-dependent interaction, with transition probability depending on the matrix element of $\mathbf{L} + 2\mathbf{S}$. Physically, the matrix element refers to an electromagnetic wave's magnetic field interacting with an atom's magnetic moment due to orbital motion and with a particle's intrinsic magnetic moment.

8.12.2 From the Wigner–Eckart Theorem

The Wigner–Eckart theorem developed in Section 8.11 provides a general formulation of selection rules. To review, the Wigner–Eckart theorem Eq. (8.48) is

$$\langle \alpha,\, j,\, m | T_q^{(k)} | \alpha',\, j',\, m' \rangle = (j'\, k\, m'\, q | j\, m)\, \langle \alpha,\, j || T_q^{(k)} || \alpha',\, j' \rangle .$$

The quantity on the left is a matrix element to be evaluated, where α, α' are other quantum numbers not involved with angular momentum. On the right, the factor $(j'\, k\, m'\, q | j\, m)$ is a *3-j* coefficient and $\langle \alpha,\, j || T_q^{(k)} || \alpha',\, j' \rangle$ is the reduced matrix element. The operator $\mathbf{T}_q^{(k)}$ is a tensor operator of rank k.

The power of the theorem comes from the general nature of $\mathbf{T}_q^{(k)}$ and from the fact that all of the dependence on m, m', q is contained in tabulated *3-j* coefficients. Selection rules therefore come from the known properties of the *3-j* coefficients without reference to the detailed nature of the system expressed in the reduced matrix element.

Consider as an example the z component of electric dipole radiation, corresponding to $k = 1$ and $q = 0$. The corresponding *3-j* coefficient is

$$(j'\, 1\, m'\, 0 | j\, m).$$

The *3-j* coefficient $= 0$ unless $j', 1, j$ satisfy the angular momentum triangle condition $j = j' + 1,\, j',\, j' - 1$ so this is the selection rule $\Delta j = \pm 1, 0$. As a corollary the coefficient is 0 if $j = 0,\, j' = 0$, and we see again that electric dipole radiation is forbidden in a transition between a state $j = 0$ and a state $j' = 0$ because angular momentum cannot be conserved when a photon is emitted.

The coefficient $(j'\, 1\, m'\, 0 | j\, m) = 0$ unless $m = m' + q = m' + 0 = m'$ so that $\Delta m = 0$ is the selection rule for allowed z transitions. Transitions corresponding to the components $x + iy$ ($q = 1$) and $x - iy$ ($q = -1$) depend on the *3-j* coefficient $(j'\, 1\, m' \pm 1 | j\, m)$. These transitions are allowed only if $m = m' \pm 1$, so the corresponding selection rule is $\Delta m = \pm 1$.

8.12.3   From Parity-Inversion

Let $\mathcal{P}$ be the inversion operator that changes the signs of spatial coordinates:

$$\mathcal{P}\{x, y, z\} \longrightarrow \{-x, -y, -z\}.$$

Consider the effect of $\mathcal{P}$ on vector $\mathbf{r}$:

$$\begin{aligned}
\mathcal{P}\mathbf{r} &= \mathcal{P}(x\hat{\mathbf{i}} + y\hat{\mathbf{j}} + z\hat{\mathbf{k}}) \\
&= (-x\hat{\mathbf{i}} - y\hat{\mathbf{j}} - z\hat{\mathbf{k}}) \\
&= -\mathbf{r}.
\end{aligned}$$

Because $\mathcal{P}$ changes the sign of $\mathbf{r}$, we say that $\mathbf{r}$ has *negative parity* $(-)$.

$$\mathcal{P}\mathbf{r} = (-)\mathbf{r}$$

As another example, consider the effect of $\mathcal{P}$ on the scalar radius r.

$$\begin{aligned}
\mathcal{P}r &= \mathcal{P}\sqrt{x^2 + y^2 + z^2} \\
&= \sqrt{(-x)^2 + (-y)^2 + (-z)^2} \\
&= \sqrt{x^2 + y^2 + z^2} \\
&= r
\end{aligned}$$

The sign of r is not changed by $\mathcal{P}$, so r has *positive parity* $(+)$.

$$\mathcal{P}r = (+)r$$

A matrix element that changes parity under inversion must equal 0 and cannot describe a physical process. Parity is *conserved* in transitions, with the exception of certain weak interactions such as beta decay. Parity is conserved for the electromagnetic forces treated in this section.

The selection rule for electric dipole radiation is an important example of parity conservation. The electric dipole moment vector

$$\boldsymbol{\rho} = q(x\hat{\mathbf{i}} + y\hat{\mathbf{j}} + z\hat{\mathbf{k}})$$

has $(-)$ parity. Consider the matrix element $\int \psi_n^* \boldsymbol{\rho} \psi_{n'} d\,\tau$. If ψ_n and $\psi_{n'}$ have the same parity, the matrix element changes sign under inversion and must equal 0. Electric dipole radiation can occur only in transitions between states of opposite parity.

Amplify this argument by considering the parity of atomic wave functions. The radial part of an atomic wave function has $(+)$ parity and need not be considered further. The angular part of any given atomic wave function is a spherical harmonic. To determine the parity of spherical harmonics, use the ladder-generating function Eq. (8.34).

$$Y_\ell^m = \underbrace{I_+ \ldots I_+}_{2\ell \text{ times}}(x - iy)^\ell$$

Equation (8.13),

$$I_\pm = \mp\hbar(x \pm iy)\frac{\partial}{\partial z} \pm \hbar z\left(\frac{\partial}{\partial x} \pm i\frac{\partial}{\partial y}\right),$$

shows that applying inversion to x, y, z leaves $I_\pm$ with unchanged (+) parity. The parity of Y_ℓ^m is determined solely by the term $(x - iy)^\ell$ that has parity $(-1)^\ell$. An atomic wave function with quantum number ℓ therefore has definite parity $(-1)^\ell$. It follows that electric dipole transitions are allowed if $\Delta\ell = \pm 1$ but forbidden if $\Delta\ell = 0$.

8.13 Brief Bios

Otto Stern (1888–1969), born in Poland, was awarded the 1943 Nobel Prize in Physics, cited "for his contribution to the development of the molecular ray method and his discovery of the magnetic moment of the proton." He was forced to leave Germany in the 1930s and took a post at the Carnegie Institute of Technology. He retired to Berkeley, California.

The German mathematicians Alfred Clebsch (1833–72) and Paul Gordan (1837–1912) developed the Clebsch–Gordan coefficients in the 1860s while both were on the faculty of the University of Giessen. Their motivation, carried out long before quantum mechanics, was research on spherical harmonics. Section 8.8 showed that spherical harmonics are basis functions for the rotation group, and the Clebsch–Gordan coefficients therefore turned out to apply to quantum mechanical angular momentum. The Wigner *3-j* coefficients were developed in the 1930s and have a higher degree of symmetry than Clebsch–Gordan coefficients.

Gordan deserves a footnote in history because, while at the University of Erlangen, his one and only graduate student was the brilliant Emmy Noether.

Julian Schwinger (1918–94), born in New York City, shared the 1965 Nobel Prize in Physics with Sin-Itiro Tomonaga and Richard Feynman, cited for "their fundamental work in quantum electrodynamics, with deep-ploughing consequences for the physics of elementary particles." A brilliant theorist, he published his first research paper at age 16. One of his most important works was calculation of the anomalous magnetic moment of the electron (1948).

$$gs \approx 2\left(1 + \frac{\alpha}{2\pi} - \ldots\right)$$

The first correction term is engraved on his tombstone in Mt. Auburn cemetery in Cambridge, Massachusetts. He taught primarily at Harvard University and UCLA.

Disclosure: As a graduate student at Harvard, the author took a course on quantum electrodynamics with Prof. Schwinger.

Summary of Chapter 8

Chapter 8 discusses representations of the rotation group and angular momentum operators, central to applications in quantum mechanics.

a) The Stern–Gerlach experiment showed that the magnetic moment of atoms, hence the angular momentum, is spatially quantized – directed only in discrete directions.

b) The Stern–Gerlach experiment stimulated the development of quantum numbers: principal quantum number n, orbital angular momentum quantum number L and its projection m_L, spin quantum number S and its projection m_S. Quantum numbers are the labels for wave functions.

c) Angular momentum in quantum mechanics is represented by operators analogous to classical angular momentum but with de Broglie momentum $\mathbf{p} = -i\hbar\nabla$. A finite angle rotation can be generated by repeated application of an angular momentum operator. The rotation is represented by a complex exponential with angular momentum operator in the argument.

d) Angular momentum operators obey nonzero commutation relations, showing that different components of angular momentum do not commute, hence are not all simultaneously measurable. Matrices for the angular momentum components can be derived from first-order rotation matrices, from which the commutation relations follow.

e) Rotations about a fixed axis form an Abelian group, which has only 1-dimensional representations $e^{\pm im\theta}$ where m is an integer.

f) Ladder operators are complex linear combinations of angular momentum components $I_\pm = I_x \pm i I_y$. They raise or lower the m value of a basis function u_m and can be used to show that $-j \leq m \leq j$, where j can be an integer or half-odd integer, giving $2j + 1$ basis functions. I^2 and I_z commute, so the basis functions can be labeled by j and by m as u_{jm}. $I_z u_{jm} = m u_{jm}$ and $I^2 u_{jm} = j(j + 1)u_{jm}$. In Dirac bracket notation $I_z |j\ m\rangle = m|j\ m\rangle$. Rotation can change m but not j, so the u_{jm} for a given j form an irreducible block in a rotation group's matrix representation.

g) For $j = 1$, the $2j + 1 = 3$ basis functions $u_m = u_1, u_0, u_{-1}$ can be used to develop the matrix representations of the components of the angular momentum operators to evaluate the series expansion of the rotation operator in exponential form. For j an integer, the matrix elements involve single-valued real angles. For j half an odd integer, the matrix elements involve double-valued angles characteristic of spinors.

h) The $u_{\ell m}$ are spherical harmonics Y_ℓ^m. They can be evaluated for a given ℓ by applying the I_+ ladder operator 2ℓ times to the lowest basis function $u_{\ell,-\ell} = (x - iy)^\ell$, or without using group theory by solving the angular part of Laplace's partial differential equation. The spherical harmonics also appear in the central column of the matrix $D^{(\ell)*}$.

i) Spin basis functions cannot be represented by continuous functions because they have discrete values and are therefore represented by discrete matrices. Matrices for the components of spin $\frac{1}{2}$ angular momentum are proportional to the Pauli spin matrices.

j) Coupling quantum-mechanical angular momenta is done with coefficients such as the Wigner *3-j* coefficients. Coupling two independent sets of angular momentum basis functions $u_m^{(\ell)}$, $v_{m'}^{(\ell')}$ symbolized by the direct (Kronecker) product $D^{(\ell)} \otimes D^{(\ell')}$ gives the overall basis functions W. The explicit example of positronium (spin $\frac{1}{2}$ of the e^-, spin $\frac{1}{2}$ of the e^+) is worked out to develop the *3-j* coefficients for $D^{(\frac{1}{2})} \otimes D^{(\frac{1}{2})}$.

k) The Wigner *3-j* coefficients obey a number of orthogonality properties.

l) The Wigner–Eckart theorem uses *3-j* coefficients to express matrix elements for given j, j' in terms of a reduced matrix element independent of m, m'.

m) Selection rules for a given type of transition, for example, electric dipole transitions, state whether the transition is allowed or forbidden depending on whether its matrix element is nonzero or not. Methods for deriving selection rules include properties of spherical harmonics, the properties of *3-j* coefficients, and parity-inversion. If the matrix element is 0, the transition is forbidden. An atomic wave function with angular momentum quantum number ℓ has parity $(-1)^\ell$, so in atoms an electric dipole transition $(-)$ parity is allowed only between states of different parity. Spin plays no role in electric dipole transitions because electric fields cannot change electron spin.

Problems and Exercises

8.1 Consider two operators **A** and **B**. Show that

$$e^{-i\mathbf{A}} e^{-i\mathbf{B}} = e^{-i(\mathbf{A}+\mathbf{B})}$$

only if **A** and **B** commute.

8.2 Show that the angular momentum operators (I_x, I_y, I_z) are Hermitian.

8.3 Consider an operator $\mathbf{U} = e^{i\theta\mathbf{G}}$ with generator **G**. If **U** commutes with a Hamiltonian **H**, use the first-order method to show that **G** also commutes with **H**. What does this imply about **G**?

8.4 Show that the ladder operators $I_\pm$ are not Hermitian.

8.5 Show that $\left[I^2, I_\pm \right] = 0$.

8.6 Find the eigenvalues of I_x.

$$I_x = \begin{pmatrix} 0 & 0 & 0 \\ 0 & 0 & -i \\ 0 & i & 0 \end{pmatrix}$$

8.7 Suppose that an irreducible representation of the rotation group is represented by 7×7 matrices.
(a) What is the value of its angular momentum quantum number j?
(b) What are the possible values of its magnetic quantum number m?

8.8 Suppose that an irreducible representation of the rotation group is represented by 4×4 matrices.
(a) What is the value of its angular momentum quantum number j?
(b) What are the possible values of its magnetic quantum number m?

8.9 Evaluate $I_x \,|j\, m\rangle$.

8.10 Evaluate $I_y \,|j\, m\rangle$.

8.11 For $j = 1$, consider the basis functions u_{1m}. Show by calculation that $I_+ u_{10}$ is proportional to the spherical harmonic Y_1^1.

8.12 Consider the basis function u_{11} for $D^{(1)}(\alpha, \beta, \gamma)$. Calculate the result of rotating u_{11} through the Euler angles $\gamma = 0$, $\beta = \frac{\pi}{4}$, $\alpha = \frac{\pi}{2}$.

8.13 Consider $I^2 = I_x^2 + I_y^2 + I_z^2$. Use Eqs. (8.27), (8.28), and (8.29) to show that $I^2 u_{jm} = j(j + 1)\, u_{jm}$.

8.14 For $\ell = 2$, develop by calculation an expression proportional to $Y_2^{-1}(\theta, \phi)$.

8.15 For $\ell = 3$, develop by calculation an expression proportional to $Y_3^{-2}(x, y)$.

8.16 a) Evaluate $e^{-i\phi I_z} \,|jm\rangle$.
b) Use the result of a) to evaluate the matrix the element $\langle jm'|e^{-i\phi I_z}|jm\rangle$.

8.17 In the HD molecule of hydrogen, one nucleus is ^{1}H (spin $\frac{1}{2}$) and the other is a deuteron ^{2}H (spin 1). What are the possible spin angular momentum states of the molecule? The two electrons are paired and assumed to be in a state $j = 0$.

8.18 In the HD molecule of hydrogen, one nucleus is ^{1}H (spin $\frac{1}{2}$) and the other is a deuteron ^{2}H (spin 1). The two electrons are paired and assumed to be in a state $j = 0$. Referring to Table F.2 in Appendix F, express the overall spin wave function $W_{-1/2}^{(\frac{1}{2})}$ in terms of the ^{1}H spin wave functions $v^{(\frac{1}{2})}$ and the ^{2}H spin functions $u^{(1)}$.

8.19 In a molecule of "heavy hydrogen," D_2, the nuclei are both deuterons (spin 1). The two electrons are paired and assumed to be in a state $j = 0$. Referring to Table F.3 in Appendix F, what is the overall spin wave function W if $u_1^{(1)}$ and $v_0^{(1)}$ are the spin functions of the individual deuterons?

8.20 In a molecule of "heavy hydrogen," D_2, the nuclei are both deuterons (spin 1). The two electrons are paired and assumed to be in a state $j = 0$. Referring to Table F.3 in Appendix F, what is the overall spin wave function W if $u_0^{(1)}$ and $v_1^{(1)}$ are the spin functions of the individual deuterons?

8.21 What angular momentum values are possible from the combination of $j = 2$ and $j = 1$? Draw sketches according to the vector model to illustrate your answers.

8.22 What angular momentum values are possible from the combination of $j = \frac{3}{2}$ and $j = \frac{1}{2}$? Draw sketches according to the vector model to illustrate your answers.

8.23 Use values from Table F.2 in Appendix F to illustrate the orthogonality relation Eq. (8.43).

8.24 Use values from Table F.3 in Appendix F to illustrate the orthogonality relation Eq. (8.44).

8.25 Consider a state $\psi_{\ell m}$ with $\ell = 2$ and $m = 1$, and a state with $\ell = 1$ and $m = 0$. Show, using spherical harmonics, whether an electric dipole transition is allowed between these states.

8.26 Consider an atom with a single electron outside closed shells. Take the z component of a magnetic dipole transition with matrix element

$$\langle \ell', m', m'_s | L_z + 2S_z | \ell, m, m_s \rangle .$$

Let

$$\ell = 1 \quad m = 0 \quad m_s = +\frac{1}{2}.$$

For what values of ℓ', m', m'_s will this be an allowed transition?

8.27 An atom's electron distribution will have an electric quadrupole moment if the distribution is a nonspherical ellipsoid. The matrix element for an electric quadrupole transition is proportional to the electric quadrupole operator $Q = x^2 + y^2 - 2z^2 = r^2 - 3z^2$.
(a) What is the parity of Q?
(b) What does the parity of Q imply about the parity of the initial and final states of an allowed electric quadrupole transition?

8.28 The quadrupole operator Q discussed in Problem 27 is a second-rank tensor $\mathbf{T}_q^{(k)}$ with $k = 2$. Based on this, what are the selection rules Δj and Δm_j for electric quadrupole transitions?

9

THE STRUCTURE OF ATOMS

9.1   Introduction

This chapter is concerned with interactions within an atom and the electron structure of atoms.

We begin with a review of the notation for chemical elements. Elements are identified by one or two letters, for example, H for hydrogen and Ag for silver. The atomic number Z of the element, the number of protons in its nucleus, is a subscript at the lower left of the letter symbol. Z is also equal to the number of electrons in a neutral atom. The mass number A, the total number of protons and neutrons in the nucleus, is a superscript at the upper left. A can symbolize different isotopes of an element. The mass number is often omitted for an element's principal (most abundant) isotope. For example $_1$H symbolizes the most abundant isotope of hydrogen $Z = 1$. Deuterium ("heavy hydrogen") is symbolized $_1^2$H. Its nucleus consists of one proton and one neutron so $A = 2$.

9.2   Zeeman: An Important Experiment (1897)

Chapter 8 began with the important Stern–Gerlach experiment that demonstrated directional quantization of angular momentum. This chapter begins with Dutch physicist Pieter Zeeman's experiments carried out more than 20 years earlier that helped lead to the key role of angular momentum in atomic structure. Zeeman's important work predated the formal discovery of the electron and demonstrated the role of electrons in atomic spectra.

9.2.1   Classical Theory of the Zeeman Effect

In the 1890s Dutch theorist Hendrik Lorentz (1853–1928) at the University of Leiden speculated that because classical theory predicts electromagnetic radiation from moving charges, atoms might contain moving charged particles responsible for spectral lines. Lorentz theorized that the motion, hence the emitted spectra, could be affected by an applied magnetic field. His former student Pieter Zeeman (1865–1943) took up this research topic after assuming a post at the University of Amsterdam.

The light source in Zeeman's first experiments was asbestos soaked in sodium chloride and heated in a Bunsen burner to generate bright yellow sodium *D* spectral lines. With the source placed between the poles of a magnet, Zeeman observed in 1897 a broadening of spectral lines when the magnet was turned on and an immediate disappearance of the broadening when the magnet was turned off. Although imperfect, this was the first observation of what is now known as the *Zeeman effect*.

Zeeman then obtained a stronger electromagnet and used an electric arc struck between cadmium electrodes as the light source. Cadmium vapor is toxic, but Zeeman lived to age 78; experiments today use a spectral lamp with cadmium vapor enclosed in a protective glass bulb. Cadmium has a bright blue spectral line at 480 nm, and Zeeman observed the line to be split by the magnetic field into a triplet of three separate spectral lines. The center line had the same wavelength as the original unperturbed line, with the two new lines equally spaced higher and lower in wavelength.

Lorentz explained the triplet using classical mechanics and classical electromagnetism, the only tools available before quantum mechanics. His model pictured an atom with a particle of charge *q* and mass *m* bound by a spherically symmetric force with force constant *k* and executing harmonic motion about equilibrium.

In a magnetic field **B** a charge of mass *m* moving with speed **v** experiences a magnetic force $q\mathbf{v} \times \mathbf{B}$. With **B** in the *z* direction, $\mathbf{B} = B\hat{\mathbf{k}}$, the equations of motion are

$$m\ddot{x} = -kx + q\dot{y}B \qquad (9.1)$$
$$m\ddot{y} = -ky - q\dot{x}B \qquad (9.2)$$
$$m\ddot{z} = -kz. \qquad (9.3)$$

A solution of Eq. (9.3) is $A\cos(\omega_0 t)$, where $\omega_0 = \sqrt{k/m}$ is the frequency of the central line. Note that ω_0 is independent of **B**. The equations predict two additional modes having frequencies $\omega_+ > \omega_0 > \omega_-$. As a trial solution for ω_+, assume

$$x = A_+ \sin(\omega_+ t) \qquad (9.4)$$
$$y = A_+ \cos(\omega_+ t). \qquad (9.5)$$

Substituting in Eq. (9.1) gives

$$-\omega_+^2 + \omega_0^2 + \left(\frac{q}{m}\right)B\omega_+ = 0. \qquad (9.6)$$

Let $\Delta\omega$ be the frequency shift $\omega_\pm = \omega_0 \pm \Delta\omega$. Using the good approximation $\omega_0 \gg \Delta\omega$ in Eq. (9.6) gives

$$-2\omega_0 \Delta\omega + \frac{q}{m} B\omega_0 = 0$$

$$\Delta\omega = \left(\frac{q}{2m}\right) B.$$

Leaving details to the problems, Eqs. (9.1) and (9.2) imply that the two shifted lines are linearly polarized perpendicular to the field. They are labeled σ (from German *senkrecht*, perpendicular). The central unshifted line is polarized parallel to the field and is labeled π. As observed along the direction of **B** through a hole drilled along the axis of a pole piece, the shifted lines are circularly polarized (one clockwise, the other counterclockwise). Zeeman's observations confirmed these predictions.

Measuring the frequency shift and the field strength allowed Zeeman to determine q/m of the active particle, later named the *electron*. His result was within 10 percent of the modern value for the electron, and he also determined that it was negatively charged. He speculated that atomic spectra are due to radiation from electrons. Shortly after Zeeman's work, British physicist J. J. Thomson (1856–1940) directly observed the electron in electric discharges in gases at low pressure.

Lorentz and Zeeman shared the 1902 Nobel Prize in Physics cited "in recognition of the extraordinary service they rendered by their researches into the influence of magnetism upon radiation phenomena."

Magnetic Fields of Sunspots

Zeeman early expressed the hope that the effect he had discovered would have application in astronomy. His hope was soon realized. In 1908 he received a letter from George Hale at the Mt. Wilson Observatory (founded by Hale) in Pasadena, California, saying that he had determined that sunspots were vortices and surmised that the rotating charges could produce magnetic fields. As proof, he told Zeeman that when he scanned his spectrometer over a sunspot, he observed Zeeman splitting of atomic hydrogen spectral lines but did not see splitting when scanning over areas free of sunspots.

9.3   Quantum Theory of the Zeeman Effect

Zeeman's experiments with the 480 nm cadmium line showed the *normal* triplet of three separate spectral lines predicted by Lorentz's classical theory. However, further experiments showed that many other lines exhibit so-called *anomalous* splittings that were not the "normal" triplet. Lorentz and other physicists of the time could not account for the "anomalous" multiplicity and observed separations. Another quantum number was needed. The complete theory of the anomalous splittings came only with the advent of quantum mechanics and the discovery of electron spin.

The bright yellow light from sodium street lamps is due to the intense D doublet spectral lines in the emission spectrum. The doublet is a closely spaced pair D_1 (589.6 nm) and D_2 (590.0 nm), separated by 0.597 nm.

The figure, to scale in wavelength, shows the calculated "anomalous" Zeeman splitting in a magnetic field of 5 T = 50 kG. The upper sketch is field off and the lower is field on. With improved apparatus, Zeeman was able to photograph the splitting. The later quantum-mechanical calculation shown in the figure agreed with his result. The analysis in Section 9.3.2 shows how to calculate the wavelength of any line in weak-field Zeeman splitting.

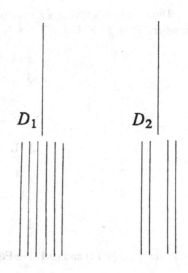

The magnetic moment μ gives a handle on angular momentum. The magnetic moment is

$$\boldsymbol{\mu} = \mu_B g_J \mathbf{J}$$
$$= \left(\frac{e\hbar}{2m_e}\right)(g_L \mathbf{L} + g_S \mathbf{S}). \qquad (9.10)$$

$\mu_B = \frac{e\hbar}{2m_e}$ is the Bohr magneton of the electron, and g_J is the *g-factor* for the state characterized by quantum number J. One unit of orbital angular momentum $\hbar$ generates one unit of magnetic moment, so $g_L = 1$. One unit of intrinsic spin generates slightly more than two units of magnetic moment, so $g_S \approx 2$.

The magnetic energy of an atom's magnetic moment $\boldsymbol{\mu}$ in a magnetic field $\mathbf{B}$ shifts energy levels by $E = -\boldsymbol{\mu} \cdot \mathbf{B}$. Let the field be in the z direction, $\mathbf{B} = B\hat{\mathbf{k}}$.

$$\mu_z = \mu_B g_J J_z$$
$$= \mu_B g_J M_J$$
$$E = (\mu_B g_J M_J)B$$

The Zeeman effect was telling physicists something about angular momentum and atomic structure.

9.3.1 Weak Field Zeeman

The magnetic field due to the orbital motion of electrons in an atom can interact with an electron's magnetic moment, causing an energy shift proportional to $\mathbf{L} \cdot \mathbf{S}$ called variously *LS coupling*, *spin-orbit coupling*, or *Russell–Saunders coupling*. In weak magnetic fields, *LS* coupling causes $\mathbf{L}$ and $\mathbf{S}$ to precess around $\mathbf{J}$, as suggested by the sketch, so L, S, J, and $J_z = M$ are good quantum numbers. The spin-orbit energy is $a\mathbf{L} \cdot \mathbf{S}$, where the parameter a depends on details of atomic structure.

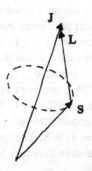

The task is to express g_J in terms of the good quantum numbers J, M, L, S. Using $\mathbf{J} = \mathbf{L} + \mathbf{S}$ and Eq. (9.10),

$$g_J \mathbf{J} = g_L \mathbf{L} + g_S \mathbf{S}$$
$$g_J \mathbf{J} \cdot \mathbf{J} = g_L \mathbf{J} \cdot \mathbf{L} + g_S \mathbf{J} \cdot \mathbf{S} \tag{9.11}$$
$$\mathbf{S} = \mathbf{J} - \mathbf{L}$$
$$\mathbf{S}^2 = \mathbf{J}^2 + \mathbf{L}^2 - 2\mathbf{J} \cdot \mathbf{L}$$
$$\mathbf{J} \cdot \mathbf{L} = \frac{1}{2}(\mathbf{J}^2 + \mathbf{L}^2 - \mathbf{S}^2) \tag{9.12}$$
$$\mathbf{L} = \mathbf{J} - \mathbf{S}$$
$$\mathbf{L}^2 = \mathbf{J}^2 + \mathbf{S}^2 - 2\mathbf{J} \cdot \mathbf{S}$$
$$\mathbf{J} \cdot \mathbf{S} = \frac{1}{2}(\mathbf{J}^2 + \mathbf{S}^2 - \mathbf{L}^2). \tag{9.13}$$

Inserting Eqs. (9.12) and (9.13) in Eq. (9.11) gives

$$g_J \mathbf{J}^2 = \frac{g_L}{2}(\mathbf{J}^2 + \mathbf{L}^2 - \mathbf{S}^2) + \frac{g_S}{2}(\mathbf{J}^2 - \mathbf{L}^2 + \mathbf{S}^2). \tag{9.14}$$

The basis functions for the irreducible representations are $u(J, L, S)$, so each term obeys $\mathbf{I}^2 u = I(I + 1)u$. With $g_L = 1$ and $g_S \approx 2$, Eq. (9.14) gives

$$g_J J(J + 1) = \frac{1}{2}(J(J + 1) + L(L + 1) - S(S + 1))$$
$$+ (J(J + 1) - L(L + 1) + S(S + 1))$$

$$g_J = 1 + \frac{J(J + 1) - L(L + 1) + S(S + 1)}{2J(J + 1)}. \tag{9.15}$$

g_J is known as the *Landé g-factor*.

9.3.2   Term Notation and Zeeman Splitting

There is a standard notation for expressing the angular momentum quantum numbers of an atom's quantum states. The orbital angular momentum L is represented by an upper case letter. For historical reasons, the first few letters and their numerical equivalents are $S \rightarrow 0$, $P \rightarrow 1$, $D \rightarrow 2$, $F \rightarrow 3$ followed in numerical and alphabetical order except that J, P, and S are omitted from the sequence. Letters beyond F are rarely needed.

A superscript at the upper left of the letter gives the spin quantum number S in the form $2S + 1$. A subscript at the lower right gives the total angular momentum quantum number J, omitted in some listings if its value is clear from the relation $\mathbf{J} = \mathbf{L} + \mathbf{S}$. The notation has the form $^{2S+1}L_J$ and is called the *term symbol*. The quantum numbers S, L, J specify a *level*. The Zeeman effect reveals that a level can be split into several states of different energies. An individual state is characterized by a complete set of quantum numbers such as S, L, J, M_J.

The term symbol for the ground level of hydrogen is $^2S_{1/2}$ with $L = 0$, $S = 1/2$, $J = 1/2$. Another example is the term symbol $^2P_{3/2}$ for the ground level of chlorine $L = 1$, $S = 1/2$, $J = 3/2$. The nearby level $^2P_{1/2}$ is ≈ 0.1 eV higher than the true ground level, due mainly to LS coupling energy.

A set of closely associated terms having nearly the same wavelengths is called a *multiplet*. The three closely spaced lines in the normal Zeeman effect are a *triplet*. Two lines forming a multiplet are a *doublet*, and a single line is a *singlet*.

The diagram (not to scale) shows the Zeeman energy levels in a constant magnetic field for the red line of cadmium at 643.8 nm, a source often used in physics labs to measure the Zeeman effect. Lines are emitted by transitions from an upper Zeeman-split $^1D_2$ level at ≈ 7.1 eV to a lower Zeeman-split $^1P_1$ level at ≈ 5.2 eV.

The upper level $^1D_2$ is characterized by $S = 0$, $L = 2$, $J = 2$, and the lower level $^1P_1$ has $S = 0$, $L = 1$, $J = 1$. The Landé factor is $g_J = 1$ for both states, a result that holds for any $S = 0$ state. In this example, S = 0 for both levels, so electron spin does

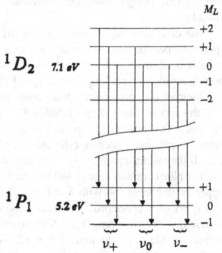

not affect the splitting, and the result is therefore a normal Zeeman triplet.

The magnetic field energy splits the D level into a multiplet of $2L+1 = 5$ states with $M_L = 2, 1, 0, -1 -2$ and splits the P level into 3 states $M_L = 1, 0, -1$ as the sketch shows. $g_J = 1$ for both multiplet states in this example, so all have the same spacing and the result is a normal Zeeman triplet. The selection rule $\Delta M_L = 0, \pm 1$ limits the number of different frequencies to three.

To calculate Zeeman line spacing, a general approach is first to calculate the frequencies of a line and an adjacent line using Landé factors, then convert frequency difference to wavelength difference. Assume a moderate field strength of $0.5\,\text{T} = 5$ kG.

$$h\nu_+ = [(M_L = +2\,g_J)_D - (M_L = +1\,g_J)_P]\,\mu_B B$$

$$h\nu_0 = [(M_L = +1\,g_J)_D - (M_L = +1\,g_J)_P]\,\mu_B B$$

$$\Delta\nu = (\nu_+ - \nu_0)$$

$$= \frac{1}{h}\,[(2)(1) - (1)(1)]\,\mu_B B = \frac{1}{h}\mu_B B$$

$$= \left(\frac{9.27 \times 10^{-24}\,\text{J}\cdot\text{T}^{-1}}{6.63 \times 10^{-34}\,\text{J}\cdot\text{s}^{-1}}\right)(0.5\,\text{T})$$

$$= 7.0 \times 10^9 \text{ s}^{-1}$$

$$\Delta\lambda = \frac{\Delta\nu}{\nu}\lambda = \frac{\Delta\nu}{c}\lambda^2$$

$$= \left(\frac{7.0 \times 10^9 \text{ s}^{-1}}{3 \times 10^8 \text{ m} \cdot \text{s}^{-1}}\right)(643.8 \times 10^{-9} \text{ m})^2$$

$$= 9.7 \times 10^{-12} \text{ m} \approx 0.01 \text{ nm}.$$

Note that only the magnetic energy difference contributes to the spacing, because the nonmagnetic energy associated with each multiplet is the same for every line and cancels.

The calculated separation is too small to be seen visually through an ordinary spectroscope. An interferometer such as a Fabry–Perot etalon or Lummer–Gehrcke plate is used instead. They create an interference pattern by using multiple internal reflections to generate long path-length differences.

Zeeman did not use an interferometer but instead relied on visual observation through a spectroscope fitted with a ruled optical grating (5670 lines/cm) as the dispersive element. From measurements of line broadening due to unresolved Zeeman-split lines, Zeeman estimated that he would be able to distinguish individual lines if they were separated by more than ≈ 0.02 nm.

The blue cadmium line at 480 nm studied by Zeeman is due to a transition from an upper excited level $^3S_1$ with $S = 1$, $L = 0$, $J = 1$ to a lower excited level $^3P_1$ with $S = 1$, $L = 1$, $J = 1$. According to theory, the separation of Zeeman-split lines increases with magnetic field strength, so Zeeman obtained an electromagnet with tapered pole pieces capable of producing 3.2 T = 32 kG. Using the new magnet, he saw the 480 nm line split into a distinct triplet.

9.3.3   Strong Field (Paschen–Back) Zeeman

In a sufficiently high magnetic field, the magnetic energy $\mu \cdot \mathbf{B}$ can be substantially greater than the LS coupling energy. In the high field regime, $\mathbf{L}$ and $\mathbf{S}$ are decoupled and precess separately about the field direction as shown. Now J is no longer a good quantum number, but the projections of $\mathbf{L}$ and $\mathbf{S}$ on $\mathbf{B}$ are the good quantum numbers M_L and M_S. If the Hamiltonian's matrix elements are evaluated using M_L and M_S, its diagonal elements are much larger than the off-diagonal elements.

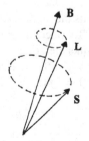

The Landé g-factor no longer applies in the high field regime. The magnetic energy is best expressed in terms of the good quantum numbers M_L and M_S.

$$(g_L M_L + g_S M_S)\mu_B B \approx (M_L + 2M_S)\mu_B B$$

The regime where the magnetic energy is much larger than the spin-orbit energy is called the *Paschen–Back effect* named after the German physicists who first observed it. The selection rules for the Paschen–Back effect are $\Delta M_S = 0$ and $\Delta M_L = 0, \pm 1$, so in the high field regime a spectral line is always split into a triplet just as in the normal Zeeman effect when $S = 0$.

How high a magnetic field is needed to enter the Paschen–Back regime? Compare the measured spin-orbit energy for the first excited state of the hydrogen atom to the magnetic energy of one Bohr magneton in a field B:

$$\text{spin-orbit energy:} \approx 7.2 \times 10^{-24} \text{ J}$$
$$\text{magnetic energy:} = \mu_B B \approx (9.3 \times 10^{-24}) \text{ J} \cdot \text{T}^{-1} \times B.$$

The LS coupling energy is 10 percent of the magnetic energy for $B \approx 10 \text{ T} = 100$ kG. Smaller fields are sufficient for some atoms.

Modern researchers do not use special terms such as normal, anomalous, and Paschen–Back, viewing them simply as different cases of the Zeeman effect.

9.4 Fine Structure

LS coupling and relativistic effects are the main contributors to *fine structure* shifting and splitting of spectral lines. This section discusses only the effect of LS coupling, the main cause of fine structure in low-Z atoms. Consider $\mathbf{J} = \mathbf{L} + \mathbf{S}$ and take, for example, $L = 2$, $S = 1$. According to the vector model, the possible values of J are 3, 2, 1. When $L > S$, as in this example, the multiplicity is $2S + 1 = 3$ and the energy level is split into three. If $L < S$, the multiplicity is $2L + 1$. If, for example, $L = 0$, $S = \frac{3}{2}$, the only possible value of J is $\frac{3}{2}$ and the multiplicity is 1.

The diagram (not to scale) shows the fine structure of the lowest energy levels in ^{1}H. The $^2S_{1/2}$ ground state is not split because $L = 0$ and $S = \frac{1}{2}$, so only the state $J = \frac{1}{2}$ is possible. The first excited state at $\approx$ 10.2 eV has $L = 1$, $S = \frac{1}{2}$, so there are two P levels, $^2P_{3/2}$ and $^2P_{1/2}$.

	eV
$^2P_{3/2}$	
	4.5×10^{-5}
$^2P_{1/2}$	10.2
$^2S_{1/2}$	0.0

Table 9.1 lists measured energy levels and multiplet splittings for ^3D states in Ca and in V^{++}. These species were chosen because they are *isoelectronic* – they have the same number of electrons.

As the data suggest, LS splitting between adjacent levels tends to increase with J. This behavior can be predicted for low-Z atoms, where J, L, and S are good quantum numbers.

Table 9.1 Measured fine structure

	J	cm^{-1}	eV
Ca	1	20335.4	2.521
	2	20349.3	$2.521 + 1.7 \times 10^{-3}$
	3	20371.0	$2.521 + 4.4 \times 10^{-3}$
V^{++}	1	18269.5	2.265
	2	18293.9	$2.265 + 3.9 \times 10^{-3}$
	3	18353.8	$2.265 + 7.4 \times 10^{-3}$

Data from NIST (National Institute of Standards and Technology) Atomic Spectra Database.

$$\mathbf{S} = \mathbf{J} - \mathbf{L}$$
$$\mathbf{S}^2 = \mathbf{J}^2 + \mathbf{L}^2 - 2\mathbf{J} \cdot \mathbf{L}$$
$$\mathbf{L} \cdot \mathbf{S} = \mathbf{L} \cdot (\mathbf{J} - \mathbf{L})$$
$$= \mathbf{J} \cdot \mathbf{L} - \mathbf{L}^2$$
$$= \frac{1}{2}(\mathbf{J}^2 - \mathbf{L}^2 - \mathbf{S}^2)$$

Because $\mathbf{J}$ commutes with $\mathbf{L}$ and $\mathbf{S}$, the matrix element is diagonal.

$$\langle |\mathbf{L} \cdot \mathbf{S}| \rangle = \frac{1}{2}[J(J+1) - L(L+1) - S(S+1)] \qquad (9.16)$$

Strictly speaking, L and S are never absolutely good quantum numbers because the spin-orbit coupling introduces off-diagonal elements in the Hamiltonian. As long as the spin-orbit coupling is small, there is only a small correction to the energy levels. In atoms of higher Z, however, first-order corrections are no longer adequate.

9.4.1 Making Estimates

As a first step in analyzing a physical system according to an assumed model, it is convenient to make a rough estimate of the magnitude with a "back of the envelope" calculation to see if the model could describe the situation or to see if an effect is large enough to include in a calculation or small enough to ignore. This section illustrates the approach by making an estimate of the spin-orbit coupling of the $n = 2$ first excited ^2P level of the hydrogen atom.

The starting point is a physical model, here the energy of an electron's magnetic moment interacting with the magnetic field produced by circulating charges. Use simple results from physics and known values of physical constants.

From the Biot–Savart law, the magnetic field produced at the center of a circular current loop of radius r due to a current I is

$$B = \frac{\mu_0 I}{2r}.$$

The current due to an electron traveling with speed v around the loop is

$$I = \frac{ev}{2\pi r}$$

$$B = \frac{\mu_0 ev}{4\pi r^2}.$$

To estimate v, use the Bohr quantization condition $m_e v r = n\hbar$.

$$v = \frac{n\hbar}{m_e r}$$

$$B = \frac{\mu_0 n e \hbar}{4\pi m_e r^3}$$

Let r be $n^2 a_0 = 4a_0$ where a_0 is the first Bohr radius, then multiply B by the Bohr magneton times $g_S \approx 2$. Use SI units and divide by e to express the result in eV:

$$
\begin{aligned}
\frac{g_S \mu_B B}{e} &= \left(2\frac{e\hbar}{2m_e}\right)\left(\frac{\mu_0 n\hbar}{4\pi m_e (4a_0)^3}\right) \\
&\approx \left(\frac{(1.6\times 10^{-19})(1.05\times 10^{-34})}{(9.1\times 10^{-31})}\right)\left(\frac{(4\pi\times 10^{-7})(2)(1.05\times 10^{-34})}{4\pi(9.1\times 10^{-31})(4\times 5.29\times 10^{-11})^3}\right) \\
&\approx 4.5\times 10^{-5} \quad \text{eV},
\end{aligned}
$$

surprisingly close to the measured value 4.5×10^{-5} eV, pretty good for a 10-minute calculation. Such rough estimates are not expected to have better than order-of-magnitude accuracy.

9.5 Example: Intermediate Field Zeeman

Consider the first excited level of $_1\text{H}$, where LS coupling splits the ^2P state into $^2P_{3/2}$ and $^2P_{1/2}$.

In a weak magnetic field, **B**, the Zeeman effect

$$\Delta E_B = \mu_B g_J B M_J$$

splits the $^2P_{3/2}$ state into four levels and splits the $^2P_{1/2}$ into two levels, as illustrated schematically in the sketch. Because of the selection rule, $\Delta M_J = 0, \pm 1$, only the six transitions shown are allowed. When the spin-orbit energy and

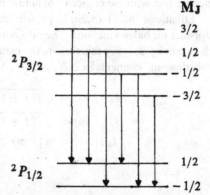

the magnetic energy are comparable, the off-diagonal terms in the Hamiltonian matrix $\mathcal{H}$ cannot be neglected.

LS coupling energy with strength factor a and magnetic energy interactions are added to the Hamiltonian.

$$\Delta\mathcal{H} = a\mathbf{L}\cdot\mathbf{S} + \boldsymbol{\mu}\cdot\mathbf{B} \tag{9.17}$$

This section diagonalizes $\Delta\mathcal{H}$ to express the energy levels as functions of B. The purpose is to exhibit the Zeeman effect over the full range from weak to strong magnetic fields with LS coupling included.

There are several ways to treat the diagonalization. Taking a simple approach, express the operators in $\Delta\mathcal{H}$ as matrices. Diagonalization gives the same results regardless of whether weak field or strong field quantum numbers are used as long as the matrix elements are calculated accordingly. Use the strong field quantum numbers M_L and M_S. In this example, $L = 1$ and $S = \frac{1}{2}$, so there are six basis functions corresponding to $M_L = +1, 0, -1$ and $M_S = +\frac{1}{2}, -\frac{1}{2}$. The matrices to be included in the Hamiltonian will therefore be 6×6.

Next express the operators $a\mathbf{L}\cdot\mathbf{S}$ and $\boldsymbol{\mu}\cdot\mathbf{B}$ as matrices. Consider first $a\mathbf{L}\cdot\mathbf{S}$. Using ladder operators $L_\pm, S_\pm$,

$$\begin{aligned}
\mathbf{L}\cdot\mathbf{S} &= L_x S_x + L_y S_y + L_z S_z \\
&= \frac{1}{4}(L_+ + L_-)(S_+ + S_-) - \frac{1}{4}(L_+ - L_-)(S_+ - S_-) + L_z S_z \\
&= \frac{1}{2}(L_+ S_- + L_- S_+) + L_z S_z.
\end{aligned}$$

The magnetic energy is

$$\begin{aligned}
\boldsymbol{\mu}\cdot\mathbf{B} &= (M_L + g_S M_S)\mu_B B \\
&= (M_L + 2M_S)\mu_B B,
\end{aligned}$$

with $g_S \approx 2$. The magnetic energy should not be written $g_J \mu_B B M_J$ because this is inconsistent with the choice of quantum numbers M_L, M_S.

Equations (8.27)–(8.29) (repeated here for convenience) show how ladder operators act on basis functions. They do not change j. I_z leaves m unchanged, and $I_\pm$ change m by ± 1. For this reason, the matrices for $L_z S_z$ and $(M_L + 2M_S)\mu_B B$ have only diagonal elements in $\Delta\mathcal{H}$, and $(L_+ S_- + L_- S_+)$ has only off-diagonal elements.

$$\begin{aligned}
I_+ |j, m\rangle &= \sqrt{j(j+1) - m(m+1)}\,|j, m+1\rangle \\
I_- |j, m\rangle &= \sqrt{j(j+1) - m(m-1)},|j, m-1\rangle \\
I_z |j, m\rangle &= m\,|j, m\rangle
\end{aligned} \tag{9.18}$$

The matrix elements for $L_\pm$ and $S_\pm$ follow from Eq. (9.18).

$$\langle j', m'|I_+|j, m\rangle = \sqrt{j(j+1) - m(m+1)}\, \delta_{j'j}\delta_{m'm+1}$$
$$\langle j', m'|I_-|j, m\rangle = \sqrt{j(j+1) - m(m-1)}\, \delta_{j'j}\delta_{m'm-1}$$
$$\langle j', m'|I_z|j, m\rangle = m\, \delta_{j'j}\delta_{m'm} \tag{9.19}$$

In this example, $j' = j = 1$ for L operators and $j' = j = \frac{1}{2}$ for S operators.

In the 3×3 matrix below left, let the elements of $L_\pm$ and L_z be labeled $\chi_{m',m}$ with $m', m = 1, 0, -1$. The 2×2 matrices below right for $S_\pm$ and S_z have elements labeled $\xi_{m',m}$ with $m', m = \frac{1}{2}, -\frac{1}{2}$.

$$\begin{pmatrix} \chi_{1,1} & \chi_{1,0} & \chi_{1,-1} \\ \chi_{0,1} & \chi_{0,0} & \chi_{0,-1} \\ \chi_{-1,1} & \chi_{-1,0} & \chi_{-1,-1} \end{pmatrix} \qquad \begin{pmatrix} \xi_{\frac{1}{2},\frac{1}{2}} & \xi_{\frac{1}{2},-\frac{1}{2}} \\ \xi_{-\frac{1}{2},\frac{1}{2}} & \xi_{-\frac{1}{2},-\frac{1}{2}} \end{pmatrix}$$

Use Eq. (9.19) to calculate the matrix elements of L_+ and S_-.

$$\langle j, m' \mid L_+ \mid j, m\rangle = \begin{pmatrix} 0 & \sqrt{2} & 0 \\ 0 & 0 & \sqrt{2} \\ 0 & 0 & 0 \end{pmatrix} \qquad \langle j, m' \mid S_- \mid j, m\rangle = \begin{pmatrix} 0 & 0 \\ 1 & 0 \end{pmatrix}$$

The matrix $\frac{a}{2}L_+S_-$ is the 6×6 direct (Kronecker) product matrix of the two matrices. This matrix is the contribution of $\frac{a}{2}L_+S_-$ to $\Delta\mathcal{H}$.

$$\begin{pmatrix} 0 & 0 & 0 & 0 & 0 & 0 \\ 0 & 0 & \frac{a}{\sqrt{2}} & 0 & 0 & 0 \\ 0 & 0 & 0 & 0 & 0 & 0 \\ 0 & 0 & 0 & 0 & \frac{a}{\sqrt{2}} & 0 \\ 0 & 0 & 0 & 0 & 0 & 0 \\ 0 & 0 & 0 & 0 & 0 & 0 \end{pmatrix}$$

Proceeding in the same way, calculate the matrix $\frac{a}{2}L_-S_+$ and the matrices aL_zS_z and $(M_L + 2M_S)\mu_B B$. The net matrix for $\Delta\mathcal{H}$ results from adding the matrices term by term.

In the result, each column is headed by M_L, M_S. The matrix elements are expressed in dimensionless form by extracting a factor of a: $\frac{\mu_B}{a}B \to CB$, where $C \equiv \frac{\mu_B}{a} \approx 1.93 \text{ T}^{-1}$ for $_1$H. Energy levels are expressed in dimensionless form as $\frac{E}{a}$.

$$\underline{\quad 1,\tfrac{1}{2} \quad 1,-\tfrac{1}{2} \quad 0,\tfrac{1}{2} \quad 0,-\tfrac{1}{2} \quad -1,\tfrac{1}{2} \quad -1,-\tfrac{1}{2} \quad}$$

$$
\begin{pmatrix}
\tfrac{1}{2}+2CB & 0 & 0 & 0 & 0 & 0 \\
0 & -\tfrac{1}{2}+0 & \tfrac{1}{\sqrt{2}} & 0 & 0 & 0 \\
0 & \tfrac{1}{\sqrt{2}} & CB & 0 & 0 & 0 \\
0 & 0 & 0 & -CB & \tfrac{1}{\sqrt{2}} & 0 \\
0 & 0 & 0 & \tfrac{1}{\sqrt{2}} & -\tfrac{1}{2}+0 & 0 \\
0 & 0 & 0 & 0 & 0 & \tfrac{1}{2}-2CB
\end{pmatrix}
$$

There are two different 1×1 diagonal terms and two different 2×2 blocks. The two diagonal terms give the energy states $M_L = 1$, $M_S = \tfrac{1}{2}$ and $M_L = -1$, $M_S = -\tfrac{1}{2}$ without further calculation.

$$\frac{E}{a} = \frac{1}{2} \pm 2CB$$

These states vary linearly with B for all B. Because of LS coupling, they now originate at $\frac{E}{a} = \frac{1}{2}$ for $B = 0$ instead of at 0.

The 2×2 submatrices have off-diagonal terms, so a quadratic secular equation must be solved to find the energies. To simplify the notation, let A_{ij} be the matrix elements of a submatrix as shown.

$$
\begin{pmatrix}
A_{11} & A_{12} \\
A_{21} & A_{22}
\end{pmatrix}
$$

The off-diagonal elements in each submatrix are equal: $A_{21} = A_{12}$. This is a general result of Theorem 10 (Section 6.3.2) for the case of real quantities. It is explicit in the $\Delta \mathcal{H}$ matrix, where $A_{21} = A_{12} = \frac{1}{\sqrt{2}}$ in both submatrices.

The secular equation in determinant form is

$$
\begin{vmatrix}
A_{11} - \frac{E}{a} & A_{12} \\
A_{12} & A_{22} - \frac{E}{a}
\end{vmatrix} = 0.
$$

For each 2×2 submatrix there are two real solutions corresponding to $\pm$.

$$\frac{E}{a} = \frac{(A_{11} + A_{22}) \pm \sqrt{(A_{11} - A_{22})^2 + 4A_{12}^2}}{2}$$

This form is convenient because it is only necessary to plug in the A_{ij}.

The diagram shows the calculated energies plotted versus B. The four states for $^2P_{3/2}$ begin at $\frac{E}{a} = \frac{1}{2}$, and the two states for $^2P_{1/2}$ begin at $\frac{E}{a} = -1$. For $_1$H, the ^2P levels in the first excited state are therefore separated by $\frac{3}{2}a$. The measured separation is 7.2×10^{-24} J $= 4.5 \times 10^{-5}$ eV so that $a \approx 3.0 \times 10^{-5}$ eV.

The short dash lines are the levels for $M_L = 1$, $M_S = \frac{1}{2}$ and for $M_L = -1$, $M_S = -\frac{1}{2}$. The long dash lines show the Zeeman states at small B becoming Paschen–Back states $M_L = 0$, $M_S = \pm\frac{1}{2}$.

The solid lines show the Zeeman states $M_L = -1$, $M_S = \frac{1}{2}$ and $M_L = 1$, $M_S = -\frac{1}{2}$ merging at large B to become a single Paschen–Back state.

The diagram shows the allowed transitions in ^{1}H from the ^2P excited level Zeeman states to the ^2S ground level Zeeman states in the strong field Paschen–Back regime $\mu_B B \gg a$. The transitions follow the selection rules $\Delta M_S = 0$ and $\Delta M_L = 0, \pm 1$. $\Delta E = 10.2$ eV is the energy difference between ^2P and ^2S when $B = 0$. The energy differences for the numbered transitions are therefore

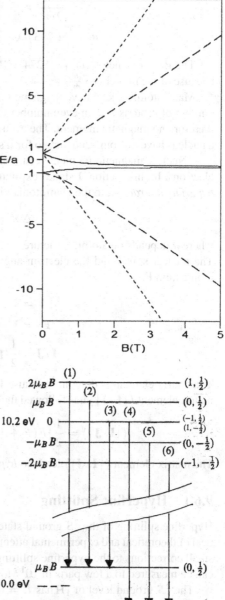

(1) & (4): $\Delta E + \mu_B B$

(2) & (5): ΔE

(3) & (6): $\Delta E - \mu_B B$

The spectral lines form a triplet, as is always the case for Paschen–Back. (2) and (5) are the central line; (1) and (4) are blue-shifted up by $\mu_B B$; (3) and (6) are red-shifted down.

9.6 Nuclear Spin and Hyperfine Structure

Both the proton p^+ and the neutron n^0 have intrinsic spin $\frac{1}{2}$ and decidedly nonclassical magnetic moments, especially for n^0 that has no measurable electric charge. The magnetic moments of nuclei are expressed as multiples of the *nuclear magneton* μ_N.

$$\mu_N = \frac{e\hbar}{2m_{p^+}} \approx 5.501 \times 10^{-27} \, \mathrm{J \cdot T^{-1}}$$

$$\mu_{p^+} \approx 2.793 \, \mu_N \qquad \mu_{n^0} \approx -1.913 \, \mu_N$$

The Bohr magneton $\mu_B \approx 9.274 \times 10^{-24} \, \mathrm{J \cdot T^{-1}}$ is considerably larger than μ_N because of the mass ratio $\frac{m_{p^+}}{m_{e^-}} \approx 1800$.

Many atomic nuclei have magnetic moments. Exceptions are nuclei with an even number of protons and an even number of neutrons; these nuclei all have spin 0 and therefore no magnetic moment. The even–even nuclei have spin 0 because protons in a nucleus have a strong tendency to form spin-cancelling pairs and so do the neutrons.

Nuclear magnetic moments can interact with the magnetic moments of atomic electrons. In this section, **I** stands explicitly for nuclear spin. The contribution of the *hyperfine interaction* to the Hamiltonian is

$$\Delta \mathcal{H} = \alpha \mathbf{I} \cdot \mathbf{J},$$

where α depends on atomic structure.

The nuclear spin **I** and the electron angular momentum **J** combine to give angular momentum **F**.

$$\mathbf{F} = \mathbf{I} + \mathbf{J}$$

$$\mathbf{F}^2 = \mathbf{I}^2 + \mathbf{J}^2 + 2\mathbf{I} \cdot \mathbf{J}$$

$$\mathbf{I} \cdot \mathbf{J} = \frac{1}{2}(\mathbf{F}^2 - \mathbf{I}^2 - \mathbf{J}^2)$$

The matrix element is diagonal because **F** commutes with $\mathcal{H}$. Our calculation of the matrix element $\langle \, | \mathbf{I} \cdot \mathbf{J} | \, \rangle$ has followed the same approach that led to **L·S** in Eq. (9.16).

$$\langle \, | \mathbf{I} \cdot \mathbf{J} | \, \rangle = \frac{1}{2}[F(F+1) - I(I+1) - J(J+1)] \qquad (9.20)$$

The matrix element $\langle \, | \mathbf{I} \cdot \mathbf{J} | \, \rangle$ leads to *hyperfine splitting* of levels with different F.

9.6.1 Hyperfine Splitting

Hyperfine splitting of the ^{2}S ground state of the hydrogen atom has received a great deal of theoretical and experimental attention. Because ^{1}H is the simplest atom, many small corrections to the hyperfine splitting can be calculated accurately. The splitting can be measured to a few parts in 10^{13} to provide a check on the calculations.

The ^2S ground level of $_1$H has $L = 0$ and $S = \frac{1}{2}$, so $J = \frac{1}{2}$. The ^{1}H nucleus is a proton with $I = \frac{1}{2}$, and the hyperfine splitting is between two levels: $F = 1$ (the higher) and $F = 0$. The energy difference is $\approx 5.9 \times 10^{-6}$ eV, usually expressed in terms of frequency $\nu \approx 1420$ MHz or wavelength $\lambda = 21$ cm.

The nucleus in deuterium $_1^2$H consists of a proton and neutron with parallel spins to give $I = 1$. For deuterium, the hyperfine levels of ^2S are $F = 1 \pm \frac{1}{2}$ so that $F = \frac{3}{2}$ and $\frac{1}{2}$, with $\nu \approx 327$ MHz.

^{1}H and Radio Astronomy

Hydrogen is the most abundant element in the universe, and in the 1940s physicists speculated that hydrogen in distant gas clouds might be emitting 1420 MHz radiation. In 1951 Edward Purcell and Harold Ewen at Harvard University successfully detected it, using an antenna in the form of a pyramid roughly 1 m on a side with the large open end pointed to the heavens. This discovery was the starting point for the ever-expanding field of radio astronomy. The appearance of the universe can be very different when viewed in different spectral regions, leading to increased understanding.

Collisions in space populate the upper $F = 1$ level of hydrogen atoms. The $F = 1$ (nuclear and electron spins parallel) and $F = 0$ (spins antiparallel) hyperfine levels in the ground state of $_1$H both have (+) parity. Electric dipole radiation has ($-$) parity and is therefore forbidden. The most probable allowed transition is then magnetic dipole radiation. The probability of spontaneous transition $F = 1$ to $F = 0$ is very small, partly because of the small energy difference. Consequently the lifetime of the $F = 1$ level is very long, $\approx 10^7$ years. 1420 MHz astronomy is nevertheless possible because there is a *lot* of hydrogen in the universe.

9.6.2 Hyperfine Structure Zeeman

Taking into account hyperfine structure, the Hamiltonian for an atom in a magnetic field B_z is

$$\mathcal{H} = \mathcal{H}_0 + \Delta\mathcal{H}$$
$$\Delta\mathcal{H} = \alpha\mathbf{I}\cdot\mathbf{J} - (g_J\mu_B J_z + g_I\mu_N I_z)B_z,$$

where $\mathcal{H}_0$ is the unperturbed Hamiltonian.

$\mathbf{F}$ does not commute with $\Delta\mathcal{H}$ for $B_z \neq 0$, but $J_z + I_z$ does commute – the axial rotation group again. The ground state of $_1$H has $J = \frac{1}{2}$, $I = \frac{1}{2}$, so possible values of $J_z + I_z$ are $+1, 0, -1$.

Weak Field

Consider the limiting case of a magnetic field so weak that the magnetic energies are much less than the hyperfine energy. For the ground state of ^{1}H, this would be a field < 100 G. The off-diagonal terms of the matrix are small compared to the diagonal elements. In this approximation, $\mathbf{J}$ and $\mathbf{I}$ precess around $\mathbf{F}$ so that I, J, F, M_F are all good quantum numbers. For the $F = 1$ level in ^{1}H, the quantum number M can be $0, \pm 1$, and for the $F = 0$ level, $M = 0$.

A rigorous calculation of the weak field Zeeman effect uses the Wigner–Eckart theorem to calculate the matrix elements, but a plausible physical picture is simpler and achieves the same result. As the sketch suggests, the precessing $\mathbf{J}$ and $\mathbf{I}$ average onto $\mathbf{F}$ according to

$$\frac{\langle\,|\,\mathbf{J}\cdot\mathbf{F}\,|\,\rangle}{\sqrt{F(F+1)}} \quad\text{and}\quad \frac{\langle\,|\,\mathbf{I}\cdot\mathbf{F}\,|\,\rangle}{\sqrt{F(F+1)}}.$$

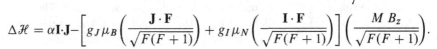

The nonrigorous magnitude $\sqrt{F(F+1)}$ for $\mathbf{F}$ is taken from the quantum mechanics relation $\mathbf{F}^2\psi = F(F+1)\psi$.

The precessing $\mathbf{F}$ averages onto $\mathbf{B}$ according to

$$\frac{\langle\,|\,\mathbf{F}\cdot\mathbf{B}\,|\,\rangle}{\sqrt{F(F+1)}} = \frac{MB_z}{\sqrt{F(F+1)}}$$

$$\Delta\mathcal{H} = \alpha\mathbf{I}\cdot\mathbf{J} - \left[g_J\mu_B\left(\frac{\mathbf{J}\cdot\mathbf{F}}{\sqrt{F(F+1)}}\right) + g_I\mu_N\left(\frac{\mathbf{I}\cdot\mathbf{F}}{\sqrt{F(F+1)}}\right) \right]\left(\frac{MB_z}{\sqrt{F(F+1)}}\right).$$

The remaining task is to evaluate $\mathbf{J}\cdot\mathbf{F}$ and $\mathbf{I}\cdot\mathbf{F}$.

$$\mathbf{F} = \mathbf{J} + \mathbf{I}$$

$$\mathbf{J}\cdot\mathbf{F} = \mathbf{J}^2 + \mathbf{I}\cdot\mathbf{J}$$

Use Eq. (9.20) for $\mathbf{I}\cdot\mathbf{J}$.

$$\langle\,|\,\mathbf{J}\cdot\mathbf{F}\,|\,\rangle = J(J+1) + \frac{1}{2}[F(F+1) - J(J+1) - I(I+1)]$$

$$= \frac{1}{2}[F(F+1) + J(J+1) - I(I+1)]$$

With a change of variables $I \leftrightarrow J$,

$$\langle\,|\,\mathbf{I}\cdot\mathbf{F}\,|\,\rangle = \frac{1}{2}[F(F+1) - J(J+1) + I(I+1)].$$

Assembling the terms,

$$\langle\,|\,\Delta\mathcal{H}\,|\,\rangle = \frac{\alpha}{2}[F(F+1) - I(I+1) - J(J+1)]$$

$$-\left(\frac{MB_z}{2F(F+1)}\right)\Big(g_J\mu_B[(F(F+1) + J(J+1) - I(I+1)]$$

$$+ g_I\mu_N[F(F+1) - J(J+1) + I(I+1)]\Big).$$

Because $\mu_N \ll \mu_B$, the term in μ_N is only a small correction. However, note that I appears in all the terms on an equal footing with F and J. The reason is that angular momentum is quantized so that $F\hbar$, $I\hbar$, and $J\hbar$ are comparable in magnitude. The magnetic moment of the nucleus is small compared to the electron's, but the spin angular momentum of the nucleus is substantial.

9.7 Multi-electron Atoms

If electrons in atoms obeyed Newtonian mechanics, all the electrons would occupy the lowest possible energy level – they would all be in the atom's ground state. The Stern–Gerlach experiment, the Zeeman effect, atomic spectra, and the Ritz combination principle all demonstrate that an atom's electrons are instead distributed among discrete energy levels. The object here is to extend the discussion to multi-electron atoms.

9.7.1 The Pauli Exclusion Principle

Soon after the development of Bohr's quantum theory, chemists and others wondered whether his ideas could help explain the chemical properties of multi-electron atoms. The nineteenth-century Russian chemist Mendeleev's periodic table had demonstrated that chemical elements tend to fall into groups with similar chemical properties. A prime example is the family of noble gases He, Ne, Ar, ..., all of which are the least chemically reactive elements. Mendeleev arranged the sequence of elements in terms of atomic mass to demonstrate recurring properties, but it was later shown that arrangement by atomic number Z is fundamental.

Regularities become apparent when elements are arranged by Z. Consider the atomic numbers of the first few noble gases: $_2$He, $_{10}$Ne, $_{18}$Ar. Z differs by 8 from one to the next. The same difference of 8 is seen for the chemically similar alkali metals: $_3$Li, $_{11}$Na, $_{19}$K. These observations led researchers to visualize electrons in atoms as occupying distinct layers called *shells*.

Pauli was struck by these empirical regularities. To explain them in a simple way, he postulated in 1925 that only one electron in an atom can have the same values of all four quantum numbers n, ℓ, m_ℓ, m_s. The exclusion principle *excludes* electrons that have identical sets. Because of this principle, atoms have finite dimensions, keeping matter from collapsing.

The four quantum numbers must satisfy certain relations: $\ell \leq n - 1$, $m_\ell = -\ell, -\ell + 1, \ldots, +\ell$, giving $2\ell + 1$ possibilities. m_s can have only two values: $\pm\frac{1}{2}$. Thus, for a given n and ℓ there are $2(2\ell + 1)$ available states that can be occupied. Electrons having the same n and ℓ are said to be *equivalent*.

In $_2$He, the two electrons in the ground state $n = 1$ necessarily have quantum numbers $\ell = 0$ and $m_s = \pm\frac{1}{2}$. No further electrons can be added to the $n = 1$ level. Fully occupied levels are said to form *closed shells*. The $n = 1$, $\ell = 0$ shell, called the K shell, becomes closed with two electrons. Their spins must be antiparallel $\boxed{\uparrow\downarrow}$ to satisfy Pauli exclusion.

The angular momentum quantum number ℓ of an electron is specified by the same code letters as before but written lower case: $s = 0$, $p = 1$, $d = 2$, $f = 3$, ... In standard notation, the *electron configuration* of a K shell is written $1s^2$, where $n = 1$ is the principal quantum number, s is angular momentum $\ell = 0$, and the superscript 2 at the upper right means that there are two electrons in this state.

Using relativistic quantum mechanics, Pauli proved the important *spin-statistics theorem*. It states that electrons and any other particles with half odd integer spin (called *fermions*) have only antisymmetric wave functions. Particles with integer spin are *bosons*. Bosons have only symmetric wave functions and are not subject to the exclusion principle. In 1995, experimenters produced an aggregate cloud of thousands of bosons in the same quantum state, having a single wave function like an individual atom. It is called the *Bose–Einstein condensate* because Einstein predicted this new state of matter in 1925, building on the ideas of Indian physicist Satyendra Bose (1894–1974).

9.8   The Helium Atom

Bohr's old quantum theory gave an excellent account of the main features of the hydrogen atom spectrum, but the helium atom with two electrons presented insuperable difficulties. Experimentally, helium has two independent sets of spectral lines; this was a puzzle in the old quantum mechanics but is explained using group theory, as we now show.

Taking into account only Coulomb forces, the Hamiltonian $\mathcal{H}$ for electrons 1 and 2 in a helium atom is

$$\mathcal{H}(r_1, r_2) = -\frac{\hbar^2}{2\mu}\nabla^2_{(1)} - \frac{\hbar^2}{2\mu}\nabla^2_{(2)} - \frac{2e^2}{4\pi\epsilon_0 r_1} - \frac{2e^2}{4\pi\epsilon_0 r_2} + \frac{e^2}{4\pi\epsilon_0 |\mathbf{r_1} - \mathbf{r_2}|},$$

where the reduced mass $\mu \approx m_e$. The third and fourth terms on the right are the electrostatic interactions of each electron with the doubly charged helium nucleus. The fifth term is the electrostatic interaction of the two electrons; the old quantum theory had no way to apply the quantization condition $mvr=n\hbar$ to two interacting electrons.

The Schrödinger wave equation for He is

$$\mathcal{H}(r_1, r_2)\Psi(r_1, r_2) = E_n\Psi(r_1, r_2).$$

Interchanging 1 and 2 leaves the Hamiltonian unchanged because electrons are indistinguishable.

$$\mathcal{H}(r_1, r_2) = \mathcal{H}(r_2, r_1)$$
$$\mathcal{H}(r_2, r_1)\Psi(r_2, r_1) = E_n\Psi(r_2, r_1)$$
$$\mathcal{H}(r_1, r_2)\Psi(r_2, r_1) = E_n\Psi(r_2, r_1)$$

It follows that

$$\Psi(r_2, r_1) = \beta\Psi(r_1, r_2).$$

Because of normalization, $\beta^2 = 1$ so that $\beta = \pm 1$. The wave functions are either symmetric (s) or antisymmetric (a).

$$\text{symmetric:} \quad \Psi_s(r_2, r_1) = +\Psi_s(r_1, r_2)$$
$$\Psi_s(r_1, r_2) = \psi(r_1, r_2) + \psi(r_2, r_1)$$
$$\text{antisymmetric:} \quad \Psi_a(r_2, r_1) = -\Psi_a(r_1, r_2)$$
$$\Psi_a(r_1, r_2) = \psi(r_1, r_2) - \psi(r_2, r_1)$$

Let $f(r)$ be the wave function of each $1s$ electron in the ground state of He. The symmetric ground state wave function is $f(r_1)f(r_2) + f(r_2)f(r_1)$. The antisymmetric ground state wave function essentially vanishes, $f(r_1)f(r_2) - f(r_2)f(r_1) = 0$, so the ground state must be symmetric in the spatial variables.

The assumed wave function is not exact. An exact wave function results from diagonalization by a similarity transformation, $\mathbf{T}\psi(\mathbf{r}_i)u(\sigma_i)\mathbf{T}^{-1}$. However, character determines irreducible representations, and character is invariant under similarity transformations. The symmetry of the inexact product wave function therefore has the same symmetry as the exact wave function.

Let $u_{\frac{1}{2}}^{(\frac{1}{2})}(1)$ and $u_{\frac{1}{2}}^{(\frac{1}{2})}(2)$ be the spin functions of the two electrons, labeled (1) and (2). The spin functions $u(1)$ and $u(2)$ commute because they refer to different sets of spin variables.

Use the *3-j* coefficients for $D^{(\frac{1}{2})} \otimes D^{(\frac{1}{2})}$ from Appendix F, Table F.1 to express the overall spin functions $W^{(S)}$ in terms of $u(1)$ and $u(2)$. The overall spin wave function $W^{(S)}$ can have only $S = 0$ and $S = 1$.

$$W_1^{(1)} = \left(u_{\frac{1}{2}}^{(\frac{1}{2})}(1) \, u_{\frac{1}{2}}^{(\frac{1}{2})}(2) \right) \qquad W_0^{(1)} = \frac{1}{\sqrt{2}} \left(u_{\frac{1}{2}}^{(\frac{1}{2})}(1) \, u_{-\frac{1}{2}}^{(\frac{1}{2})}(2) + u_{-\frac{1}{2}}^{(\frac{1}{2})}(1) \, u_{\frac{1}{2}}^{(\frac{1}{2})}(2) \right)$$

$$W_{-1}^{(1)} = \left(u_{-\frac{1}{2}}^{(\frac{1}{2})}(1) \, u_{-\frac{1}{2}}^{(\frac{1}{2})}(2) \right) \qquad W_0^{(0)} = \frac{1}{\sqrt{2}} \left(u_{\frac{1}{2}}^{(\frac{1}{2})}(1) \, u_{-\frac{1}{2}}^{(\frac{1}{2})}(2) - u_{-\frac{1}{2}}^{(\frac{1}{2})}(1) \, u_{\frac{1}{2}}^{(\frac{1}{2})}(2) \right)$$

According to the Pauli principle, the wave functions of electrons must be antisymmetric, taking into account both spatial coordinates and spin variables. The ground state $1s^2$ is symmetric in spatial coordinates, so the spin part must be antisymmetric. Inspection of the overall spin functions W shows that only $W_0^{(0)}$ is antisymmetric. Hence the ground state of He has spin wave function $S = 0$ with the electron spins antiparallel. $\boxed{\uparrow\downarrow}$. It is called a *singlet state* because its multiplicity is $2S + 1 = 1$.

In the first excited state of He, one electron is in the $1s$ ground state and the other is in the first excited state $2s$. If their spins are antiparallel, this would be a singlet state symmetric in spatial coordinates and antisymmetric in spin coordinates. However, there is another excited state also, with one electron in $1s$ and the other in $2s$. The wave function of this state is antisymmetric in spatial coordinates, hence necessarily symmetric in spin variables. The three spin functions $W^{(1)}$ are all symmetric. This excited state therefore has $S = 1$ with the electron spins parallel $\boxed{\uparrow\uparrow}$. It is called the *triplet state* because $2S + 1 = 3$.

All the energy levels of He form two distinct sets, singlet and triplet. He in the singlet states is called *parahelium*, and He in the triplet states is called *orthohelium*, analogous to the discussion of positronium Section 8.10.1. Electric dipole transitions are forbidden between singlet and triplet states because electric fields cannot change spin states. Magnetic dipole radiation is allowed but has low probability. The measured lifetime for the transition from the first excited state in orthohelium to the parahelium ground state is ≈ 8000 s. The triplet excited state is said to be *metastable* because its lifetime is orders of magnitude longer than the lifetime of an excited state where electric dipole transitions are allowed.

9.8.1 Perturbation Theory

The Hamiltonian for He does not have a closed form solution in terms of known functions. Physicists and mathematicians have developed numerical methods for obtaining approximate solutions to such problems.

The simplest approximation method is called *perturbation theory*, where a term is treated as a correction to an exact Hamiltonian. As an illustration of perturbation theory, write the Hamiltonian for He as $\mathcal{H} = \mathcal{H}_0 + \mathcal{V}$, where the electron–electron interaction $\mathcal{V}$ is treated as a perturbation.

$$\mathcal{H} = \mathcal{H}_0 + \frac{e^2}{4\pi\epsilon_0|\mathbf{r}_1 - \mathbf{r}_2|} \equiv \mathcal{H}_0 + \mathcal{V}$$

$\mathcal{H}_0$ is the sum of two hydrogen atom Hamiltonians and has an exact solution as the product of hydrogen atom wave functions with exact energy E_0. For the first excited state $1s$, $2s$ of He, the wave functions for $\mathcal{H}_0$ can be written

$$u_{12} = f_{1s}(1)g_{2s}(2)$$
$$u_{21} = f_{1s}(2)g_{2s}(1),$$

where f, g are hydrogen atom wave functions. Let the wave function for the first excited state of He be ψ_0.

$$\psi_0 = c_1 u_{12} + c_2 u_{21}$$
$$\mathcal{H}_0\psi_0 = E_0\psi_0$$
$$(\mathcal{H}_0 + \mathcal{V})\psi_0 \approx (E_0 + E_1)\psi_0$$
$$\mathcal{V}\psi_0 = E_1\psi_0$$

$E_0 + E_1$ is the approximate energy of the $1s$, $2s$ excited state.

$$\mathcal{V}(c_1 u_{12} + c_2 u_{21}) = E_1(c_1 u_{12} + c_2 u_{21})$$

Combining with $\langle u_{12}|$ from the left, the matrix elements are

$$c_1 \langle u_{12} | \mathcal{V} | u_{12} \rangle + c_2 \langle u_{12} | \mathcal{V} | u_{21} \rangle = c_1 E_1$$

because $\langle u_{12}|u_{12}\rangle = 1$ and $\langle u_{12}|u_{21}\rangle = 0$. Similarly,

$$c_1 \langle u_{21} | \mathcal{V} | u_{12} \rangle + c_2 \langle u_{21} | \mathcal{V} | u_{21} \rangle = c_2 E_1.$$

Write the matrix elements in simplified notation as

$$c_1 \mathcal{V}_{11} + c_2 \mathcal{V}_{12} = c_1 E_1 \tag{9.21}$$
$$c_1 \mathcal{V}_{21} + c_2 \mathcal{V}_{22} = c_2 E_1.$$

Because electrons are indistinguishable,

$$\langle u_{12} \mid \mathcal{V} \mid u_{12} \rangle = \langle u_{21} \mid \mathcal{V} \mid u_{21} \rangle$$
$$\mathcal{V}_{11} = \mathcal{V}_{22}$$
$$\langle u_{12} \mid \mathcal{V} \mid u_{21} \rangle = \langle u_{21} \mid \mathcal{V} \mid u_{12} \rangle$$
$$\mathcal{V}_{12} = \mathcal{V}_{21}.$$

The secular equation for E_1 is

$$\begin{vmatrix} \mathcal{V}_{11} - E_1 & \mathcal{V}_{12} \\ \mathcal{V}_{12} & \mathcal{V}_{11} - E_1 \end{vmatrix} = 0$$

with solutions

$$E_1 = \mathcal{V}_{11} \pm \mathcal{V}_{12}.$$

Using $E_1 = \mathcal{V}_{11} + \mathcal{V}_{12}$ in Eq. (9.21) gives

$$c_1(-\mathcal{V}_{12}) + c_2(\mathcal{V}_{12}) = 0$$
$$c_1 = c_2.$$

The wave function $\psi_0 = c_1(u_{12} + u_{21})$ is spatially symmetric and therefore corresponds to a singlet state. For $E_1 = \mathcal{V}_{11} - \mathcal{V}_{12}$, the solution is $c_1 = -c_2$, and the wave function is spatially antisymmetric and corresponds to a triplet state.

Calculations using atom wave functions show that $\mathcal{V}_{12} > 0$, so the energy of the first excited singlet state is higher by $2\mathcal{V}_{12}$ than the energy of the first excited triplet state. The sketch (to scale, energies in eV) shows the measured energies.

20.6 ———   ——— 19.8

orthohelium
(triplet)

Perturbation calculations are most accurate if the perturbation is much smaller than the unperturbed $E_1 \ll E_0$. For higher accuracy, perturbation theory can be extended to more terms. The corrections are progressively smaller but more difficult to calculate.

0.0 ———

9.9 The Structure of Multi-electron Atoms

parahelium
(singlet)

Quantitative calculations of the detailed structure of atoms with many electrons is a job for supercomputers. This section is therefore largely qualitative.

9.9.1   Electron Configurations and the States of Atoms

Given the electron configuration of a level, the problem is to determine the atom's possible states. This section is an introduction to how group theory attacks this problem.

Take as an example the electron configuration $1s2p^2$. The s electron contributes zero to the spatial angular momentum and each of the p electrons has $\ell = 1$. According to the vector model, the contribution to orbital angular momentum from sp^2 can be $L=0,1,2$. Orbital angular momentum states are therefore restricted to S,P,D.

The vector model sketch for the two electrons (labeled 1 and 2) makes plausible that the $L = 0$ S states and $L = 2$ D states have symmetric orbital angular momentum states and hence require an antisymmetric spin function according to the Pauli principle. The $L = 1$ P state is antisymmetric and requires a symmetric spin function.

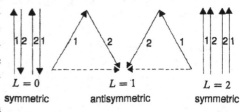

Consider the possible spin states. The three electrons in sp^2 can produce only two distinct spin values: $S = \frac{1}{2}$ with spins $\boxed{\uparrow\uparrow\downarrow}$ spin superscript $2S+1 = 2$, and $S = \frac{3}{2}$ with spins $\boxed{\uparrow\uparrow\uparrow}$ spin superscript $2S + 1 = 4$. The symmetric spin function required by the P state is fulfilled by the $S = \frac{3}{2}$ spin function $\boxed{\uparrow\uparrow\uparrow}$ with $2S + 1 = 4$. ^4P is therefore one of the allowed states of the sp^2 electron configuration. States ^4S and ^4D are fully symmetric and are not possible states because they do not obey Pauli exclusion.

Developing the spin functions for $S = \frac{1}{2}$ is more complicated, but group theory can help. Because electrons are indistinguishable, their permutations commute with the Hamiltonian, so permutation is a good quantum number.

$$\begin{pmatrix} 1 & 2 & 3 \\ 1 & 2 & 3 \end{pmatrix} \quad \begin{pmatrix} 1 & 2 & 3 \\ 2 & 1 & 3 \end{pmatrix} \quad \begin{pmatrix} 1 & 2 & 3 \\ 1 & 3 & 2 \end{pmatrix} \quad \begin{pmatrix} 1 & 2 & 3 \\ 3 & 2 & 1 \end{pmatrix} \quad \begin{pmatrix} 1 & 2 & 3 \\ 3 & 1 & 2 \end{pmatrix} \quad \begin{pmatrix} 1 & 2 & 3 \\ 2 & 3 & 1 \end{pmatrix}$$

$$\mathbf{P}_1 \qquad\quad \mathbf{P}_2 \qquad\quad \mathbf{P}_3 \qquad\quad \mathbf{P}_4 \qquad\quad \mathbf{P}_5 \qquad\quad \mathbf{P}_6$$

Conveniently, the permutation group $\mathbf{S}_3$ of three objects is isomorphic to the familiar **32** group. The symmetric group $\mathbf{S}_3$ therefore also has three classes, hence three irreducible representations. Because of the isomorphism, $\mathbf{S}_3$ and **32** are equivalent and have the same character table. The characters and matrices of irreducible representations 1 and A are all 1-dimensional.

Spin $\frac{1}{2}$ states of sp^2 are the three permutations $\boxed{\uparrow\uparrow\downarrow}\ \boxed{\uparrow\downarrow\uparrow}\ \boxed{\downarrow\uparrow\uparrow}$. This reducible representation is of dimension 3, so the character of $\mathbf{P}_1$ is 3. The permutations in the class $\{\mathbf{P}_2, \mathbf{P}_3, \mathbf{P}_4\}$ all leave one spin unchanged, so the character of this class is 1.

	$\{\mathbf{P}_1\}$	$\{\mathbf{P}_2, \mathbf{P}_3, \mathbf{P}_4\}$	$\{\mathbf{P}_5, \mathbf{P}_6\}$
1	1	1	1
A	1	-1	1
Γ	2	0	-1

The permutations in the class $\{\mathbf{P}_5, \mathbf{P}_6\}$ leave no spin unchanged, so the character is 0. Proofs are left to the problems.

The character set 3, 1, 0 equals the sum of characters $1 + \Gamma$. There is therefore a symmetric $S = \frac{1}{2}$ spin state, which matches with the antisymmetric spatial P state to give the allowed state ^2P. The 2-dimensional Γ representation provides two $S = \frac{1}{2}$ spin states, leading to the remaining allowed states ^2S and ^2D. All four of the allowed sp^2 states ^2S, ^2P, ^4P, ^2D have been observed in singly ionized $_4$Be. The sp^2 states that do not obey Pauli exclusion are not observed.

Slater Determinant Wave Functions

The He atom has only two electrons and requires only simple algebra to write symmetric and antisymmetric wave functions. How can we write an antisymmetric wave function for more than two electrons?

In 1929, John C. Slater (1900–76) realized that determinants are perfectly adapted to describe the antisymmetric wave functions of multi-electron systems. Consider, for example, a configuration with three electrons labeled 1, 2, 3 and having wave functions $\psi_\alpha, \psi_\beta, \psi_\gamma$. The Greek subscripts symbolize a wave function with a set of the four quantum numbers n, ℓ, m_ℓ, m_s.

The Slater determinant wave function for three electrons is

$$\Psi(1, 2, 3) = \frac{1}{\sqrt{6}} \begin{vmatrix} \psi_\alpha(1) & \psi_\alpha(2) & \psi_\alpha(3) \\ \psi_\beta(1) & \psi_\beta(2) & \psi_\beta(3) \\ \psi_\gamma(1) & \psi_\gamma(2) & \psi_\gamma(3) \end{vmatrix}.$$

For N electrons there are $N!$ permutations, so the normalization factor is $\frac{1}{\sqrt{N!}}$.

According to the properties of determinants, exchanging two rows or two columns changes the sign, showing that $\Psi(1, 2, 3)$ is antisymmetric. Further, if two rows are the same, $\Psi(1, 2, 3) = 0$, a statement of Pauli's exclusion principle.

Evaluate the determinant in cofactors of the first row so as to treat electrons equally.

$$\begin{aligned} \sqrt{6}\,\Psi(1, 2, 3) = {} & \psi_\alpha(1)\psi_\beta(2)\psi_\gamma(3) - \psi_\alpha(1)\psi_\beta(3)\psi_\gamma(2) \\ & + \psi_\alpha(2)\psi_\beta(3)\psi_\gamma(1) - \psi_\alpha(2)\psi_\beta(1)\psi_\gamma(3) \\ & + \psi_\alpha(3)\psi_\beta(1)\psi_\gamma(2) - \psi_\alpha(3)\psi_\beta(2)\psi_\gamma(1) \end{aligned}$$

Changing the $(-)$ signs to $(+)$ produces a symmetric wave function that can apply to bosons.

9.9.2   Electron Configuration and the Building-Up Principle

Applying the exclusion principle sequentially to atoms of the elements beginning with $_1$H develops the ideas of shells and their ability to accommodate electrons, an approach called the *building-up principle* (German *Aufbauprinzip*).

Chemical reactions involve the binding of electrons between atoms. An atom must be ionized to take part in an ionic bond, and one of its electrons must have an energy not far below the ionization potential to join in a covalent bond. The ionization potential of $_2$He is 24.6 eV above the K shell ground state, so it is no wonder that $_2$He is not known to join in a stable chemical bond. The ability to contribute an electron to a bond is not the only aspect of chemical reactivity. In an ionic bond, one species contributes an electron to the other, forming an electric bond $(+)(-)$. The ability to attract and accommodate electrons is another factor important in determining reactivity.

Continue with $_3$Li. According to the Pauli principle, two of the electrons in $_3$Li fill the K shell $\boxed{\uparrow\downarrow}$, so the third electron must go into the next higher n level. The electron configuration of the ground state is therefore $1s^22s$, and the spectroscopic label is ^2S. The figure shows measured energy levels in Li for $n = 1$ to 5. It is an example of a *Grotrian diagram* named after W. Grotrian who in 1928 published energy level diagrams in this style. The solid line at 5.39 eV is the energy required to remove the $2s$ electron, a moderate amount of energy that makes Li a reactive element.

All the excited states of Li are doublets ($2S + 1 = 2$) because they involve only one electron. As n increases, the spacing between levels becomes smaller, as also shown by the Balmer series of H. There is an unlimited number of levels approaching the ionization limit.

Going from $_3$Li to $_4$Be, the fourth electron completes the 2s shell. The ground state configuration is thus $1s^22s^2$ with label ^1S. Any atom with only closed shells has ground state ^1S.

All the electrons in a closed shell are paired. $\boxed{\uparrow\downarrow}$ A closed shell has $L = 0$ and $S = 0$, so a closed shell is spherically symmetric and need not be considered when developing an atom's term symbol.

Going to five electrons in $_5$B, the fifth electron is outside the $n = 1$ and $n = 2$ closed shells, so the configuration is $1s^22s^22p$. With $L = 1$ and $S = \frac{1}{2}$, the ground state is ^2P. Because of spin-orbit coupling, the ground state consists of two states, $J = \frac{1}{2}$, $J = \frac{3}{2}$, with nearly the same energy.

$_6$C has two equivalent p electrons, $1s^22s^22p^2$. By the vector model, the p electrons can have angular momentum $L = 0, 1, 2$ corresponding to states S, P, D. If both electrons have $m_\ell = 1$ the electrons must have antiparallel spins because of the

Pauli principle, corresponding to a ^1D state with $L = 2$, $S = 0$. Similarly, if both electrons have $m_\ell = 0$, the spins are again antiparallel and the state is ^1S.

To determine the multiplicity of the P state, suppose that one electron has $m_\ell = 1$ and the other $m_\ell = 0$ to give an antisymmetric orbital state. All spin states $\boxed{\uparrow}\,\boxed{\uparrow}$ $\boxed{\uparrow}\,\boxed{\downarrow}$ $\boxed{\downarrow}\,\boxed{\uparrow}$ $\boxed{\downarrow}\,\boxed{\downarrow}$ are therefore allowed by the Pauli principle. The spin states $\boxed{\uparrow}\,\boxed{\downarrow}$ $\boxed{\downarrow}\,\boxed{\uparrow}$ are identical because electrons are indistinguishable. The states belong to $M_S = +1$, 0, -1, so $S = 1$ and the state is ^3P, which is the observed ground state.

Hund's Rules

The building-up principle cannot by itself specify which state ^1S, ^3P, ^1D is the actual ground state of $_6$C. In 1927, German physicist Friedrich Hund (1896–1997) stated three rules to solve this problem. Starting with the most important, they are as follows:

(1) In a configuration, the state with greatest multiplicity lies lowest.

(2) If several states have the same multiplicity, the state with largest L lies lowest.

(3) If a shell is half-filled or less, the state with lowest J has the lowest energy. If the shell is more than half-filled, the state with largest J lies lowest.

From rule (1), the ground state of $_6$C is ^3P as verified by experiment. From rule (2), ^1D is lower than ^1S. The observed energies are 1.26 eV for ^1D compared to 2.68 eV for ^1S.

According to rule (1), electrons fill a shell in order with one electron in every m_ℓ level and begin to pair only when all the m_ℓ levels are occupied. The physical basis for rule (1) is that unpaired electrons experience less shielding from the nuclear charge by other electrons and hence lie lower because they experience a higher electrostatic attraction.

Table 9.2 lists measured properties of the first ten elements.
Data from NIST Atomic Spectra Database.

The (+) summation signs in electron configurations separate closed shells from the shell being filled. Because closed shells do not affect the filling, only the newly added electrons need to be shown in a configuration, along with the symbol of the previous element that has closed shells. As an example, the configuration of C can be written [He] $2s^2 2p^2$.

As Table 9.2 demonstrates, ionization limits reach a maximum when a shell becomes filled, as shown by He, Be, Ne.

Note from Table 9.2 that the term symbols of B and F are the same. This is a general result; if a shell can accommodate N electrons, the N' electron in the shell has the same ground state as the N − N' electron. The reason is that N − N' + N' = N so that the electron configurations of these two electrons taken together are the same as for a closed shell with $S = 0$ and $L = 0$. These electrons must therefore have the same S and L to compensate.

Table 9.2 $_1$H to $_{10}$Ne

element	ground state configuration	term symbol	first excited state (eV)	ionization limit (eV)
$_1$H	$1s$	^2S	10.2	13.6
$_2$He	$1s^2$	^1S	19.8	24.6
$_3$Li	$1s^2 + 2s$	^2S	1.85	5.39
$_4$Be	$1s^2 + 2s^2$	^1S	2.73	9.32
$_5$B	$1s^2 + 2s^2 + 2p$	^2P	3.55	8.30
$_6$C	$1s^2 + 2s^2 + 2p^2$	^3P	1.26	11.3
$_7$N	$1s^2 + 2s^2 + 2p^3$	^4S	2.38	14.5
$_8$O	$1s^2 + 2s^2 + 2p^4$	^3P	1.97	13.6
$_9$F	$1s^2 + 2s^2 + 2p^5$	^2P	12.7	17.4
$_{10}$Ne	$1s^2 + 2s^2 + 2p^6$	^1S	16.6	21.6

The K shell for $n = 1$ is closed at $_2$He and accommodates 2 electrons. The $n = 2$ L shell closes at $_{10}$Ne and accommodates an additional $2 + 6 = 8$ electrons. Its electron configuration is [Ne] $3s^2 3p^6$, with ground state ^1S as expected for an atom with fully closed shells.

The outer shell of F is $2p^5$, just one electron short of a closed shell. The small radius of the F atom has electrons relatively close to the nucleus, hence strongly attracted. This combination makes F the most reactive of all elements, even forming compounds with noble gas Xe.

The $n = 3$ M shell has a closed subshell at $_{18}$Ar that can accommodate an additional $2 + 6 = 8$ electrons; its electron configuration is [Ne] $3s^2 3p^6$. Something unexpected happens for $_{36}$Kr. Its configuration is [closed shells] $4s^2 4p^6$ as predicted, but Kr has 36 electrons instead of the $2 + 8 + 8 + 8 = 26$ expected. The reason is that at higher Z there are many closely spaced levels, and from Ar to Kr, electrons can populate d levels. The configuration of $_{29}$Cu, for example, is [Ar] $3d^{10} 4s$, with a complete closed M shell (10 electrons in d levels), accounting for the additional 10 electrons in $_{36}$Kr. The closed shell filling order from $_1$H to $_{36}$Kr is therefore $2 - 8 - 8 - 18$.

The building-up principle can be relied upon only for low-Z atoms, but nevertheless it gives insight as to why elements in the periodic table have regularly recurring properties.

Summary of Chapter 9

Chapter 9 discusses the Zeeman effect, its relation to quantum numbers, important interactions in atoms, and the electron structure of atoms. The placement of elements in the periodic table follows from the Pauli exclusion principle.

a) In standard notation for chemical elements, the charge number Z (the number of protons in the nucleus) is written at the lower left of the element's chemical symbol and the mass number A (total number of protons and neutrons) is at the upper left. An example is $^{40}_{18}$Ar with $Z = 18$, $A = 40$, the most abundant isotope of argon (99.6 percent abundance).

b) Well before the advent of Bohr's atomic model, physicists speculated that electrons in atoms were in motion and might therefore respond to an applied magnetic field. Classical equations of motion for a charged particle in a magnetic field predict three frequencies of rotation interpreted as the frequency of a central spectral line independent of magnetic field and a splitting into two spectral lines symmetrically on either side. The frequency of the shift from the central line is predicted to be proportional to magnetic field $\delta\omega = \left(\frac{q}{2m}\right)B$.

c) Pieter Zeeman's experiments verified the predictions and yielded a measurement of $\frac{q}{m}$ for the circulating particles (later named electrons).

d) According to quantum theory, a magnetic moment is $\mu = \left(\frac{e\hbar}{2m_e}\right)(g_L\mathbf{L} + g_S\mathbf{S})$ where $\mathbf{L}$ is orbital angular momentum ($g_L = 1$) and $\mathbf{S}$ is spin angular momentum ($g_S \approx 2$). They are quantized.

e) Magnetic fields from orbiting electrons can interact with an electron's magnetic moment to produce spin-orbit coupling, an energy shift proportional to $\mathbf{L} \cdot \mathbf{S}$. In a sufficiently weak external magnetic field, $\mathbf{L}$ and $\mathbf{S}$ remain coupled to give a net angular momentum $\mathbf{J} = \mathbf{L} + \mathbf{S}$, and the resulting magnetic moment is $\mu = \mu_B g_J \mathbf{J}$, where g_J is the Landé g factor.

f) Notation for expressing the quantum numbers of an atom's energy levels uses a capital letter for the orbital angular momentum L according to the scheme $S = 1$, $P = 2$, $D = 3$, $F = 4$, ... The spin quantum number S is written in the form $2S + 1$ at the upper left, and quantum number J is at the lower right if needed: $^{2S+1}L_J$. It is called the term symbol. Observed Zeeman lines result from a transition from an upper Zeeman split level to a lower Zeeman split level, according to selection rules.

g) In sufficiently high fields (Paschen–Back regime), $\mathbf{L}$ and $\mathbf{S}$ are decoupled and precess independently around $\mathbf{B}$ so that now M_L and M_S are good quantum numbers. The magnetic energy in this regime is therefore $\approx (M_L + 2M_S)\mu_B B$.

h) The spin-orbit coupling energy $a\mathbf{L} \cdot \mathbf{S}$ can shift and split spectral lines. A rough estimate of the magnitude of spin-orbit coupling energy can be made from the laws of classical physics together with the Bohr quantization condition.

i) For the Zeeman effect in intermediate fields, there are no good quantum numbers and it is necessary to diagonalize the Hamiltonian matrix to derive secular equations for the energies. To write the nondiagonal Hamiltonian, express all the operators in terms of a chosen set of quantum numbers; for example, M_L and M_S. The calculated energies at low B show weak-field Zeeman splitting with shifts due to spin-orbit energy. At high fields, some levels merge to develop the Paschen–Back triplet.

j) Hyperfine structure is due to interaction of the nuclear magnetic moment with atomic electron magnetic moments. Although nuclear magnetic moments are much smaller than electron moments, the nucleus has quantized angular momentum of the same order as electron angular momentum, so hyperfine splitting is readily observable. In the ground state of the hydrogen atom, there are two hyperfine levels, $F = 1$ and $F = 0$, separated by 1420 MHz, the basis for radio astronomy of hydrogen.

k) In a weak field, $\mathbf{I}$ and $\mathbf{J}$ are coupled: $\mathbf{F} = \mathbf{I} + \mathbf{J}$. $\mathbf{I}$ and $\mathbf{J}$ precess about $\mathbf{F}$, and $\mathbf{F}$ precesses about the magnetic field. I, J, F, M_F are good quantum numbers.

l) Because electrons are indistinguishable, the Hamiltonian for He remains the same if the two electrons 1 and 2 are interchanged. Helium's wave functions are basis functions for the permutation group of order 2 and are therefore either symmetric or antisymmetric in the spatial variables. The ground state $1s^2$ must be symmetric because the antisymmetric wave function essentially vanishes if both electrons have the same spatial wave functions. The spin wave functions are constructed from $D^{(\frac{1}{2})} \otimes D^{(\frac{1}{2})}$. According to the Pauli exclusion principle, the overall wave function must be antisymmetric, so in the ground state the two electrons must have opposite spins. In this way, all the states of helium are divided into singlet and triplet states. Electric dipole transitions between singlet and triplet states are forbidden.

m) Bohr's theory has no way of dealing with multi-electron atoms because of the electron–electron interaction term. Without this term, the Hamiltonian reduces to the sum of two hydrogenic Hamiltonians having known solutions. The energy difference between the singlet and triplet excited states $1s2s$ can be estimated by treating the electron–electron interaction as a perturbation.

n) To account for the observed ordering of atoms in the periodic table, Pauli's exclusion principle postulates that no two electrons in an atom can have the same set of quantum numbers n, ℓ, m_ℓ, m_s. A consequence is that electrons in atoms are arranged in shells having limited occupancy.

o) Species with half-integer spin are fermions and have antisymmetric wave functions that can be described by determinants. Species with integer spin are bosons and have symmetrical wave functions. Only a limited number of fermions can occupy the same shell, but the possible number of bosons in the same state is unlimited.

p) In an electron shell of given n and ℓ, m_ℓ has $2\ell + 1$ possible values and m_s has two possible values, so the maximum occupancy of the shell is $2(2\ell + 1)$. At maximum occupancy, the shell is closed and additional electrons must go into higher shells. This picture, called the building-up principle, accounts qualitatively for the shell structure of multi-electron atoms and can predict the spectroscopic label of a state.

q) At larger n, there is an increased number of closely spaced levels and the building-up principle is no longer a reliable guide. Hund's rules provide a series of criteria to determine which of several allowed states has the lowest energy.

Problems and Exercises

9.1 (a) Show that according to the classical theory of the Zeeman effect (Section 9.2.1) a possible motion is circular motion, either ccw or cw, in the $x - y$ plane perpendicular to the magnetic field direction.
(b) Of the two possible circular motion, ccw and cw, which one corresponds to the higher frequency ω_+?
(c) For the possible classical oscillation modes, what is the appearance of the electron motion as observed perpendicular to the magnetic field direction?

9.2 (a) From Eq. (9.15) the Landé g-factor is

$$g_J = 1 + \frac{J(J+1) - L(L+1) + S(S+1)}{2J(J+1)}. \tag{9.15}$$

Show that g_J can be written in the equivalent form

$$g_J = \frac{3}{2} + \frac{(S-L)(S+L+1)}{2J(J+1)}.$$

(b) Consider the case $L = S$. Evaluate the sum of the g_J over the allowed values of J.

(c) Consider the case $L = 3, S = 1$. Evaluate the sum of the g_J over the allowed values of J.

9.3 Show that the Landé g-factor = 1 for $S = 0$.

9.4 Here is a list of several atoms and their ground states. For each case, determine J, L, S and show that the vector model is satisfied.
(a) neon (Ne): $^1S_0$
(b) silver (Ag): $^2S_{1/2}$
(c) iron (Fe): $^5D_4$

9.5   Here is a list of several atoms and their ground states. For each case, determine J, L, S and show that the vector model is satisfied.
       (a) beryllium (Be): ${}^1S_0$
       (b) aluminum (Al): ${}^2P_{1/2}$
       (c) platinum (Pt): ${}^3D_3$

9.6   (a) The ground state of iron (Fe) is ${}^5D_4$. What is the Landé g-factor of this state?
       (b) One of the excited states of Fe is ${}^3F_2$. What is the Landé g-factor of this state?

9.7   An excited state of mercury (Hg) is ${}^3P_J$, where the possible values of J are $2, 1, 0$.
       (a) What is the Landé g-factor of the state ${}^3P_2$?
       (b) What is the Landé g-factor of the state ${}^3P_1$?
       (c) What can you say regarding the Landé g-factor of the state ${}^3P_0$?

9.8   Consider transitions from a ${}^3S_1$ state to a lower ${}^3P_1$ state. In a magnetic field of 3.2 T the transitions show weak-field Zeeman splitting. Calculate the span in wavelength of the Zeeman pattern from the longest wavelength allowed transition to the shortest wavelength transition.

9.9   The visible spectrum of sodium is dominated by two closely spaced bright yellow lines at 589.0 nm and 589.6 nm. The line at 589.0 nm is due to a transition from a ${}^2P_{3/2}$ level to a lower ${}^2S_{1/2}$ level. How many different Zeeman lines would it exhibit in a relatively weak magnetic field?

9.10   Consider transitions from a ${}^2D$ state to a $2P$ state in the strong field Paschen-Back regime. List all allowed transitions and show that there are only three different spectral lines.

9.11   Beginning in the 1920s, Russian physicist Pyotr Kapitza or Kapitsa (1894–1984, Nobel laureate in physics 1978) measured the Paschen–Back effect to an accuracy of 1 percent to 3 percent in various atoms. He generated magnetic fields higher than 10 T (100 kG) for a few milliseconds by short-circuiting a generator across a coil. A spark discharge from a bank of capacitors excited the sample for about a microsecond, short compared to the slow decay of the maximum field. A line separation of 0.42 nm was measured for the 404.6 nm line of Hg, a line due to the transition from a ${}^3S_1$ level at 7.7 eV to a ${}^3P_0$ level at 4.7 eV. (The term notations are for zero magnetic field.) Estimate the magnetic field in this experiment.

9.12   Taking into account only fine structure due to LS coupling, calculate the energy difference between a level characterized by J and the neighboring level $J - 1$. The result is known as the *Landé interval rule* for J.

9.13   Consider hyperfine splitting in the absence of a magnetic field. Calculate the energy difference between a level characterized by F and the neighboring level $F - 1$. The result is known as the *Landé interval rule* for F.

9.14 The ground state of sodium $^2S_{1/2}$ is subject to hyperfine splitting into two levels with a separation of 1.771×10^9 Hz.
(a) The atomic structure factor α introduced in Section 9.6 is 885.8×10^6 Hz for the ground state, as established both experimentally and theoretically. From the data, what is the nuclear spin of the sodium nucleus?
(b) From the result in part (a), what are the F values of the hyperfine levels in the ground state of sodium?

9.15 Hyperfine splitting in ^{1}H is 1420×10^6 Hz in the absence of a magnetic field where I, J, F, M_F are all good quantum numbers. In a moderately strong field **I** and **J** are decoupled so that M_I and M_J are good quantum numbers.
(a) Estimate the field required to decouple **I** and **J** and by calculating the field that marks the magnetic energy equal to the hyperfine energy.
(b) In a magnetic field required that decouple **I** and **J**, how many levels are there?
Which of the levels, labeled (M_I, M_J), very linearly with B?

9.16 Consider the spin states of three electrons $\boxed{\uparrow\uparrow\downarrow}\ \boxed{\uparrow\downarrow\uparrow}\ \boxed{\downarrow\uparrow\uparrow}$. Treating these as basis functions for S_3, develop the 3×3 matrix representation of P_2 and show that its character is 1.

9.17 Consider the spin states of three electrons $\boxed{\uparrow\uparrow\downarrow}\ \boxed{\uparrow\downarrow\uparrow}\ \boxed{\downarrow\uparrow\uparrow}$. Treating these as basis functions for S_3, develop the 3×3 matrix representation of P_5 and show that its character is 0.

9.18 Show that the Slater determinant wave function for two electrons is antisymmetric.

9.19 The ground state of $_8$O has two $1s$ electrons, two $2s$, and four $2p$.
(a) What is the complete electron configuration in standard notation?
(b) What are the possible values of J?
(c) What value of J do Hund's rules predict for the ground state?

9.20 The ground state of $_{13}$Al has two $3s$ and one $3p$ electron outside closed shells.
(a) What is the complete electron configuration in standard notation?
(b) What is a possible spectroscopic term symbol for the ground state?

10

PARTICLE PHYSICS

10.1 Introduction

The first nine chapters have laid out how group theory developed from a branch of pure mathematics in the nineteenth century to an essential tool for describing atomic phenomena. Group theory was then put on the shelf and taken down from time to time to help with calculations in atomic and molecular physics.

Group theory experienced a rebirth in the 1930s, when it was seen to be applicable to understanding the fundamental structure of matter. This chapter is an introduction to these ideas. Particle physics is a very extensive subject, so the first part of the chapter provides necessary background.

10.1.1 The Fundamental Forces

Only a few fundamental forces are currently known.

(1) **Strong Interaction** This is the strongest of the known forces and holds nuclei together against the electric repulsion of the protons. Its range of interaction is only a few fm (1 fm = 1 femtometer = 1×10^{-15} m). The radius of a proton is $\approx$ 0.8 fm.

(2) **Electromagnetic Force** Electric and magnetic forces were shown in the nineteenth century to be two aspects of the more general electromagnetic interaction. The electric repulsion of protons in a nucleus is only about 1 percent of the strong nuclear force. The range is infinite, but not thought to play a significant role in cosmology because the universe seems to be electrically neutral. However, electromagnetism plays a major role in our everyday world. A block on a table experiences an upward contact force due to atoms in the table exerting electric repulsion on neighboring atoms in the block.

(3) **Weak Interaction** This force is weaker than the strong force and electromagnetism. Its range of interaction is less than 10^{-3} fm. The interaction of an electron (a point particle) with a nucleus takes place by this force. One example is when a neutron emits an electron and becomes a proton, a process called β decay. If neutrinos (ν) or antineutrinos ($\bar{\nu}$) take part in a reaction, the weak nuclear force must be involved.

(4) **Gravity** Gravity is very weak compared to the other fundamental forces, but because of its infinite range it is important in cosmology perhaps along with other suspected but as yet undiscovered cosmological forces. Gravity is the main subject of general relativity, but despite nearly a century of effort it has not been possible to unite gravity and quantum theory.

In 1935, Japanese theorist Hideki Yukawa (1907–81) developed a theory of interactions. In his model, forces are due to transfer of bosons (integer spin) between particles. A few years later, pi mesons π^+, π^-, π^0 were observed experimentally and were found to have the properties Yukawa's theory predicted for bosons. Yukawa was the first Japanese Nobel laureate in physics (1949), cited "for his prediction of the existence of mesons on the basis of theoretical work on nuclear forces."

The uncertainty principle $\Delta E \Delta t \approx \hbar$ gives a qualitative estimate of the mass of a mediating boson. If the time is very short, energy conservation does not apply, and the energy change can be sufficient to account for the mass-energy $M_0 c^2$ of the boson. Moving at roughly the speed of light, the boson travels the range r of the force in a time $\frac{r}{c}$ so that

$$M_0 c^2 \approx \frac{\hbar c}{r} \tag{10.1}$$

If the range is short, the mass of the boson is correspondingly large.

In particle physics, energies are often expressed in GeV = 10^9 eV = 1.60×10^{-10} J. The rest mass of a particle is stated in units of energy/c^2 or often just as energy. The mass of a proton is, for example, 0.938 GeV and the mass of an electron is 0.51 MeV.

Electromagnetism has infinite range, so the mediating spin-1 photon must have zero mass, as observed. Taking the range of the strong nuclear force to be 2.0 fm, Eq. (10.1) estimates the pion mass to be 150 MeV, close to the experimental masses π^+, π^- (140 MeV) and π^0 (135 MeV). The range of the weak nuclear force is not known exactly, but it is much shorter than the range of the strong nuclear force, so its bosons are massive: W^+, W^- (80 GeV) and the neutral Z (91 GeV).

10.2 Natural Units

Ordinary units from macroscopic physics such as the kilogram–meter–second–ampere in the SI system do not fit the scale of atoms and particles, where masses and sizes are small and speeds and energies are high. It has therefore become conventional in atomic physics and particle physics to use a units system called *natural units* to fit the scale of quantum phenomena better.

By definition, the following quantities are assigned value 1 in natural units:

reduced Planck's constant $\hbar = 1$ speed of light $c = 1$

electron mass $m_e = 1$ vacuum permittivity $\epsilon_0 = 1$.

In SI units,

$$\hbar = 1.05 \times 10^{-34} \, \text{J} \cdot \text{s} \quad c = 3.00 \times 10^8 \, \text{m} \cdot \text{s}^{-1}$$

$$m_e = 9.11 \times 10^{-31} \, \text{kg} \quad \epsilon_0 = 8.85 \times 10^{-12} \, \text{SI units}.$$

10.2.1   Converting between Systems of Units

To convert between different systems of units, for example, between SI units and natural units, the starting point is to recognize the difference between *dimensions* and *units*. Dimensions express the fundamental attributes of a physical quantity in terms of mass (dimension M), length (dimension L), time (dimension T), and electric current (dimension I). As an example, velocity has the dimensions LT^{-1}. The corresponding *units* are derived by assigning values to the dimensions; for instance, meters/second for velocity.

The dimensions of any quantity can be expressed in terms of M, L, T, I by referring to a relation it satisfies. The dimensions of a quantity are always the same regardless of the physics; energy has the same dimensions in both nonrelativistic and relativistic mechanics. For convenience, here are the dimensions of the fundamental natural units, using Maxwell's notation of square brackets to express dimensions:

$$[\hbar] : L^2 M T^{-1} \quad [c] : LT^{-1}$$

$$[m_e] : M \quad [\epsilon_0] : L^{-3} M^{-1} T^4 I^2.$$

As a simple example, $[\hbar]$ is obtained from the relation $E = h\nu$ so that $[h]$ has the same dimensions as $\frac{[mv^2]}{[\nu]} \rightarrow L^2 M T^{-1}$.

To obtain numerical conversion factors, the trick is to write the dimensions of a quantity in terms of the dimensions of fundamental units. Here is the conversion to natural units of the SI unit second.

$$[s] = [\hbar]^\alpha [c]^\beta [m_e]^\gamma$$

$$T^1 = (L^{2\alpha} M^\alpha T^{-\alpha})(L^\beta T^{-\beta})(M^\gamma)$$

$$= L^{(2\alpha+\beta)} M^{(\alpha+\gamma)} T^{-(\alpha+\beta)}$$

$$2\alpha + \beta = 0$$

$$\alpha + \gamma = 0$$

$$-(\alpha + \beta) = 1$$

Solving,

$$\alpha = 1 \quad \beta = -2 \quad \gamma = -1.$$

The result for 1 second in natural units is therefore

$$\frac{\hbar}{m_e c^2} = \frac{1}{(1)(1)^2} = 1,$$

and in SI units it is

$$\frac{\hbar}{m_e c^2} = \frac{1.05 \times 10^{-34}}{(9.11 \times 10^{-31})(3.00 \times 10^8)^2}$$
$$= 1.28 \times 10^{-21}.$$

In other words, 1 second in natural units is 1.28×10^{-21} SI second. This is a convenient value for particle physics and atomic physics because a particle traveling near the speed of light travels 3.8×10^{-13} m in this time, about 100 times the diameter of a proton and about 1/100 the Bohr radius.

Further examples of unit conversions are left to the problems.

10.3 Isospin

In beginning courses, protons (p^+) and neutrons (n^0) are considered to be fundamental particles. We shall see later in this chapter that this picture is not correct, but we begin with how group theory was originally applied to p^+ and n^0.

10.3.1 The Nucleon Isospin Doublet

It is fair to say that the application of group theory to particle physics began in 1932 when Heisenberg considered the similarity of protons (p^+) and neutrons (n^0). They have nearly equal masses (0.938 GeV for p^+, 0.940 GeV for n^0), they both have spin 1/2, they play much the same role in nuclei, and they experience the strong nuclear force. In practice, p^+ and n^0 are easily distinguishable by the electric charge of the p^+, but in a nucleus the Coulomb force is only about 1 percent of the strong interaction.

Taking these similarities into account, researchers viewed p^+ and n^0 as two states of a *nucleon* and assigned a new quantum operator symbolized **T**. Here is where group theory enters particle physics. The essential insight was to view **T** as analogous to quantum angular momentum but describing "rotations" in abstract isospin space and not in physical space. Because of the experimental fact that the strong nuclear interaction is the same for any pair p^+p^+, p^+n^0, n^0n^0, the Hamiltonian must be independent of **T**. Hence **T** commutes with the Hamiltonian and is therefore a good quantum number. p^+ and n^0 can therefore be viewed as two states of **T** = 1/2 forming a doublet with isospin quantum number +1/2, and they can also be interpreted as basis functions for the 2-dimensional irreducible representation of SU(2) introduced in Section 7.5.

$$p^+ = \begin{pmatrix} 1 \\ 0 \end{pmatrix} \qquad n^0 = \begin{pmatrix} 0 \\ 1 \end{pmatrix}$$

Table 10.1 Some useful mass-energies

Species	Mass-energy (MeV)
Mass unit	931.494
m_{p^+}	938.272
m_{n^0}	939.565
m_{e^-}	0.511

Because of the difference in p^+ and n^0 masses, nucleon isospin is an approximate symmetry, but it holds to good accuracy for the strong interaction.

The isospin formalism follows in detail the results for quantum angular momentum operators developed in Chapter 8, and there is no need to repeat them here. The main differences are in notation and interpretation. The isospin operator T_3 is analogous to the spin operator σ_z. If w_+ is the eigenfunction for p^+ and w_- for n^0,

$$T_3 w_+ = \left(\tfrac{1}{2}\right) w_+ \qquad T_3 w_- = -\left(\tfrac{1}{2}\right) w_-.$$

By the analogy with quantum angular momentum, isospin also has ladder operators T_+ and T_-, where

$$T_- w_+ = w_- \qquad T_+ w_- = w_+.$$

The ladder operator T_- converts p^+ to n^0, and T_+ converts n^0 to p^+.

Binding Energy and Mirror Nuclei

One line of evidence for the charge independence of the strong nuclear force comes from a consideration of nuclear binding energy. The *binding energy (BE)* of nucleons in a nucleus is the difference between the mass of a nucleus and the mass of its separated constituent nucleons. Consider, for example, the binding energy of ${}^{15}_{7}\text{N}$, which has $7p^+$ and $8n^0$.

$$BE = 7m_{p^+} c^2 + 8m_{n^0} c^2 - M_N c^2$$
$$BE/c^2 = 7m_{p^+} + 8m_{n^0} - M_N,$$

where M_N is the mass of the nitrogen-15 nucleus. Binding energy is stated in terms of the mass-energy of the constituents, typically in units of MeV. If atomic masses are used instead of nuclear masses, the electrons should be included among the constituents. For low-Z atoms, omitting the binding energy of the electrons makes only a small correction to the binding energy.

Table 10.1 shows some mass-energy values useful for calculating nuclear binding energies.

The atomic mass unit in the table is for the modern unified atomic mass scale (u) based on the ${}^{12}\text{C}$ mass $\equiv 12.00000$ u.

Credits: Data in Tables 10.1, 10.2, 10.3, 10.4, 10.6, 10.7, 10.9 are taken from the 2020 Review of Particle Physics, cited as P. A. Zyla et al. (Particle Data Group), *Prog. Theor. Exp. Phys.* **2020**, 083C01 (2020).

The binding energies of *mirror nuclei* demonstrate the charge independence of the strong force. If a nucleus has X protons and X' neutrons, its mirror has X' protons and X neutrons so $^{15}_{8}$O is a mirror to $^{15}_{7}$N. The following table shows the calculated *BE* for each using the data in Table 10.1 and the experimental atomic masses.

$$^{15}_{7}\text{N}: \; 15.000109 \; u = 13,972.5 \; \text{MeV}$$
$$7\,p^{+} + 8\,n^{0} + 7\,e^{-} = 14,088.0 \; \text{MeV}$$
$$BE = 115.5 \; \text{MeV}$$

$$^{15}_{8}\text{O}: \; 15.003065 \; u = 13,975.3 \; \text{MeV}$$
$$8\,p^{+} + 7\,n^{0} + 8\,e^{-} = 14,087.2 \; \text{MeV}$$
$$BE = 111.9 \; \text{MeV}$$

The difference of 3.6 MeV between the binding energies is largely accounted for by the Coulomb potential energy due to the additional repulsive proton in $^{15}_{8}$O, which decreases the *BE*. The binding energies of the mirror nuclei are thus nearly equal, showing that p^{+} and n^{0} experience essentially the same strong force.

The mass of $^{15}_{8}$O is greater than the mass of $^{15}_{7}$N so there is mass-energy available to permit decay. A $^{15}_{8}$O nucleus can capture an electron to become a nucleus of $^{15}_{7}$N, a process with a lifetime of 122 s.

10.3.2 The Deuteron

A deuteron is the nucleus of "heavy hydrogen," ^{2}H, and consists of a proton and a neutron. In particle physics, it is symbolized d or D.

Assuming that the strong nuclear force is charge independent, each nucleon can be assigned isospin quantum number 1/2. *3-j* coefficients apply to isospin because isospin obeys the same relations as quantum angular momentum. The isospin of the deuteron is therefore $D^{(\frac{1}{2})} \otimes D^{(\frac{1}{2})}$, just as for positronium in Section 8.10.1.

The proton and neutron both have intrinsic spin 1/2, and Table F.1 in Appendix F shows that $D^{(\frac{1}{2})} \otimes D^{(\frac{1}{2})}$ can have two possible isospin values: T = 1 in a symmetric isospin triplet state $W^{(1)}$ or T = 0 in an antisymmetric singlet state $W^{(0)}$. The deuteron therefore can have nuclear spin I = 1 or I = 0. The total isospin and nuclear spin wave functions must be antisymmetric, giving only the two choices: I = 1, T = 0 or I = 0, T = 1. The former is the observed ground state, as further supported by the deuteron's electric quadrupole moment. According to the Wigner–Eckart theorem, there can be an electric quadrupole moment only if intrinsic spin is ≥ 1.

Table 10.2 Pion isospin triplet

Pion	Isospin	Mass (MeV)
π^+	+1	139.6
π^0	0	134.8
π^-	−1	139.6

Table 10.3 Sigma isospin triplet

Sigma	Isospin T_3	Mass (MeV)
Σ^+	+1	1,189
Σ^0	0	1,193
Σ^-	−1	1,197

10.3.3 The Pion Isospin Triplet

The three pions π^+, π^-, and π^0 take part in the strong interaction, so **T** commutes with the Hamiltonian. They all have intrinsic spin 0, and they have nearly the same mass, 135–140 MeV, all of which suggests that they are an isospin triplet analogous to the p^+ - n^0 isospin doublet. The pions can be identified as basis functions for $T = 1$ so that $T_3 = 1, 0, -1$, with the assignments shown in Table 10.2.

10.3.4 The Sigma Isospin Triplet

The three Sigma particles Σ^+, Σ^0, Σ^- all have intrinsic spin $\frac{1}{2}$, and all have masses within a few MeV of 1,190 MeV (see Table 10.3). They are identified as an isospin triplet $T = 1$.

10.3.5 The Delta Isospin Quadruplet

The nucleon isospin doublet wave functions transform like the $D^{(\frac{1}{2})}$ representation of SU(2), and the pion isospin triplet wave functions transform like its $D^{(1)}$ representation. Nature has, in addition, provided the Delta quadruplet that transforms like $D^{(\frac{3}{2})}$ (see Table 10.4).

The Δ quadruplet particles are very short-lived, $\approx 5 \times 10^{-24}$ s, and their masses are not known with high accuracy. Their decay modes are to nucleons and pions, strongly suggesting they have a rightful place among the other particles that experience the strong nuclear interaction.

Table 10.4 Delta isospin quadruplet

Delta	Isospin T_3	Mass (MeV)
Δ^{++}	3/2	1,232
Δ^{+}	1/2	1,232
Δ^{0}	−1/2	1,232
Δ^{-}	−3/2	1,232

10.4 Cross Section

In 1911 Ernest Rutherford and his students bombarded a thin gold foil with 4.87 MeV α particles ($^4_2$He^{++}) emitted by the radioactive decay of ^{226}Ra. They were amazed that some of the α particles underwent a nearly 180° backscatter, as amazing as if a hurtling cannon ball bounced back from a sheet of paper. Their *scattering* experiments were the first to demonstrate that atoms have a massive nucleus. Ever since, scattering has been an essential tool for studying interactions and the structure of matter.

Suppose that a foil of thickness Δx has $N\Delta x$ scattering targets per unit area, where N is the number of targets per unit volume. Let I_0 be the *flux* of bombarding particles, measured as the number incident on the target per unit area per unit time. A bombarding particle is scattered if it is deviated from its path by interactions with the target. The flux of scattered particles Δn is

$$\Delta n = \sigma_{tot} I_0 N\Delta x,$$

where σ_{tot} is the *total cross section*. σ_{tot} is a function of bombarding energy and has the dimensions of area. A common (nonstandard) unit for expressing nuclear cross sections is the *barn* (b)=10^{-24} cm^2, named because nuclear cross sections of this size are "big as a barn." Many cross sections are much smaller, so the units mb = 10^{-27} cm^2 and μb = 10^{-30} cm^2 are often used. The area associated with σ_{tot} is not necessarily a geometric area but must be interpreted as a measure of the range of the interaction forces. Scattering experiments tell a lot about interactions.

The total cross section is proportional to the square of the reaction matrix element. The reaction matrix element is the matrix element of $\mathcal{H}$ between the wave function of the initial reactants and the wave function of the final products.

$$\sigma_{tot} \propto \left| \langle \psi_{reactants} \mid \mathcal{H} \mid \psi_{products} \rangle \right|^2 \tag{10.2}$$

10.4.1 Pion–Nucleon Scattering

As an application of isospin, consider the elastic scattering of pions incident on nucleons according to the reactions

(a) $\quad \pi^+ + p^+ \to \pi^+ + p^+ \quad$ cross section σ^+

(a') $\quad \pi^- + p^+ \to \pi^- + p^+ \quad$ cross section σ^-.

For elastic scattering, the incident π energy should be less than a few hundred MeV, because at higher energies inelastic scattering can produce additional particles.

Isospin is assumed to be conserved for reactions (a) and (a'). The isospin portions of the wave functions are

$$u_1^{(1)} \;\rightarrow\; \pi^+ \qquad u_{-1}^{(1)} \;\rightarrow\; \pi^- \qquad v_{1/2}^{(\frac{1}{2})} \;\rightarrow\; p^+.$$

The isospin wave functions W for these reactions are $D^{(1)} \otimes D^{(\frac{1}{2})}$. The *3-j* coefficients are tabulated in F, Table F.2. For elastic scattering such as (a) and (a'), the reactants and products both have the same isospin wave functions:

(a) $\quad u_1^{(1)} v_{1/2}^{(\frac{1}{2})} = W_{3/2}^{(\frac{3}{2})}.$

(a') $\quad u_{-1}^{(1)} v_{1/2}^{(\frac{1}{2})} = \sqrt{\tfrac{1}{3}}\, W_{-1/2}^{(\frac{3}{2})} - \sqrt{\tfrac{2}{3}}\, W_{-1/2}^{(\frac{1}{2})}.$

For (a'), the term in $W^{(\frac{1}{2})}$ is small for an incident pion energy less than a few hundred MeV and can be neglected. $\mathscr{F}$ represents the matrix elements of $\mathscr{H}$ from terms not involving isospin.

$$\left\langle W_{3/2}^{(\frac{3}{2})} \,\middle|\, \mathscr{H} \,\middle|\, W_{3/2}^{(\frac{3}{2})} \right\rangle = (1) \times \mathscr{F}$$

$$\left\langle \sqrt{\tfrac{1}{3}}\, W_{-1/2}^{(\frac{3}{2})} \,\middle|\, \mathscr{H} \,\middle|\, W_{-1/2}^{(\frac{3}{2})} \right\rangle = \sqrt{\tfrac{1}{3}} \times \mathscr{F}$$

$$\sigma^+ = (1) \times \mathscr{F}^2$$

$$\sigma^- = \left(\frac{1}{3}\right) \times \mathscr{F}^2$$

The isospin prediction is thus $\sigma^+ \approx 3\sigma^-$. In the diagram (vertical scale in mb), the data points show measured cross sections for σ^+. The two dashed curves represent boundaries for $3 \times$ the measured σ^- cross sections. The data support the isospin prediction.

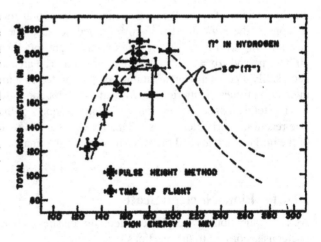

Reprinted figure from J. Ashkin *et al.* Phys. Rev. **96**, 1104 (1954). Copyright 1954 by the American Physical Society.

Table 10.5 Particle-antiparticle comparison

Property	Particle	Antiparticle
Mass m	m	m
Electric charge Q	Q	$-Q$
Magnetic moment μ	μ	$-\mu$
Isospin T_3	T_3	$-T_3$

10.5 Antiparticles

Every subatomic particle has a corresponding *antiparticle*. We saw an example in the discussion of positronium in Section 8.10.1 formed by the temporary union of an electron and its antiparticle, the positron. It is largely a matter of convention which is called the particle and which the antiparticle, but some names such as electron, proton, and neutron are historically established. Table 10.5 compares the properties of a particle and its antiparticle. Additional properties will be discussed later.

If a particle is electrically charged, the symbol for its antiparticle is the same but with reversed charge. For example, the symbol for an electron is e^-, and for its antiparticle it is e^+. If a particle does not carry electric charge, the symbol for its antiparticle is the particle's symbol with an overbar. For example, the antineutron is symbolized $\bar{n}$. Some particles such as the photon and the π^0 are their own antiparticle.

10.6 The Lagrangian

In Chapter 5 on matrix mechanics and in Chapter 6 on wave mechanics, calculations were based on the Hamiltonian $\mathcal{H}$. Particle physics theory is instead expressed in terms of the Lagrangian $\mathcal{L}$, the predecessor of the Hamiltonian. In classical mechanics, $\mathcal{H}$ and $\mathcal{L}$ and their equations of motion are

$$\text{Hamiltonian } \mathcal{H} = \text{kinetic energy} + \text{potential energy}$$

$$\dot{q}_i = \frac{\partial \mathcal{H}}{\partial p_i} \qquad \dot{p}_i = -\frac{\partial \mathcal{H}}{\partial q_i}$$

$$\text{Lagrangian } \mathcal{L} = \text{kinetic energy} - \text{potential energy}$$

$$0 = \frac{d}{dt}\left(\frac{\partial \mathcal{L}}{\partial \dot{q}_i}\right) - \frac{\partial \mathcal{L}}{\partial q_i}, \tag{10.3}$$

where q_i is a coordinate and p_i its conjugate momentum. Equation (10.3) for $\mathcal{L}$ is called the *Euler–Lagrange* equation. It can be written succinctly as $\dot{p}_i = \frac{\partial \mathcal{L}}{\partial q_i}$.

When Eq. (10.3) is applied to problems in classical mechanics (pendulum, mass on a spring, etc.), it is equivalent to using Newton's laws, but with the advantage that the geometric complications of evaluating force vectors is avoided because $\mathcal{L}$ involves only scalars.

10.6.1   Stationary Action

Unlike the rather ad hoc nature of Newton's laws, the Euler–Lagrange equation stems from a fundamental principle: $\mathcal{L}$ satisfies the principle of *stationary action*.

If a particle travels from point a to point b (perhaps acted upon by forces), the *action S* in two space variables is

$$S = \int_a^b \mathcal{L}(x, \dot{x}, y, \dot{y}, t)\, ds. \tag{10.4}$$

The integrand is formally the *Lagrangian density* but is often referred to as just the Lagrangian.

The action obeys an *extremal* principle for the path actually followed:

$$\delta \int_a^b \mathcal{L}\, ds \rightarrow \text{minimum or maximum or saddle point.} \tag{10.5}$$

Explanations are in order. Equation (10.4) is not to be interpreted as an ordinary definite integral of a given function. It is an example of a *functional* that has a value for all paths from a to b. Of all those values, Eq. (10.5) says that the actual path followed is the one that makes S unchanged to first order for any path that is close to the extremal path and also has endpoints a and b, as suggested by the sketch.

The situation is analogous to the problem in ordinary calculus of finding the maximum or minimum of a function $f(x)$. The slope of the function is 0 at the extremum $x = x_0$, so its Taylor's series expansion about x_0 is

$$f(x) = f(x_0) + 0 + \tfrac{1}{2}(x - x_0)^2 \left.\frac{d^2 f}{dx^2}\right|_{x_0} + \ldots,$$

with $f(x) \approx f(x_0)$ correct to first order.

10.6.2   The Lagrangian and Invariance (Noether's Theorem)

As just discussed, the Lagrangian obeys an extremal principle of stationary action. This section derives a plausible derivation of another important property, one that leads to a conservation law, as stated in *Noether's theorem*. The brilliant mathematician Emmy Noether derived a comprehensive version of this theorem in 1915 using Lie groups.

Consider a 1-dimensional Lagrangian $\mathcal{L}(x, \dot{x})$ that is independent of time. Change x by a small amount $\epsilon \mathcal{P}(x)$, where $\mathcal{P}$ will be chosen later to represent the particular transformation of interest.

$$\delta x \approx \epsilon \mathcal{P}(x)$$

$$\mathcal{L}(x', \dot{x}') \approx \mathcal{L}(x, \dot{x}) + \epsilon \frac{d\mathcal{L}}{dx}$$

$$\frac{d\mathcal{L}}{dt} = \frac{\partial \mathcal{L}}{\partial x} \mathcal{P} + \frac{d\mathcal{L}}{\partial \dot{x}} \frac{d\mathcal{P}}{dt}$$

Assume that $x(t)$ is along a path that satisfies the Euler–Lagrange equation Eq. (10.3), a path that causes no change in $\mathcal{L}$. In this case the system is said to have a *continuous symmetry*.

$$\frac{d\mathcal{L}}{dt} = \frac{d}{dt}\left(\frac{\partial \mathcal{L}}{\partial \dot{x}} \mathcal{P} \right)$$

$$0 = \frac{d}{dt}\left(\frac{\partial \mathcal{L}}{\partial \dot{x}} \mathcal{P} \right) \qquad (10.7)$$

In Eq. (10.7) the quantity in parentheses is constant in time – it is a conserved quantity. It is called the *Noether charge*, where the term "charge" is interpreted broadly to mean any conserved quantity.

To illustrate the meaning of $\mathcal{P}$, consider a 1-dimensional motion with the coordinates translated in time. Determine the operator $\mathcal{P}$ for time translation

$$\delta x \approx \epsilon \mathcal{P}(x)$$

$$x(t') \approx x(t) + \epsilon \frac{dx}{dt}$$

$$\delta x = \epsilon \frac{dx}{dt}$$

$$= \epsilon \mathcal{P}$$

so that

$$\mathcal{P} = \frac{dx}{dt} = \dot{x}. \qquad (10.8)$$

Consider a 1-dimensional Lagrangian.

$$\mathcal{L} = \tfrac{1}{2} m \dot{x}^2 - V(x)$$

Using Eqs. (10.7) and (10.8),

$$0 = \frac{d}{dt}\left(m\dot{x}(\dot{x}) - \tfrac{1}{2} m\dot{x}^2 + V(x) \right)$$

$$= \frac{d}{dt}\left(\tfrac{1}{2} m\dot{x}^2 + V(x) \right)$$

$$= \frac{d\mathcal{H}}{dt}.$$

The Hamiltonian $\mathcal{H}$ (the classical energy) is therefore a constant of the motion, if $\mathcal{H}$ and $\mathcal{L}$ have no explicit dependence on time. Energy is conserved because of symmetry under time translation.

Noether proved general theorems that if a certain symmetry makes $\mathcal{L}$ invariant, there is a corresponding conservation law, with one conserved quantity for each symmetry. Today, her theorems play a central role in theoretical physics, and especially in particle physics, where they help identify conserved quantities.

10.7   Gauge Theory

The object of this section is to introduce the essential ideas of *gauge theory*, using the example of electrodynamics, the simplest case. Gauge theory deals with quantum fields and their accompanying particles and is one of the foundations of the Standard Model.

In preparation for the example, here is a summary of notation for the vector calculus operator ∇:

$$\nabla \times \mathbf{F} = \text{ curl of vector } \mathbf{F}$$
$$\nabla \cdot \mathbf{F} = \text{ divergence of vector } \mathbf{F}$$
$$\nabla \xi = \text{ gradient of scalar } \xi$$
$$\nabla(fg) = (\nabla f)g + f\nabla g$$
$$\nabla fg = (\nabla f)g.$$

Consider two of Maxwell's equations in the absence of electric currents.

$$\nabla \times \mathbf{E} = -\frac{\partial \mathbf{B}}{\partial t} \qquad \nabla \times \mathbf{B} = \frac{\partial \mathbf{E}}{\partial t}$$

Expressing $\mathbf{E}(\mathbf{r}, t)$ and $\mathbf{B}(\mathbf{r}, t)$ in terms of the electric scalar potential $\phi(\mathbf{r}, t)$ and the magnetic vector potential $\mathbf{A}(\mathbf{r}, t)$ gives

$$\mathbf{E} = -\nabla\phi - \frac{\partial \mathbf{A}}{\partial t} \tag{10.9}$$

$$\mathbf{B} = \nabla \times \mathbf{A}. \tag{10.10}$$

Equations (10.9) and (10.10) are left unchanged by the simultaneous transformations

$$\mathbf{A}' = \mathbf{A} + \nabla \beta \tag{10.11}$$

$$\phi' = \phi - \frac{\partial \beta}{\partial t}, \tag{10.12}$$

where β is an arbitrary positive or negative scalar. The set of Eqs. (10.11) and (10.12) is an example of a *gauge transformation*.

A coordinate transformation, $\mathbf{r} \rightarrow \mathbf{r}'$, transforms the fields to their values at a new point $\mathbf{r}'$. In contrast, a gauge transformation brings us back to the electric and magnetic fields $\mathbf{E}(\mathbf{r}, t)$, $\mathbf{B}(\mathbf{r}, t)$ at the original location.

There are many different gauge transformations in electromagnetism, including the Lorenz gauge (not *Lorentz*) that puts the potentials in relativistic form. All gauge transformations lead to the same field equations. Choice of gauge can lead to simplifications in calculations, but there is much more to gauge theory. To set the stage, we go back in history.

Although gauge transformations in electromagnetism had been known for a long time, they were considered only to be mathematical properties having no particular physical importance. The situation began to change in 1918, when German mathematician Hermann Weyl (1885–1955) tried to unite general relativity and electromagnetism. His attempt was unsuccessful, but in the development of his theory he made use of *gauge invariance* of Lagrangians. About 10 years after his struggles with general relativity, Weyl and others realized that gauge invariance was instead eminently suited to combining electromagnetism and the newly developed quantum mechanics. As Weyl stated in a 1929 publication, "the electromagnetic field is a necessary accompaniment of the matter wave field and not of gravitation."

Over the next several decades, Weyl's seminal ideas were extended by others into a well-developed *gauge theory* that is now the foundation of the Standard Model of particle physics. The term "gauge" has no particular technical meaning. Weyl used this concept to mean a spatial coordinate scaling factor and referred to it by the German *Eichen*, meaning a measuring standard.

Weyl expressed his scaling factor gauge as a real exponential, but other workers realized that in quantum mechanics, gauge should be a complex exponential. When applied to wave functions in quantum mechanics, the gauge transformation, Eqs. (10.11) and (10.12), needs the additional equation,

$$\psi' = e^{i \frac{q}{\hbar} \beta} \psi. \tag{10.13}$$

Such complex exponentials are called *phase factors*, and it would have been more appropriate to use the term *phase transformation* instead of gauge transformation. The reason for choosing this particular phase factor will soon become apparent.

Because $\psi^* \psi$ is the observable probability density by Born's Rule (Section 6.6.1), it must remain the same under transformation by any phase factor $e^{i\gamma}$:
$\psi'^* \psi' = e^{-i\gamma} e^{i\gamma} \psi^* \psi = \psi^* \psi$.

Any two gauges commute. These properties remind us of the 1-dimensional unitary Abelian group U(1) introduced in Section 7.5.1. U(1) is the group associated with the electromagnetic field, just as SU(2) is associated with isospin.

If a gauge is constant, it has the same value everywhere and is said to be *global*. However, if a gauge depends on space-time coordinates $\gamma = \gamma(\mathbf{r}, t)$, it is a *local* gauge that varies with location. An arbitrary local gauge $e^{i\gamma}\psi$ is not a solution of Schrödinger's equation because the derivatives in Schrödinger's equation act on γ and change the equation by introducing additional terms.

10.7.1   Local Gauge Invariance of Schrödinger's Equation

This section will show that the choice of local gauge in Eq. (10.13) keeps Schrödinger's equation unchanged so that $e^{i\frac{q}{\hbar}\beta}\psi$ is a valid solution. In this case we say that Schrödinger's equation is *locally gauge invariant*, a property that leads to a remarkable result and reconnects us to group theory.

It is well known from classical physics that for a particle with charge q in a magnetic field, the particle's momentum must include the vector potential $\mathbf{A}(\mathbf{r}, t)$,

$$\mathbf{p}_{gen} = -i\hbar\nabla - q\mathbf{A},$$

so that Schrödinger's equation must be written

$$i\hbar\frac{\partial\psi}{\partial t} = \frac{\mathbf{p}_{gen}^2}{2m}\psi + q\phi\,\psi$$

$$i\hbar\frac{\partial\psi}{\partial t} = \tfrac{1}{2m}(-i\hbar\nabla - q\mathbf{A})^2\psi + q\phi\,\psi$$

$$= \tfrac{1}{2m}(i\hbar\nabla + q\mathbf{A})\cdot(i\hbar\nabla + q\mathbf{A})\psi + q\phi\,\psi$$

$$= \tfrac{1}{2m}[-\hbar^2\nabla^2\psi + i\hbar q\nabla\mathbf{A}\psi + 2i\hbar q\mathbf{A}\nabla\psi + q^2\mathbf{A}^2\psi] + q\phi\psi. \quad (10.14)$$

Now apply the gauge transformation equations, Eqs. (10.11), (10.12), (10.13), to Schrödinger's equation, Eq. (10.14).

$$i\hbar\frac{\partial\left(e^{i\frac{q}{\hbar}\beta}\psi\right)}{\partial t} = \tfrac{1}{2m}[-\hbar^2\nabla^2\left(e^{i\frac{q}{\hbar}\beta}\psi\right) + i\hbar q\nabla(\mathbf{A} + \nabla\beta)e^{i\frac{q}{\hbar}\beta}\psi$$

$$+ 2i\hbar q(\mathbf{A} + \nabla\beta)\nabla(e^{i\frac{q}{\hbar}\beta}\psi) + q^2(\mathbf{A} + \nabla\beta)^2 e^{i\frac{q}{\hbar}\beta}\psi] + q\left(\phi - \frac{\partial\beta}{\partial t}\right)e^{i\frac{q}{\hbar}\beta}\psi$$

Evaluate the left-hand side.

$$i\hbar\frac{\partial\left(e^{i\frac{q}{\hbar}\beta}\psi\right)}{\partial t} = -\hbar\frac{q}{\hbar}e^{i\frac{q}{\hbar}\beta}\frac{\partial\beta}{\partial t}\psi + i\hbar e^{i\frac{q}{\hbar}\beta}\frac{\partial\psi}{\partial t}$$

$$= -qe^{i\frac{q}{\hbar}\beta}\left(\frac{\partial\beta}{\partial t}\right)\psi + i\hbar e^{i\frac{q}{\hbar}\beta}\frac{\partial\psi}{\partial t}$$

Evaluate the right-hand side term by term.

$$-\hbar^2\nabla^2\left(e^{i\frac{q}{\hbar}\beta}\psi\right) = -\hbar^2\nabla\left[\nabla\left(e^{i\frac{q}{\hbar}\beta}\psi\right)\right]$$

$$= -i\hbar q\left[(\nabla^2\beta)\psi + 2\nabla\beta\nabla\psi\right]e^{i\frac{q}{\hbar}\beta}\psi$$

$$+ q^2(\nabla\beta)^2 e^{i\frac{q}{\hbar}\beta}\psi - \hbar^2(\nabla^2\psi)e^{i\frac{q}{\hbar}\beta}$$

$$i\hbar q\nabla(\mathbf{A} + \nabla\beta)e^{i\frac{q}{\hbar}\beta}\psi = i\hbar q\left(\nabla\mathbf{A} + \nabla^2\beta\right)e^{i\frac{q}{\hbar}\beta}\psi$$

$$2i\hbar q(\mathbf{A} + \nabla\beta)\nabla\left(e^{i\frac{q}{\hbar}\beta}\psi\right) = -2q^2\left[\mathbf{A}\nabla\beta + (\nabla\beta)^2\right]e^{i\frac{q}{\hbar}\beta}\psi$$

$$+ 2i\hbar q\left[\mathbf{A}\nabla\psi + \nabla\beta\nabla\psi\right]e^{i\frac{q}{\hbar}\beta}$$

$$q^2\left[\mathbf{A} + \nabla\beta\right]^2 e^{i\frac{q}{\hbar}\beta}\psi = q^2\left[\mathbf{A}^2 + 2\mathbf{A}\nabla\beta + (\nabla\beta)^2\right]e^{i\frac{q}{\hbar}\beta}\psi$$

Combining, all terms involving β cancel, including the common factor $e^{i\frac{q}{\hbar}\beta}$. The end result is identical to the original Schrödinger's equation, Eq. (10.14); Schrödinger's equation is invariant under this local gauge transformation. Although the gauge factor does not appear in Schrödinger's equation, it nevertheless contributes additional degrees of freedom.

Gauge invariance imposes substantial constraints on the fields, and especially on the potentials. An important conclusion is that Schrödinger's equation cannot exhibit gauge invariance unless the potentials are the ones that satisfy Eqs. (10.11) and (10.12). Local gauge invariance of Schrödinger's equation *requires* an accompanying field, in this case the electromagnetic field. And with this field there is a "particle" – the massless photon, a *gauge boson* that mediates the electromagnetic force.

The theory combining relativistic quantum mechanics and electromagnetism is called Quantum ElectroDynamics (QED). QED was used by Schwinger and others to calculate the value of the electron's magnetic moment g-factor. The result is in agreement with the experimental value to ≈ 1 part in 10^{11}.

10.7.2 Lagrangian, Gauge Theory, and Particle Physics

Over the past 60 years, modern particle physics has generated many books and many reams of research papers. All we can do at this level is outline the major features of the theory.

Particle physics theory is based on Lagrangians written according to special relativity so that calculated results are independent of the reference frame. Such Lagrangians are said to be *Lorentz invariant*. Special relativity unites space and time and expresses physical quantities in a 4-dimensional form known as *four-vectors*, a mathematical structure introduced by Einstein. A four-vector has four components labeled $0, 1, 2, 3$. Component 0 is a scalar, and $1, 2, 3$ is a vector. Examples of four-vectors are (taking $c = 1$): $[t, x, y, z]$ and $[E, \mathbf{p}]$. The union of energy and momentum in one four-vector is convenient for problems in mechanics, particularly in particle collisions.

The scalar product of two four-vectors is invariant under a Lorentz transformation. The scalar product is defined as

$$[X_0', \mathbf{A}] \cdot [X_0, \mathbf{B}] = X_0' X_0 - \mathbf{A} \cdot \mathbf{B}.$$

Take as an example the scalar product of $[E, \mathbf{p}]$ with itself.

$$\begin{aligned} [E, \mathbf{p}] \cdot [E, \mathbf{p}] &= E^2 - \mathbf{p}^2 \\ &= m_0^2 \end{aligned}$$

a well-known result from special relativity, where m_0 is rest mass, a constant and hence Lorentz invariant.

Particle physics theory is a field theory, and its Lagrangians are phrased in terms of fields. The electron (spin $\frac{1}{2}$) is a fermion field, pions (spin 1) are boson fields.

Most important, particle physics theory is a gauge theory. The reason for this importance is that the symmetry principle of local gauge invariance requires the existence of certain fields and interactions. The simplified example in Section 10.7.1 showed that the gauge invariance of Schrödinger's equation predicts electromagnetic fields.

10.8   All Those Particles – the Particle Zoo

The decays of naturally radioactive substances such as Ra occur with a decrease in their mass-energy. However, if energy is added to a system by means of kinetic energy, many reactions become possible, which is the reason for constructing particle accelerators. As of 2020, the largest accelerator so far is the Large Hadron Collider (LHC) located below ground at CERN (European Center for Nuclear Research) near Geneva on the border between France and Switzerland. The LHC is a ring 8.6 km (5.3 miles) in diameter with thousands of superconducting magnets to keep charged particles in the circular path. The LHC is designed to allow head-on collisions between bunches of particles (usually protons) circulating in opposite directions in the ring to maximize collision energy in the center-of-mass system, up to 13 TeV (1 TeV = 10^{12} eV). Current (2020) proposals for new accelerators include a much larger circular machine at CERN, nearly 32 km (20 miles) in diameter, designed to reach collision energies of 100 TeV.

Decades of work by theorists and by experimenters using accelerators have led to the discovery of dozens of hitherto unknown particles and measurement of their properties. Some decay rapidly, with lifetimes as short as 10^{-23} s and others with lifetimes as long (comparatively) as 10^{-10} s. As Enrico Fermi (1901–54, Nobel laureate in physics 1938) remarked, "If I could remember the names of all these particles, I'd be a botanist."

Our familiar world is made up of only a small number of different particles – photons, nucleons, electrons, neutrinos, and a few others that might exist here briefly in collision with an energetic cosmic ray from nature's cosmic accelerator. The short-lived particles are worthy of study because they likely existed at the very formation of the universe and may have influenced its evolution. In addition, understanding nature is one of the highest goals of humankind.

10.9   The Quark Model

By the middle of the twentieth century it was accepted that atoms are made up of e^-, p^+, n^0, then considered to be *elementary or fundamental* matter particles, all fermions

Table 10.6 Quark flavors

Name	Symbol	Charge Q	Mass (MeV)
Up	u	+2/3	2.2
Down	d	−1/3	4.7
Charm	c	+2/3	1,270
Strange	s	−1/3	92
Top	t	+2/3	172,800
Bottom	b	−1/3	4,180

with spin $\frac{1}{2}$ and having no internal structure. Scattering experiments are consistent with e^- being a structureless *point* elementary particle.

Advances from the 1960s and after have greatly changed this understanding. In 1969, energetic electrons with de Broglie wavelength smaller than a proton were scattered on a hydrogen target at the Stanford Linear Accelerator. The result exhibited peaks due to deep inelastic scattering, substantially different from the elastic scattering expected if the proton were a structureless ball of charge. The experimenters described their results as showing that the electron was scattering "from pointlike constituents within the proton." These experiments showed that p^+ and n^o are not elementary particles – they have internal structure.

A model of the proton's internal structure had been proposed earlier (in 1964) by American theorist Murray Gell-Mann (1929–2019, Nobel laureate in physics 1969) and independently by American physicist George Zweig (born 1937 in Moscow). Gell-Mann gave the name *quark* to the pointlike constituents, and today the quark model is an essential component of the Standard Model. Their quark model was not widely accepted at first, partly because physicists are cautious about adopting experimentally unsupported theories that look like "playing with numbers."

The initial quark model due to Gell-Mann and Zweig had only three different types of quarks: "up" u, "down" d, and "strange" s, later extended to six. There are six corresponding antiquarks. The properties of the six quarks are called *flavors* (see Table 10.6). The quark model uses common words such as "up," "down," "top" to label different quarks. These are only labels defined without necessary reference to their everyday meaning.

As suggested by Table 10.6, quarks are divided into three pairs called first, second, and third generation. Quarks are believed to be pointlike, and they are all fermions with intrinsic spin $\frac{1}{2}$. They are considered to interact via all the fundamental forces, although there is no experimental evidence for gravitational interaction. Only the weak force can change the flavor of a quark.

Quarks have fractional electric charge, never before seen for any particle. According to the quark model, the internal structure of a proton p^+ consists of three quarks:

Table 10.7 Leptons

Name	Symbol	Charge Q	Mass (MeV)
Electron	e^-	-1	0.51
Electron neutrino	ν_e	0	$< 1.1 \times 10^{-6}$
Muon	μ^-	-1	105.7
Muon neutrino	ν_μ	0	< 0.19
Tau	τ^-	-1	1777
Tau neutrino	ν_τ	0	< 18.2

two up quarks and one down quark uud. According to Table 10.6, they add to $+\frac{2}{3} + \frac{2}{3} - \frac{1}{3} = +1$ as expected for p^+. The antiproton p^- consists of $\bar{u}\bar{u}\bar{d}$. Charged antiparticles have the opposite sign of electric charge from the corresponding particle, so the electric charge of p^- according to the quark model is $-\frac{2}{3} - \frac{2}{3} + \frac{1}{3} = -1$. n^0 has quark content udd, giving electric charge 0 as required.

Quarks are bound within baryons or mesons by massless bosons called *gluons*. Curiously, the force exerted by gluons is weak at short distances but *increases* as quarks move apart, so quarks cannot escape, preventing individual quarks from being observed.

The top quark t was the last quark to be discovered, in p^+-p^- collisions at Fermilab 1992–95. t is the most massive known elementary particle, so massive that it decays (via the weak force) almost immediately after it is created into a W boson and a quark (most probably a b quark). Its lifetime is so short ($\approx 5 \times 10^{-25}$ s) that it decays before it can combine with other quarks to form particles. For this reason it is the only quark whose existence has actually been observed, making it an object of continuing study.

10.9.1 Classes of Elementary Particles

Several terms commonly used in particle physics are defined in terms of the quark model.

Baryon is a particle consisting of three quarks. Baryons are fermions (matter particles) with intrinsic spin half odd integer. The only observed baryon spin values are $\frac{1}{2}$ (like the nucleon doublet) or $\frac{3}{2}$ (like the Delta quadruplet), the only possible values from three spin $\frac{1}{2}$ quarks. Baryons take part in the strong interaction, but their internal quarks are actually responsible for the interactions.

Meson is a particle having an internal structure of a quark–antiquark pair. For example, π^+ is an up quark and a down antiquark $u\bar{d}$. π^0 is a combination $\frac{1}{\sqrt{2}}(u\bar{u} - d\bar{d})$. Mesons are bosons (force carriers) with integer intrinsic spin. The only observed meson spin values are 0 and 1, as expected from two spin $\frac{1}{2}$ quarks.

Table 10.8 The elementary particles

Type	Name and symbol
Fermions (matter)	**quarks** (spin 1/2): up (u) charm (c) top (t) down (d) strange (s) bottom (b) **leptons** (spin 1/2): electron (e^-) muon (μ^-) tau (τ^-) **lepton neutrinos** (spin 1/2): electron (ν_e) muon (ν_μ) tau (ν_τ)
Bosons (forces)	**massless** (spin 1): photon, gluon (g) **massive** (spin 1): W boson (W), Z boson (Z) **massive** (spin 0): Higgs

Lepton The electron e^-, muon μ^-, and tau τ^- are leptons and have similar properties except for mass (see Table 10.7). They all have electric charge -1 and they are all fermions with spin $\frac{1}{2}$. They are elementary particles with no quark structure, so they do not interact via the strong force, only with electromagnetism and the weak force. Each is accompanied by its own particular *neutrino lepton*, nearly massless neutral fermions with spin $\frac{1}{2}$.

Hadron is a general term for particles with quark content. Hadrons are, therefore, particles that interact via the strong force. Hadrons include baryons and mesons but not leptons.

Table 10.8 lists the known elementary particles. The photon, the gluon, the Z boson, and the Higgs are their own antiparticles.

10.9.2 The Charm Quark

Theorists argued that quarks should come in pairs like u-d, but in the original model strange quark s had no partner. Here is how its partner, charm quark c, was discovered.

In the early 1970s, Samuel Ting (born 1936) and his group used a proton accelerator at Brookhaven to collide GeV protons with a beryllium target. The initial object of their experiment was to investigate QED over short ranges by producing neutral mesons that are "heavy photons" with the quantum properties of ordinary photons but that are massive rather than massless. Three heavy photons were already known, with masses 0.76 GeV, 0.78 GeV, and 1.02 GeV. Like energetic photons, heavy photons decay by photoproduction into a e^--e^+ pair or a μ^--μ^+ pair.

The detected e^--e^+ pairs exhibited an unusually sharp peak at 3.1 GeV, as shown in the figure. The hatched and open traces correspond to two different detector conditions. Ting named the new heavy photon (mass 3.1 GeV) the J particle.

At the same time, Burton Richter (1931–2018) and his group at the Stanford accelerator were colliding e^--e^+ head on. The collision produced the same new neutral meson in the inverse of photoproduction. The meson then decayed to a detectable e^--e^+ pair, exhibiting the sharp peak at 3.1 GeV. Richter named the new meson Ψ. Because of the simultaneous discovery at Brookhaven and at Stanford, the particle is now called the J/Ψ. Ting and Richter shared the Nobel Prize in Physics 1976, cited for "their pioneering work in the discovery of a heavy elementary particle of a new kind."

A puzzle remained – why is the peak so narrow? A narrow peak implies that the meson state is relatively long-lived because of the complementarity of time and energy. Many mesons are, in contrast, very short-lived with peaks hundreds of times wider than for the J/Ψ. Mesons made from u, d, s cannot account for the narrow J/Ψ width.

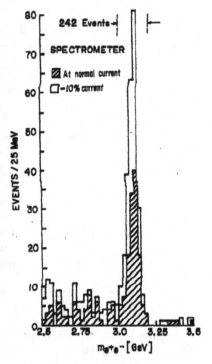

Reprinted figure with permission from S. Ting *et al.* Phys. Rev. Lett. **33**, 1404 (1974). Copyright 1974 by the American Physical Society.

Theorists proposed a new property they called *charm C*. The charm quark c has quantum number $C = +1$ and $\bar{c}$ has -1. J/Ψ is a $c\bar{c}$ bound state so that J/Ψ has $+1 - 1 = 0$ charm. A principal mechanism that hinders decay is the need to satisfy a conservation law. Assuming that charm is conserved, it follows that J/Ψ decay cannot occur by changing only one of its quarks because this would leave J/Ψ with charm +1 or −1 forbidden by conservation of charm. The decay of J/Ψ requires c and $\bar{c}$ to change flavor at the same time, a relatively slow process, hence a narrow peak.

The J/Ψ has 0 charm, so it was desirable to have particles with charm $\neq 0$ to study single charm. One answer is D mesons, where members of the D family all have a single c in their quark content (see Table 10.9).

The subscript s means that the quark content includes a strange quark. This family has become a workhorse for studying strong and weak interactions and searching for "new physics" not predicted by the Standard Model.

Table 10.9 D mesons

Symbol	Quarks	Mass (MeV)
$D^{\pm}$	$c\bar{d}$	1,869.5
D^0	$c\bar{u}$	1,864.8
$D_s^{\pm}$	$c\bar{s}$	1,969.0

10.9.3 Color Charge

The quark model for Δ^{++} is uuu. At least two of the quarks must have the same spin component, contrary to the Pauli exclusion principle. The solution to this problem is the same as for electron spin – introduce new quantum numbers.

The new property of quarks has the fanciful name *color*. Every quark has one of the quantum "numbers" red (r), green (g), blue (b), and every $\bar{q}$ has one of the quantum numbers antired $\bar{r}$, antigreen $\bar{g}$, antiblue $\bar{b}$. The colors are called "charge" in the broad sense of Noether that a charge is any intrinsic conserved quantity. In this case, the fanciful names do have a modest relation to everyday meanings because the quark content of baryons or mesons must add to colorless. Thus, a baryon (quark content qqq) must have $r + g + b$ = colorless, with each quark having a different color. A meson $q\bar{q}$ must have color-anticolor such as $r + \bar{r}$ = colorless.

The inclusion of color in the Standard Model generates a nonlinear gauge theory called Quantum ChromoDynamics (QCD), where color plays a role similar to electric charge in QED. The field Lagrangians for QED and QCD are identical except that the QCD Lagrangian has an additional factor related to color charge. QCD includes gluons, which are massless spin 1 bosons that mediate the force that keeps quarks confined inside hadrons. Gluons carry color and can interact via the strong interaction not only with quarks but also with one another, over a range roughly the size of a proton. Quarks and gluons are the only particles that carry net color charge.

The application of group theory to color charge is discussed in Section 10.11.3.

10.10 Conservation Laws and Quantum Numbers

There are fundamental conservation laws in classical physics such as conservation of energy E, linear momentum $\mathbf{p}$, angular momentum $\mathbf{L}$, electric charge Q. These laws also hold in particle physics when interpreted in the language of quantum mechanics.

They hold in both classical and quantum physics because they stem from symmetry principles that are universal.

E: symmetry under time translation – $\dot{\mathcal{H}}$ commutes with the Hamiltonian $\mathcal{H}$ (energy)
p: symmetry under space translation
L: symmetry under rotation
Q: symmetry under U(1) gauge invariance

Interactions in particle physics bring new quantum numbers and new conservation laws. Some of the conservation laws hold for all the principal interactions – strong, electromagnetism, weak – but others hold only for some interactions. Conservation laws are valuable for determining if reactions are possible or forbidden.

One clue is particle lifetime. Reactions that occur via the strong interaction have short lifetimes $\approx 10^{-23}$ s. If decay is by way of electromagnetism, the lifetime is $\approx 10^{-20}$ s, and if by the weak interaction, the lifetime is typically orders of magnitude longer, $\approx 10^{-10}$ s. If the strong interaction is forbidden by a conservation law, the reaction proceeds by electromagnetism, the next strongest. If conservation laws also forbid electromagnetism, the reaction can proceed only by the slow-acting weak interaction. If conservation laws also forbid the weak interaction, the reaction cannot occur.

Baryon Number B is conserved for all interactions. $B = +\frac{1}{3}$ for quarks and $-\frac{1}{3}$ for antiquarks. All baryons (three quarks) therefore have $B = \frac{1}{3} + \frac{1}{3} + \frac{1}{3} = +1$, while mesons (quark-antiquark) have $B = \frac{1}{3} - \frac{1}{3} = 0$.

One experimental support for conservation of baryon number is the failure to detect p^+ decay, such as

$$p^+ \rightarrow \pi^0 + e^+$$

that would be allowed by charge conservation and mass-energy but is forbidden by conservation of baryon number. p^+ decay has not been observed despite years of experimental search that have so far set a lower limit of 3.6×10^{29} years on the lifetime. For comparison, the age of the universe is estimated to be $\approx 15 \times 10^9$ years. Theories that attempt to unify all fundamental forces into one grand primordial interaction allow for proton decay, but the question remains open.

Lepton Number L was originally thought to be conserved separately for each of the three lepton flavors: e^-, μ^-, τ^-. Leptons e^-, μ^-, τ^- and their neutrinos all have lepton number +1, and all antileptons have -1. In the 1960s several experiments to detect ν_e from nuclear reactions in the Sun measured only about one-third the flux predicted by a reliable solar model. The solution to the problem was that ν_e was changing flavor on its journey, and the detectors were designed to detect only ν_e. The Standard Model predicts neutrinos to be massless, and they would therefore travel at the speed of light and have zero relativistic time interval to change flavor. When detectors that can detect all flavors were constructed, excellent agreement with the solar model was found. The conclusion is that lepton neutrinos must have some mass, however small. Therefore, only lepton number that includes all relevant flavors is conserved.

Strangeness Number S is -1 for each s and $+1$ for each $\bar{s}$. S is conserved by the strong and electromagnetic forces but not by the weak interaction, which can change the s flavor. Table 10.10 lists several hadrons with strangeness S and their quark content, electric charge Q, baryon number B, and isospin component I_3. These numbers obey the relation

$$Q = I_3 + \tfrac{1}{2}(B + S).\tag{10.17}$$

Equation (10.17) is known as the *Gell-Mann Nishijima relation*. It was derived independently by Gell-Mann and Japanese theorist Kazuhiko Nishijima (1926–2009). They and others were also responsible for introducing strangeness.

Table 10.10 Some hadrons with strangeness

Name	Symbol	Quarks	Q	B	I_3	S
Omega	Ω^-	sss	-1	$+1$	0	-3
Xi	Ξ^0	uss	0	$+1$	$1/2$	-2
	Ξ^-	dss	-1	$+1$	$-1/2$	-2
Lambda	Λ^0	uds	0	$+1$	0	-1
Sigma	Σ^+	uus	$+1$	$+1$	1	-1
	Σ^0	uds	0	$+1$	0	-1
	Σ^-	dds	-1	$+1$	-1	-1
Kaon	K^+	$u\bar{s}$	$+1$	0	$+1$	$+1$
	K^0	$d\bar{s}$	0	0	0	$+1$
	$\bar{K}^0$	$\bar{d}s$	0	0	0	-1
	K^-	$\bar{u}s$	-1	0	-1	-1

The lifetimes of Σ baryons illustrate the effect of interactions. The $\Sigma^\pm$ lifetime for decay into nucleons and pions is relatively long, $\approx 10^{-10}$ s, because the decay must be via the weak interaction to change quark flavor. The Σ^0 decays much faster, $\approx 10^{-20}$ s, because it can decay to Λ^0 requiring no change of quark flavor. The lifetime of Σ^0 is three or four orders of magnitude longer than the typical lifetime for strong interactions because the decay requires a change in I from 1 to 0, causing decay to proceed by the electromagnetic interaction, which is not as strong as the strong interaction.

Λ^0 has been called a "heavy neutron" because, like n^0, it has 0 charge, spin $\tfrac{1}{2}$, and one of its decay products is p^+. Its quark content is uds, where a d in n^0 udd has been replaced by s. The Λ^0 mass is 1,116 MeV, somewhat higher than the n^0 mass, 940 MeV. Short-lived atomic nuclei have been observed, with a Λ^0 taking the place of a n^0.

Charm Number C, discussed in Section 10.9.2, assigns $+1$ to charm quark c and -1 to $\bar{c}$. Charm is conserved in strong and electromagnetic interactions, but not in weak interactions that can change quark flavor. For each of the baryons in Table 10.9, there is a corresponding "charmed" baryon in which one s is replaced by its partner c.

This can change quantum numbers and always changes Q because s has charge $-\frac{1}{3}$ and c has charge $+\frac{2}{3}$. As an example, Λ^0 has $Q = 0$, and charmed lambda Λ_c^+ has $Q = +1$. Subscript c labels the charmed version.

There are additional quantum numbers such as parity, but we have enough material now to illustrate how conservation laws apply to particle reactions.

10.10.1   Examples: Particle Reactions

This section uses conservation laws to analyze three assumed reactions to determine if they are allowed or forbidden.

$$n^0 \rightarrow p^+ + e^- + \bar{\nu}_e$$

A good place to start analysis by conservation laws is to check whether electric charge Q and baryon number B are conserved, because these laws hold for all interactions. In this example of n^0 decay, $Q = 0$ on both sides of the reaction and is conserved. On the left $B = +1$ and on the right $B = +1 + 0 + 0 = +1$, so B is also conserved. On the right, lepton number $L_e = +1$ for the e^- and -1 for the $\bar{\nu}_e$ to give $+1 - 1 = 0$ in agreement with lepton number 0 for n^0 on the left.

Quark content is udd for n^0 and uud for p^+. Changing quark flavor occurs only by the weak interaction. The presence of $\bar{\nu}$ in the reaction equation also shows that the weak interaction is involved. n^0 decay according to the reaction equation is therefore allowed, but slow.

$$K^- + p^+ \rightarrow \Omega^- + K^+ + K^0$$

$Q = 0$ on both sides and is conserved. p^+ and Ω^- are both baryons so $B = +1$ on both sides and is conserved. Look now at strangeness S. Quark content on the left is $\bar{u}s + uud$ with $S = -1$. On the right quark content is $sss + u\bar{s} + d\bar{s}$ to give $S = -3 + 1 + 1 = -1$, so S is conserved. This reaction has been used to produce Ω^-.

$$\Lambda^0 \rightarrow e^- + \pi^+ + \bar{\nu}_e$$

$Q = 0$ on both sides and so is conserved. Electron lepton number is 0 on the left and $+1 - 1 = 0$ on the right so is conserved. Λ^0 on the left is a baryon $B = +1$, but there are no baryons on the right, $B = 0$. This reaction cannot proceed by any interaction.

10.11   Group Theory and Particle Physics

It is fitting to end this book by describing how group theory and representations bring order to the particle zoo and support the Standard Model.

The discussion of isospin in Section 10.3 demonstrated how p^+ and n^0 can be represented as basis functions of SU(2) to good approximation for the strong interaction.

$$p^+ = \begin{pmatrix} 1 \\ 0 \end{pmatrix} \qquad n^0 = \begin{pmatrix} 0 \\ 1 \end{pmatrix}$$

Isomorphism with SU(2) thus allows immediate application of angular momentum relations such as *3-j* coefficients to particle physics, as illustrated by the example of the deuteron (Section 10.3.2).

The quark model initially developed by Gell-Mann and by Zweig included only three quarks u,d,s. As basis functions of a possible symmetry group for the strong interaction, they are written

$$ u = \begin{pmatrix} 1 \\ 0 \\ 0 \end{pmatrix} \qquad d = \begin{pmatrix} 0 \\ 1 \\ 0 \end{pmatrix} \qquad s = \begin{pmatrix} 0 \\ 0 \\ 1 \end{pmatrix}. \tag{10.18} $$

SU(2) can accommodate only two basis functions. Gell-Mann and Zweig therefore turned to the group SU(3).

SU(2) and SU(3) are both Special Unitary Lie groups SU(N) and have much in common. To review, SU(2) is related to the traceless (Pauli) matrices σ_k.

$$ \sigma_1 = \begin{pmatrix} 0 & 1 \\ 1 & 0 \end{pmatrix} \qquad \sigma_2 = \begin{pmatrix} 0 & -i \\ i & 0 \end{pmatrix} \qquad \sigma_3 = \begin{pmatrix} 1 & 0 \\ 0 & -1 \end{pmatrix} $$

The commutators of the Pauli matrices define the Lie algebra su(2). Exponentiated Pauli matrices (with a factor of $\frac{i}{2} \to e^{\frac{i}{2}\sigma_k}$) generate the elements of SU(2) in the infinite number of its irreducible representations. SU(2) has one Casimir operator J^2 with eigenvalue $j(j+1)$ that commutes with all the matrices. The irreducible representations of SU(2) are therefore labeled with $j = \frac{n-1}{2}$, as we saw for the rotation group SO(3). The fundamental representation $n = 2$ of SU(2) is $D^{(j)} = D^{(\frac{1}{2})}$.

10.11.1 SU(3)

The fundamental irreducible representation of SU(3) requires eight 3×3 matrices plus the identity. The generators listed here (Cartan–Weyl) are linear combinations of matrices introduced by Gell-Mann.

$$ T_0 = \tfrac{1}{2}\begin{pmatrix} 1 & 0 & 0 \\ 0 & -1 & 0 \\ 0 & 0 & 0 \end{pmatrix} \quad Y = \tfrac{1}{3}\begin{pmatrix} 1 & 0 & 0 \\ 0 & 1 & 0 \\ 0 & 0 & -2 \end{pmatrix} \quad T_+ = \begin{pmatrix} 0 & 1 & 0 \\ 0 & 0 & 0 \\ 0 & 0 & 0 \end{pmatrix} \quad T_- = \begin{pmatrix} 0 & 0 & 0 \\ 1 & 0 & 0 \\ 0 & 0 & 0 \end{pmatrix} $$

$$ U_+ = \begin{pmatrix} 0 & 0 & 0 \\ 0 & 0 & 1 \\ 0 & 0 & 0 \end{pmatrix} \quad U_- = \begin{pmatrix} 0 & 0 & 0 \\ 0 & 0 & 0 \\ 0 & 1 & 0 \end{pmatrix} \quad V_+ = \begin{pmatrix} 0 & 0 & 1 \\ 0 & 0 & 0 \\ 0 & 0 & 0 \end{pmatrix} \quad V_- = \begin{pmatrix} 0 & 0 & 0 \\ 0 & 0 & 0 \\ 1 & 0 & 0 \end{pmatrix} \tag{10.19} $$

There are two diagonal matrices T_0, Y and three pairs of ladder operators $T_\pm, U_\pm, V_\pm$.

Commutators define the algebra and are easily calculated from the matrices. For example,

$$[T_+, T_-] = T_+T_- - T_-T_+$$

$$= \begin{pmatrix} 0 & 1 & 0 \\ 0 & 0 & 0 \\ 0 & 0 & 0 \end{pmatrix} \begin{pmatrix} 0 & 0 & 0 \\ 1 & 0 & 0 \\ 0 & 0 & 0 \end{pmatrix} - \begin{pmatrix} 0 & 0 & 0 \\ 1 & 0 & 0 \\ 0 & 0 & 0 \end{pmatrix} \begin{pmatrix} 0 & 1 & 0 \\ 0 & 0 & 0 \\ 0 & 0 & 0 \end{pmatrix}$$

$$= \begin{pmatrix} 1 & 0 & 0 \\ 0 & 0 & 0 \\ 0 & 0 & 0 \end{pmatrix} - \begin{pmatrix} 0 & 0 & 0 \\ 0 & 1 & 0 \\ 0 & 0 & 0 \end{pmatrix} = \begin{pmatrix} 1 & 0 & 0 \\ 0 & -1 & 0 \\ 0 & 0 & 0 \end{pmatrix}$$

$$= 2T_0.$$

$2T_0$ acts like the isospin operator I_3. Applied to the basis functions Eq. (10.18),

$$2T_0 \,|u\rangle = (1)\,|u\rangle \qquad 2T_0\,|d\rangle = (-1)\,|d\rangle \qquad 2T_0\,|s\rangle = (0)\,|s\rangle\,.$$

The ladder operator T_+ raises d to u, and T_- lowers u to d. From Eqs. (10.18) and (10.19),

$$T_+\,|d\rangle = \begin{pmatrix} 0 & 1 & 0 \\ 0 & 0 & 0 \\ 0 & 0 & 0 \end{pmatrix} \begin{pmatrix} 0 \\ 1 \\ 0 \end{pmatrix} = \begin{pmatrix} 1 \\ 0 \\ 0 \end{pmatrix} \qquad T_-\,|u\rangle = \begin{pmatrix} 0 & 0 & 0 \\ 1 & 0 & 0 \\ 0 & 0 & 0 \end{pmatrix} \begin{pmatrix} 1 \\ 0 \\ 0 \end{pmatrix} = \begin{pmatrix} 0 \\ 1 \\ 0 \end{pmatrix}.$$

As a general rule, SU(N) has N-1 Casimir operators. SU(3) has $3-1=2$ Casimir operators, so its irreducible representations $D^{(pq)}$ have two integer labels, p and q ($p, q \geq 0$). The dimension of $D^{(pq)}$ is

$$\dim D^{(pq)} = \tfrac{1}{2}(p+1)(q+1)(p+q+2). \tag{10.20}$$

The two Casimir operators of SU(3) are both multiples of the 3×3 identity matrix, as required by *Schur's lemma*. Schur's lemma states that any matrix that commutes with all the matrices of an irreducible representation must be a multiple of the group's identity matrix. The two Casimirs are independent because they are derived by two different mathematical paths, one involving products of two group matrices and the other involving products of three.

T_0 and Y are diagonal and commute with each other, but not with the other matrices. Because $[T_0, Y] = 0$, the SU(3) irreducible representations and their basis functions can be labeled simultaneously by the eigenvalues t and y.

$$T_0\,|t, y\rangle = t\,|t, y\rangle \qquad Y\,|t, y\rangle = y\,|t, y\rangle$$

From these relations, the basis functions, Eq. (10.18), are written in Dirac notation as

$$u = |\tfrac{1}{2}, \tfrac{1}{3}\rangle \qquad d = |-\tfrac{1}{2}, \tfrac{1}{3}\rangle \qquad s = |0, -\tfrac{2}{3}\rangle\,.$$

10.11.2   Octet and Decuplet Particle Multiplets

Data on baryons and mesons suggest that particles with similar properties tend to form natural groups of 8 (octets) and groups of 10 (decuplets). Gell-Mann's introduction of strangeness S led to striking pictorial groupings when particles are plotted versus I_3 and S, as demonstrated in this section.

The key idea for the application of group theory to particle physics is that particles are isomorphic to the basis functions of irreducible representations. The dimension of an irreducible representation therefore tells how many different particles occupy the corresponding multiplet. If a particle belongs to a particular irreducible representation, all of its partners with the same symmetry must belong to the same representation because they can be transformed into one another by group operations.

Determining the dimension of an irreducible representation is important because it is the number of particles in the symmetry group. There are several ways of determining the dimension without diagonalizing matrices. Some of the methods are algebraic but lengthy and complicated. A diagrammatic method called *Young diagrams* (originally *Young tableaux*) is widely used. Construction of the diagrams is based on a set of strict non-algebraic rules that require considerable explanation and illustration. All of these methods are outside the scope of this text.

Consider mesons, which have a quark content $q\bar{q}$. With three quarks u, d, s and three antiquarks $\bar{u}$, $\bar{d}$, $\bar{s}$ there are nine distinct pairs, some with combinations like $\pi^0 = \frac{1}{2}(u\bar{u} - d\bar{d})$. The representation is reducible. Notation for reductions uses Kronecker products $\otimes$ (Section 2.8) and Kronecker sums $\oplus$ (Section 2.9). Applied to $q\bar{q}$ combinations,

$$\mathbf{3} \otimes \bar{\mathbf{3}} = \mathbf{8} \oplus \mathbf{1}. \tag{10.21}$$

The meson multiplets are an octet and a singlet. As a check, the two sides should agree numerically if $\otimes$ is replaced by $\times$ and $\oplus$ by $+$, giving in this case $3 \times 3 = 9$ and $8 + 1 = 9$. The appearance of $\mathbf{8}$ in reductions caused Gell-Mann to call it the *Eightfold Way*, a term taken from Buddhist teachings (right speech, right conduct, etc.).

Because intrinsic spins are in a different space from SU(3) symmetry, spin is an additional label for an irreducible representation. Hence there are two meson octets, one for spin 0 particles and another for spin 1. The figure shows the spin 0 octet. It is plotted versus S and I_3 (equivalently t and y). In the plot, S, I_3, Q are connected by the Gell-Mann Nishijima relation Eq. (10.17) and take on the values $+1, 0, -1$. Baryon number $B = 0$ for mesons, so Eq. (10.17) becomes

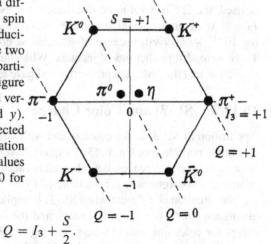

$$Q = I_3 + \frac{S}{2}.$$

The neutral η meson has spin 0 hence $I_3 = 0$. Because its quark content contains a $s\bar{s}$ pair, it also has $S = 0$, so it is located at the center of the plot along with π^0.

The basis function for the 1-dimensional singlet is the neutral spin 1 η' meson. Its quark content is completely symmetrical.

$$\eta' = \tfrac{1}{2}(u\bar{u} + d\bar{d} + s\bar{s})$$

It is more massive than the η meson and can decay into η.

Baryons have quark content qqq, so with three quarks u, d, s there is a 27×27 reducible representation. It reduces to

$$\mathbf{3} \otimes \mathbf{3} \otimes \mathbf{3} = \mathbf{10} \oplus \mathbf{8} \oplus \mathbf{8} \oplus \mathbf{1}.$$

Spin $\tfrac{1}{2}$ baryons are basis functions for an octet, and spin $\tfrac{3}{2}$ baryons are basis functions for the 10×10 decuplet shown in the figure. In the figure, baryons with superscript * are excited states, with greater mass than the ground state. For example, Σ^{*+} has a mass of 1380 MeV compared to Σ^+ with 1190. With their greater mass energy, the excited states decay quickly.

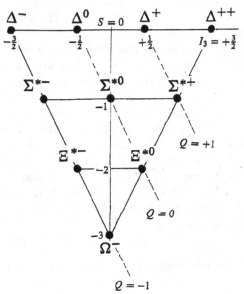

The decuplet played an historically important role. When the decuplet was first devised, the Ω^- had yet to be discovered, but the decuplet predicted $S = -3$ and $I_3 = 0$. With quark content sss, the Ω^- was predicted to be comparatively long lived ($\approx 10^{-10}$ s observed) because of strangeness conservation for strong interactions. There were also predictions of its mass. When the Ω^- was observed experimentally in 1964, it had the predicted properties and gave major support to the quark model.

10.11.3 SU(3) and Color Charge

Application of SU(3) to the quark model with three flavors helped classify the particles in the particle zoo, but SU(3) for quark flavor is only an approximate symmetry because of mass differences. It works well because u, d, s have relatively small masses compared to hadrons or to c, t, b (Table 10.6).

The situation is different when SU(3) is applied to the color gluons of QCD. All gluons are spin 1 bosons with 0 mass, and their other properties are also the same except for color quantum numbers, making SU(3) an exact symmetry in QCD and one of the foundations of the Standard Model.

There are three colors and three anticolors, which taken in pairs suggests there should be $3 \times 3 = 9$ gluons. Color–anticolor gluon content is analogous to the $q\bar{q}$

content of mesons, so SU(3) applies to color symmetry in the same way. The 9×9 representation for color charge is therefore reducible the same way as for three quarks, Eq. (10.21), to give an octet and a singlet: $3 \otimes \bar{3} = 8 \oplus 1$.

The octet irreducible representation has gluon basis functions constructed from color–anticolor pairs, as in the following set:

$$\frac{1}{\sqrt{2}}(r\bar{g} + g\bar{r}) \qquad\qquad \frac{1}{\sqrt{2}}(r\bar{b} + b\bar{b}) \qquad \frac{1}{\sqrt{2}}(b\bar{g} + g\bar{b})$$

$$\frac{1}{\sqrt{2}}(r\bar{r} - b\bar{b}) \qquad\qquad \frac{-i}{\sqrt{2}}(r\bar{b} - b\bar{r}) \qquad \frac{-i}{\sqrt{2}}(r\bar{g} - g\bar{r})$$

$$\frac{-i}{\sqrt{2}}(b\bar{g} - g\bar{b}) \qquad\qquad \frac{1}{\sqrt{6}}(r\bar{r} + b\bar{b} - 2g\bar{g}). \qquad\qquad (10.22)$$

They are linearly independent, so no gluon in the irreducible representation can be expressed as a combination of others. Group operations can transform this set into other linearly independent sets.

Quarks emit and absorb gluons while maintaining the colorless requirement of the hadron. Here is an example for the three quarks with three colors r, g, b in a baryon. Suppose that the r quark emits a gluon $\frac{1}{\sqrt{2}}(r\bar{b} + b\bar{r})$. This removes r from the quark and $\bar{b}$ from the quark's colorless $b\bar{b}$ pair, leaving the quark with color b. Now suppose the gluon interacts with a quark b. The quark absorbs $r\bar{b}$ from the gluon to give a colorless $b\bar{b}$ pair and leaving the quark as r. This example began with r and b quarks and ended with b and r, so the baryon remains colorless.

The singlet irreducible representation is linearly independent of the octet gluons and has basis function

$$\frac{1}{\sqrt{3}}(r\bar{r} + g\bar{g} + b\bar{b}).$$

This gluon cannot change the color of any quark. If it exists, it could not produce any observable effects. Only the octet gluons can produce observable reactions, seen as "jets" of hadrons produced from showers of q, $\bar{q}$, and gluons in energetic particle collisions. Collisions at extremely high energies have produced quark-gluon plasmas that are believed to re-create events immediately following the Big Bang creation of the universe.

10.12 Concluding Remark

This is an introductory text on group theory and its applications. Summarizing research work in particle physics from the past 60 years in a few dozen pages is an impossible task. Nevertheless, we made a start by discussing the groups U(1), SU(2), and SU(3), which are the only groups that play a role in the Standard Model. Our hope is that this summary treatment enables the student to move with understanding to more advanced books and articles.

Summary of Chapter 10

Chapter 10 discusses the application of group theory to modern particle physics. Necessary background material is included.

a) In order of decreasing strength, the known fundamental forces are strong interaction, electromagnetism, weak interaction, and gravitation.

b) In particle physics it is common to use units suited to the particle regime. The widely used system of natural units sets $\hbar = 1$, $c = 1$, $m_e = 1$, $\epsilon_0 = 1$. An algebraic method for converting between unit systems is demonstrated.

c) Taking p^+ and n^0 to be two states of a nucleon, the two nucleon states are described by the isospin operator $\mathbf{T}$ analogous to the spin $\frac{1}{2}$ angular momentum operator. The nucleon states are basis functions for SU(2) in abstract space. This picture is an approximation but holds to good accuracy for strong interactions at moderate energies.

d) One support for the isospin model comes from the binding energy of nuclei, where electric charge makes only a minor contribution to the strong interaction between nucleons. The deuteron nucleus is another example of the analogy between isospin and angular momentum, where Wigner *3-j* coefficients are applied to isospin.

e) The analogy between isospin and angular momentum is continued. Particles with similar properties such as mass form sets with isospin numbers T_3. Examples are the p^+-n^0 doublet $T_3 = \pm\frac{1}{2}$ and the π^+, π^0, π^- triplet $T_3 = \pm 1, 0$.

f) Isospin can be applied to moderate-energy pion-nucleon scattering, using *3-j* coefficients to estimate the ratio of matrix elements for similar reactions.

g) Every particle has its own antiparticle of equal mass. The photon, like some other particles, is its own antiparticle. If a particle has electric charge or any other electromagnetic property, the values for the antiparticle are the same magnitude but opposite in sign. The isospin component T_3 for the antiparticle also has the opposite sign. Uncharged antiparticles are symbolized with an overbar.

h) Particle physics is a field theory based on Lagrangians in relativistic form. The importance of the Lagrangian stems from its special properties such as stationary action. According to Noether's theorem, an operation that makes the Lagrangian invariant gives rise to conservation laws.

i) Particle physics theory is a gauge theory based on Lie group symmetry. U(1) gauge symmetry gives rise to electromagnetism from quantum mechanics. All interactions in particle physics are considered to arise from some form of gauge theory.

j) Collisions in powerful accelerators have generated a great number of short-lived particle species. A main object of particle physics has been to organize the particle zoo, and in this effort group theory has played a central role.

k) Experiment and theory in the 1960s showed that p^+ is not an elementary particle but has an inner structure of point-like particles called quarks. There are six known quarks, each with its own special properties or "flavors." Each quark has a corresponding antiquark. Quarks are spin $\frac{1}{2}$ fermions and carry fractional electric charge, either $+\frac{2}{3}$ or $-\frac{1}{3}$. Their masses range from a few MeV to over 100 GeV. Massless bosons called gluons bind quarks with a force that increases with separation so that individual quarks cannot escape from a nucleon or other composite particle.

l) Baryons are defined according to their quark content as particles made from three quarks. Baryons must therefore have either spin $\frac{1}{2}$ or $\frac{3}{2}$. Mesons are particles with a color–anticolor pair, so their spin must be 0 or 1. Hadron is a general term for any particle with quark content. Leptons all have charge -1. Leptons have no quark content and engage only in the weak interaction.

m) Existence of the charm quark c was inferred from the properties of the J/Ψ meson, which was theorized to have $c\bar{c}$ content. The exceptionally narrow J/Ψ decay peak (long lifetime) is explained by conservation of charm.

n) Quarks (and gluons) must carry a quantum property called color to satisfy the Pauli exclusion principle for the three quarks in a baryon. Color charge comes in three flavors, and each quark in a baryon must have a different color. In the Standard Model, color is a fundamental symmetry included in quantum chromodynamics (QCD) gauge theory.

o) Particle physics quantum numbers obey conservation laws that can determine whether reactions are allowed or forbidden. Baryon number is conserved in all reactions. Charm and strangenesss are conserved only in strong and electromagnetic interactions. Lepton number is conserved in weak interactions if all types of lepton neutrinos are taken into account.

p) The Gell-Mann Nishijima relation expresses a particle's electric charge as a linear function of isospin component I_3, baryon number B, and strangeness S.

q) The Standard Model is based on three Lie groups: U(1), SU(2), SU(3). U(1) gives the gauge theory for electromagnetism, SU(2) gives the isospin model, and SU(3) gives the u, d, s model and the fundamental color charge symmetry.

r) The fundamental representation of the SU(3) Lie group consists of eight 3×3 matrices. Two of the matrices are diagonal and commute with each other but not with the other matrices. Their eigenvalues t and y are labels for the SU(3) basis functions. SU(3) has $3 - 1 = 2$ Casimir operators that also provide two labels for the basis functions.

s) The link between particle physics and group theory is that particles are isomorphic to basis functions for irreducible representations. It is therefore important to determine how many irreducible representations are contained in a multiplet, but the methods are too lengthy to include in this text.

t) In the three-quark model u, d, s, there are nine possible pairs to represent $q\bar{q}$ mesons. The reducible representations are reduced to octet and singlet irreducible representations. There are two meson octets, corresponding to spin 0 and spin 1. A characteristic hexagonal diagram arises when known mesons are plotted versus strangeness S and isospin I_3.

u) The 27×27 reducible representation for the three-quark baryon model reduces to decuplet 10×10 irreducible representation among others. There is a decuplet for spin $\frac{1}{2}$ and one for spin $\frac{3}{2}$. Known particles fit into a characteristic diagram when plotted versus strangeness S and isospin I_3. Vacancies in the diagram predict particles to be discovered.

Problems and Exercises

10.1   Radium-226 decays with a half-life of 1,620 years to radon-222 plus a helium nucleus (α particle) according to the nuclear reaction equation

$$^{226}_{88}\text{Ra} \rightarrow ^{222}_{86}\text{Rn} + ^{4}_{2}\text{He}.$$

The atomic masses are

$$^{4}_{2}\text{He} \quad 4.002603 \text{ u}$$
$$^{226}_{88}\text{Ra} \quad 226.025408 \text{ u}$$
$$^{222}_{86}\text{Rn} \quad 222.017576 \text{ u}.$$

Use the data in Table 10.1 to calculate the energy in MeV released in the reaction. How much of this energy appears as kinetic energy of the α particle?

10.2   Consider an α particle, which is the nucleus of $^{4}_{2}\text{He}$ atomic mass 4.002603 u.
(a) Use the data in Table 10.1 to calculate the binding energy per nucleon.
(b) Release of an α particle from a nucleus is a common mode of decay for many radioactive heavy nuclei. Explain why the α decay of $^{226}_{88}\text{Ra}$ is more likely than decay by release of a p^+. The atomic mass of $^{225}_{87}\text{Fr}$ is 225.025572.

10.3   Express the natural unit of length in terms of the SI unit meter.

10.4   Express the natural unit of energy in terms of the SI unit joule and also in terms of eV.

10.5   Express the natural unit of electric charge e in terms of the SI unit coulomb. *Hint:* use the dimensionless fine structure constant $= \frac{e^2}{4\pi\epsilon_0 \hbar c}$.

10.6 Express time in natural units (combined with GeV) in terms of the SI unit second.

10.7 A consequence of Yukawa's theory was a potential energy (in natural units),

$$\frac{e^{-mr}}{r},$$

where m is the mass of the mediating boson and $r \approx 1.5$ fm is a measure of the range of the strong interaction. What value in MeV is predicted for the mass of the boson?

10.8 Making reasonable assumptions, use isospin to estimate the ratio of the total cross sections for the processes

$$\pi^- + p^+ \to \pi^- + p^+$$
$$\pi^- + p^+ \to \pi^o + n^o.$$

10.9 Making reasonable assumptions, use isospin to estimate the ratio of the total cross sections for observable pion production processes

$$p^+ + p^+ \to \pi^+ + d^+$$
$$n^o + p^+ \to \pi^o + d^+.$$

10.10 Making reasonable assumptions, use isospin to estimate the ratio of the total cross sections for observable interactions of pions and deuterons

$$\pi^+ + d^+ \to p^+ + p^+$$
$$\pi^- + d^+ \to n^o + n^o,$$

where d is a deuteron.

10.11 In special relativity, the relation between energy E, momentum p, and rest mass m_0 is

$$E^2 = p^2 + m_0^2,$$

written in natural units $c = 1$.
(a) Insert factors of c as necessary to write this relation in ordinary SI units.
(b) Consider a photon ($m_0 = 0$) with energy E. From your result, what is the momentum of the photon in SI units?

10.12 Suppose a mass m is thrown upward with initial velocity v_0 at an angle θ in a constant gravitational field.
(a) Write the Lagrangian for this situation.
(b) Use your Lagrangian to find the equations of motion, and show that they are the same as from Newton's laws.

10.13 The neutron is made of u and d. What is its composition according to the quark model?

10.14 Considering only u, d, s, and c, how many different neutral mesons are possible?

10.15 The quark content of π^+ is $u\,\bar{d}$. Based on this, what is the quark content of π^-?

10.16 The Λ^0 particle can be produced in pion-nucleon collisions according to the reaction

$$\pi^- + p^+ \rightarrow \Lambda^0 + K^o.$$

Show that the conservation laws for electric charge, baryon number, lepton number, and strangeness are all satisfied.

10.17 μ^- decays according to

$$\mu^- \rightarrow e^- + \nu_\mu + \bar{\nu}_e.$$

Show that the conservation laws for electric charge, baryon number, and lepton number are all satisfied.

10.18 The decay

$$\Xi^- \rightarrow \Sigma^- + \pi^o$$

proceeds only by the weak interaction. What conservation law prevents decay by the strong interaction?

10.19 The decay

$$\Lambda^0 \rightarrow p^+ + \pi^-$$

proceeds only by the weak interaction. What conservation law prevents decay by the strong interaction?

Appendix A

Character Tables from Class Sums

This appendix shows how to calculate the characters of the irreducible representations of a group using algebra and the product table without knowing explicit matrices.

The **32** group serves as an illustration. Every group has an irreducible identity representation 1. In 1 every group member **R** has the 1-dimensional matrix representation $D^{(1)}(\mathbf{R}) = (1)$ with corresponding character $\chi^{(1)}(\mathbf{R}) = 1$. This gives us the first line of the group's character table.

The number of irreducible representations of a group is equal to the number of classes, which can be found from the product table as in Section 2.7.3. Members of the same class all have the same character. Knowing the number of irreducible representations of a group and using the relation

$$\sum_{\beta}(h_{\beta})^2 = n$$

from Theorem 8, where h_{β} is the dimension of the β irreducible representation, it may be possible to predict the dimensions of the irreducible representations. For the **32** group of order $n = 6$ and with three classes the dimensions of the irreducible representations are 1, 1, 2. It follows that the character of an irreducible representation of **E** is equal to the dimension of the irreducible representation, giving the first column of the character table for the **32** group.

Collecting results, we now have a portion of the group's character table:

	{E}	{A, B, C}	{D, F}
1	1	1	1
A	1		
Γ	2		

We can calculate a complete character table using the product table and the concept of *class sums*. A class sum is the sum of the group members in a class. Let C_i be the sum of the group members in the ith class. The **32** group has three classes, and the class sums are

$C_1 = \mathbf{E}$ $C_2 = \mathbf{A} + \mathbf{B} + \mathbf{C}$ $C_3 = \mathbf{D} + \mathbf{F}$

Now multiply one class sum by another. The products can be evaluated using the product table. For example,

$$\begin{aligned} C_2 C_3 &= (\mathbf{A} + \mathbf{B} + \mathbf{C})(\mathbf{D} + \mathbf{F}) \\ &= \mathbf{AD} + \mathbf{AF} + \mathbf{BD} + \mathbf{BF} + \mathbf{CD} + \mathbf{CF} \\ &= \mathbf{C} + \mathbf{B} + \mathbf{A} + \mathbf{C} + \mathbf{B} + \mathbf{A} \\ &= 2C_2. \end{aligned}$$

The product of two class sums is a sum of class sums, as the example illustrates. In general,

$$C_i C_j = \sum_k c_{ijk} C_k.$$

In the example $C_2 C_3 = 2 C_2$ so that

$c_{231} = 0$ $c_{232} = 2$ $c_{233} = 0$

Here are some of the class sum coefficients for the **32** group.

$$\begin{array}{lllll} c_{121} = 0 & c_{131} = 0 & c_{221} = 3 & c_{231} = 0 & c_{331} = 2 \\ c_{122} = 1 & c_{132} = 0 & c_{222} = 0 & c_{232} = 2 & c_{332} = 0 \\ c_{123} = 0 & c_{133} = 1 & c_{223} = 3 & c_{233} = 0 & c_{333} = 1 \end{array}$$

There are many others, such as c_{111}, that are not listed.

Every member in a class has the same character (Section 2.7.3), so it is plausible that the relation for the product of class sums can be written as products of characters. Without giving a proof, let $\chi_i^{(\alpha)}$ be the character of a member in the class sum i in the α irreducible representation and let h_i be the number of group members in C_i. The character of a class sum matrix is therefore $h_i \chi_i^{(\alpha)}$. In general,

$$h_i \chi_i^{(\alpha)} h_j \chi_j^{(\alpha)} = h_\alpha \sum_k c_{ijk} h_k \chi_k^{(\alpha)}, \tag{2.11}$$

where h_α is the dimension of the α irreducible representation. (The factor h_α arises in the course of the actual proof.)

Knowing the class sum coefficients, Eq. (2.11), can be used to write more than enough equations to determine unknown characters. Here is an example using the irreducible representation Γ of the **32** group:

$$(2\chi_3^{(\Gamma)})(2\chi_3^{(\Gamma)}) = 2\sum_k c_{33k}h_k\chi_k^{(\Gamma)}$$

$$4\chi_3^{(\Gamma)}\chi_3^{(\Gamma)} = 2\times\left((1)c_{331}\chi_1^{(\Gamma)} + 2c_{333}\chi_3^{(\Gamma)}\right)$$

$$= 2\times\left((1)(2)\chi_1^{(\Gamma)} + (2)(1)\chi_3^{(\Gamma)}\right)$$

$$\left(\chi_3^{(\Gamma)}\right)^2 = \chi_1^{(\Gamma)} + \chi_3^{(\Gamma)} = 2 + \chi_3^{(\Gamma)},$$

using the known value $\chi_1^{(\Gamma)} = 2$. A solution is $\chi_3^{(\Gamma)} = -1$, in agreement with the known character table. $\chi_3^{(\Gamma)} = 2$ is also a solution, but it is not consistent with $\sum_R |\chi^{(\alpha)}(\mathbf{R})|^2 = n$ from Theorem 7.

The **32** group has three classes, so knowing $\chi_1^{(\Gamma)} = 2$ and $\chi_3^{(\Gamma)} = -1$ use Theorem 7 and sum over classes to find $\chi_2^{(\Gamma)}$.

$$\sum_\beta h_\beta \left|\chi_\beta^{(\Gamma)}\right|^2 = (2)^2 + 3\left|\chi_2^{(\Gamma)}\right|^2 + 2(-1)^2 = 6,$$

which gives the correct result $\chi_2^{(\Gamma)} = 0$.

	{E}	{A, B, C}	{D, F}
1	1	1	1
A	1		
Γ	2	0	−1

Class sums could be used to complete the **32** group's character table, but in this example it is easier to use the orthogonality of the rows as expressed in Theorem 6. Taking the scalar product of row A and row Γ,

$$(1)(2) + 3\chi_2^{(A)}(0) + 2\chi_3^{(A)}(-1) = 0,$$

which gives $\chi_3^{(A)} = 1$. The scalar product of row A and row 1 is then

$$(1)(1) + 3\chi_2^{(A)}(1) + 2(1)(1) = 0,$$

from which $\chi_2^{(A)} = -1$, completing the character table.

	{E}	{A, B, C}	{D, F}
1	1	1	1
A	1	−1	1
Γ	2	0	−1

Appendix B

Born–Jordan Proof of the Quantization Condition

This advanced material is included for completeness. Some of the mathematical development may be of interest, particularly the definition of the delta function and its Fourier transform.

In 1925, Born and Jordan published a derivation of the quantization condition, Eq. (5.13), based on Heisenberg's concepts of a few months earlier. Later researchers have criticized the Born–Jordan derivation as lacking in rigor. The proof presented here is plausible and largely follows the Born–Jordan approach but with intermediate steps added for clarity.

Born and Jordan began by calculating the action $\int p\,dq$ using classical Fourier series for coordinate q and conjugate momentum p, where q and p are ordinary continuous functions of time. Their time derivatives are denoted using Newton's dot notation $\dot{q}$, etc. In the last steps, they follow Heisenberg by expressing a continuous derivative in terms of discrete matrices, converting q and p to the matrices $\mathbf{q}$ and $\mathbf{p}$.

$$J = \int_0^{\frac{1}{\nu}} p\dot{q}\,dt$$

$$q = \sum_{n=-\infty}^{\infty} q_n e^{2\pi i n\nu t}$$

$$\dot{q} = 2\pi i\nu \sum_n n q_n e^{2\pi i n\nu t}$$

$$p = \sum_{m=-\infty}^{\infty} p_m e^{2\pi i m\nu t}$$

$$J = 2\pi i\nu \int \sum_n \sum_m n p_m q_n e^{2\pi i(n+m)\nu t}\,dt \qquad (B.1)$$

The integral over t in Eq. (B.1) is the Fourier transform of the *delta function* $\delta(t)$, a result from a theorem for the product of two classical Fourier series.

$$\delta(\xi) = \frac{1}{\mathcal{T}} \int_{-\infty}^{\infty} e^{2\pi i \xi v t} \, dt$$

$$= v \int_{-\infty}^{\infty} e^{2\pi i \xi v t} \, dt$$

The delta function $\delta(\xi)$ is the Fourier transform of a constant (here 1) and is different from the discrete Kronecker delta δ_{ij}. According to the discussion of reciprocal domains in Section 4.4, a function spread over the whole time domain must be a "spike" in the frequency domain. Put another way, a signal must extend over all time if it has only a single frequency. Limiting the signal in time generates a range of frequencies.

The delta function has the following properties:

$$\delta(\xi) = 1 \quad \text{for} \quad \xi = 0$$
$$\delta(\xi) = 0 \quad \text{for} \quad \xi \neq 0.$$

Evaluating the time integral in Eq. (B.1) in terms of the delta function,

$$J = 2\pi i v \int \sum_n \sum_m n p_m q_n e^{2\pi i (n+m)vt} \, dt$$

$$= 2\pi i \sum_n \sum_m n p_m q_n \delta(n+m).$$

The delta function $\delta(n+m) = 1$ only for $n+m = 0$. The time dependence has disappeared, and the sum over m has only the term $m = -n$:

$$J = 2\pi i \sum_n n q_n p_{-n}.$$

Next, Born and Jordan take the derivative with respect to J to give

$$1 = 2\pi i \sum_n n \frac{\partial (q_n p_{-n})}{\partial J}. \tag{B.2}$$

Although not shown explicitly in their paper, they evidently use a chain rule,

$$\frac{\partial}{\partial J} = \frac{\partial n}{\partial J} \frac{\partial}{\partial n}.$$

Using the quantum condition $J = nh$ in the chain rule, Eq. (B.2) becomes

$$1 = \frac{2\pi i}{h} \sum_n n \frac{\partial (q_n p_{-n})}{\partial n}. \tag{B.3}$$

The classical part of the calculation is complete at this point. Now Born and Jordan switch to matrix mechanics by expressing the derivative with respect to n in discrete terms:

$$\frac{\partial(q_n p_{-n})}{\partial n} \rightarrow \frac{1}{n} \left[p(n, n+k) q(n+k, n) - q(n, n-k) p(n-k, n) \right].$$

Inserting the discrete derivative in Eq. (B.3) gives

$$1 = \frac{2\pi i}{h} \sum_n \sum_k \left[p(n, n+k) q(n+k, n) - q(n, n-k) p(n-k, n) \right].$$

Each value of n gives an equation involving matrix multiplication:

$$\sum_k (p_{nk} q_{kn} - q_{nk} p_{kn}) = \frac{h}{2\pi i}$$

$$\mathbf{pq} - \mathbf{qp} = \frac{h}{2\pi i} \mathbf{1}$$

$$= -i\hbar \mathbf{1}.$$

Appendix C

Weyl Derivation of the Heisenberg Uncertainty Principle

The material in this appendix is above the level of the text but is included for completeness and for the sake of interested students.

As the measure of uncertainty, Heisenberg and most later workers used the *standard deviation* tool from statistics. Suppose that a physical quantity $\mathfrak{A}$ is measured N times to give N values $\alpha_1, \alpha_2, \ldots, \alpha_N$. The average value $\bar{\alpha}$ is

$$\bar{\alpha} = \frac{1}{N} \sum_{n=1}^{N} \alpha_n.$$

The standard deviation $\Delta\alpha$ and its square are

$$\Delta\alpha = \sqrt{\sum_{n=1}^{N} (\alpha_n - \bar{\alpha})^2}$$

$$(\Delta\alpha)^2 = \sum_{n=1}^{N} (\alpha_n - \bar{\alpha})^2.$$

The standard deviation is a measure of how far the experimental values depart from the average. Because of the square, values higher and lower than the average both make positive contributions. Values far from the average make a larger contribution. If the experimental values are all clustered near the average, the standard deviation (hence the uncertainty) is small.

Heisenberg stated that the uncertainty relation can be derived entirely from the mathematical principles of quantum theory without recourse to the wave nature of matter. His derivation of the uncertainty relation is rigorous but lengthy. Presented here is a shorter proof due to Hermann Weyl (1885–1955), a German-American mathematician at the University of Göttingen and the Institute for Advanced Study in Princeton.

Weyl used a *probability amplitude* $\psi(x)$ where $\psi^*(x)\psi(x)dx = \left|\psi^2(x)\right|dx$ represents the probability that x lies between x and $x + dx$.

$\psi(x)$ obeys boundary conditions $\psi(x) \to 0$ for $|x| \to \infty$. The normalization condition is

$$\int_{-\infty}^{\infty} \psi^*(x)\psi(x)dx = 1.$$

The average values and squared standard deviations are

$$\bar{x} = \int_{-\infty}^{\infty} \psi^*(x)x\psi(x)dx$$

$$(\Delta x)^2 = \int_{-\infty}^{\infty} (x - \bar{x})^2 \left|\psi(x)\right|^2 dx$$

$$\overline{p_x} = \int_{-\infty}^{\infty} \psi^*(x)p_x\psi(x)dx$$

$$(\Delta p_x)^2 = \int_{-\infty}^{\infty} \psi^*(x)(p_x - \overline{p_x})^2\psi(x)dx.$$

The order of the factors takes into account that p_x does not commute with a function of x, but x commutes with any function of x.

It is convenient to let $\bar{x} = 0$ and $\overline{p_x} = 0$. The final result is not affected because these conditions simply reflect a change of variables in x and in ψ. Then

$$(\Delta x)^2 = \int_{-\infty}^{\infty} \psi^*(x)x^2\psi(x)dx$$

$$= \int_{-\infty}^{\infty} (x\psi^*(x))(x\psi(x))dx. \tag{C.1}$$

The momentum operator is $p_x = -i\hbar\frac{\partial}{\partial x}$.

$$\overline{p_x} = -i\hbar \int_{-\infty}^{\infty} \psi^*(x)\frac{d\psi(x)}{dx}dx$$

$$(\Delta p_x)^2 = \int_{-\infty}^{\infty} \psi^*(x)p_x^2\psi(x)dx$$

$$= -\hbar^2 \int_{-\infty}^{\infty} \psi^*(x)\frac{d^2\psi(x)}{dx^2}dx$$

Integrating by parts gives

$$(\Delta p_x)^2 = -\hbar^2 \left(\psi^* \frac{d\psi}{dx} \right) \Bigg|_{-\infty}^{\infty} + \hbar^2 \int_{-\infty}^{\infty} \left(\frac{d\psi^*(x)}{dx} \right) \left(\frac{d\psi(x)}{dx} \right) dx$$

$$= 0 + \int_{-\infty}^{\infty} \left(\frac{d\psi^*(x)}{dx} \right) \left(\frac{d\psi(x)}{dx} \right) dx. \qquad (C.2)$$

Multiply Eq. (C.1) by Eq. (C.2). Weyl then applied the *Schwartz inequality* to give

$$(\Delta x)^2 (\Delta p_x)^2 = \hbar^2 \int_{-\infty}^{\infty} (x\psi^*(x))(x\psi(x))dx \int_{-\infty}^{\infty} \left(\frac{d\psi^*(x)}{dx} \right) \left(\frac{d\psi(x)}{dx} \right) dx$$

$$\geq \frac{\hbar^2}{4} \left(\int_{-\infty}^{\infty} \psi^*(x)\psi(x)dx \right)^2$$

$$\geq \frac{\hbar^2}{4} \times (1)$$

using the normalization condition. Thus,

$$\Delta x \Delta p_x \geq \frac{\hbar}{2}.$$

Appendix D

EPR Thought Experiment

The derivation of the EPR paradox in this appendix closely follows the approach of the 1935 paper.

The EPR model considers two systems labeled I and II in known states prior to time t. The systems are allowed to interact until they separate at time t'. They then propagate according to Eq. (6.20) and move so far apart that interactions between them are negligible. Specifically, they move a distance d apart so that the time $\frac{d}{c}$ for light to travel from I to II is longer than the time for any measurement; this is called a *space-like* separation.

Let α_I be a set of variables that describe system I, and let β_{II} be variables that describe system II. Expand the system's wave function $\Psi(\alpha_I, \beta_{II})$ in terms of system I wave functions $u(\alpha_I)$ with system II wave functions as expansion coefficients $\psi(\beta_{II})$.

$$\Psi(\alpha_I, \beta_{II}) = \sum_{n=-\infty}^{\infty} \psi_n(\beta_{II}) u_n(\alpha_I) \tag{D.1}$$

Suppose that operator $\mathbf{A}$ having eigenvalues $a_1, a_2, \ldots$ is now measured on system I. The wave function collapses to a single term $n = k$.

$$\mathbf{A}\Psi = a_k \Psi_k$$
$$= a_k \psi_k(\beta_{II}) u_k(\alpha_I) \tag{D.2}$$

The crux of the argument is to suppose that the system's wave function $\Psi(\alpha_I, \beta_{II})$ has instead been expressed in eigenfunctions v of operator $\mathbf{B}$ having eigenvalues $b_1, b_2, \ldots$. In terms of the $v(\alpha_I)$ and expansion coefficients $\phi(\beta_{II})$, the expansion in Eq. (D.1) is now written

$$\Psi(\alpha_I, \beta_{II}) = \sum_{n=-\infty}^{\infty} \phi_n(\beta_{II}) v_n(\alpha_I). \tag{D.3}$$

Let system 1 be measured by operator **B** to give eigenvalue b_j. After the collapse,

$$\mathbf{B}\Psi = b_j \Psi_j$$
$$= b_j \phi_j (\beta_{II}) v_j (\alpha_I). \qquad \text{(D.4)}$$

According to Eq. (D.2), a measurement of system I by **A** leaves system II in the state ψ_k. According to Eq. (D.4), a measurement of system I by **B** leaves system II in the state ϕ_j. System II seems to have two different wave functions even though, as EPR argue, "no real change can take place in the second system in consequence of anything that may be done to the first system" (because they have a space-like separation). They argue further that even if **A** and **B** do not commute, measurements on system I allow eigenvalues for **A** and **B** in system II to be predicted simultaneously with certainty even though system I and system II do not interact. Simultaneous measurement of noncommuting quantities is contrary to matrix mechanics and Heisenberg's uncertainty principle. EPR emphasized this apparent failure of quantum mechanics by saying, "No reasonable definition of reality could be expected to permit this."

Appendix E

Photon Correlation Experiment

To help understand the photon polarization experiments discussed in the text, consider a realizable experiment with the apparatus shown in the schematic diagram. There are two identical measuring stations **A** and **B** in spacelike separation. A source of entangled photons **S** is located halfway between the stations.

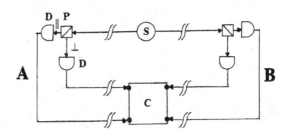

The experimental apparatus closely follows the plan of the EPR thought experiment (*Gedankenexperiment*) and is a photon analog of an experiment using Stern–Gerlach magnets to measure dichotomic particle spin $\pm\frac{1}{2}$. Each station records photon polarizations $\updownarrow$ ($\parallel$) and $\leftrightarrow$ ($\perp$) with respect to a fixed lab axis using two-channel polarizers **P** on the beam axis. The polarizer splits the incoming beam into $\parallel$ and $\perp$ components and directs them to separate light detectors **D** that generate an electrical pulse when a photon is detected. As the sketch illustrates, pulses are sent to a coincidence counter **C** that registers a count if any two pulses both arrive in the narrow time window of 20 ns. Cables are equal length to maintain the timing.

In each station the polarizer and the two light detectors are constructed as a single unit, called a *polarimeter*, and can be rotated about the beam axis. The experiment consists of measuring the accumulated counts in each coincidence counter as a function of the relative angle between the A and B polarimeters.

304

The fourfold coincidence counter accumulates separate coincidence counts N depending on which of A's inputs receives a pulse simultaneous with one of B's inputs.

$$N(A_\updownarrow B_\updownarrow) \qquad N(A_\leftrightarrow B_\leftrightarrow) \qquad N(A_\updownarrow B_\leftrightarrow) \qquad N(A_\leftrightarrow B_\updownarrow)$$

For a given setting $\hat{\mathbf{a}}, \hat{\mathbf{b}}$ of the axes, the correlation measured in a single run is

$$C_{cor}(\hat{\mathbf{a}}, \hat{\mathbf{b}}) = \frac{N(A_\updownarrow B_\updownarrow) + N(A_\leftrightarrow B_\leftrightarrow) - N(A_\updownarrow B_\leftrightarrow) - N(A_\leftrightarrow B_\updownarrow)}{N(A_\updownarrow B_\updownarrow) + N(A_\leftrightarrow B_\leftrightarrow) + N(A_\updownarrow B_\leftrightarrow) + N(A_\leftrightarrow B_\updownarrow)}$$

Alice does not know beforehand the polarization of an incoming photon – only her measurement will tell. If she measures $|\updownarrow\rangle$, then Bob will also instantaneously measure $|\updownarrow\rangle$ for his photon of the pair. Evidence for entanglement is that Bob simultaneously measures the same value for his photon as Alice had measured for hers. The measurements are correlated despite the great distance between Alice and Bob because they are one inseparable quantum state.

Appendix F

Tables of Some *3-j* Coefficients

Following are tables of some *3-j* coefficients. In the tables, reading down a column gives W in terms of the uv (compare Eq. (8.42)),

$$W_M^{(J)} = \sum_m (jj'm\,M-m|JM)\,u_m^{(j)}v_{M-m}^{(j')}, \tag{F.1}$$

so the column entries are *3-j* coefficients.

Reading across a row gives uv in terms of the W:

$$u_m^{(j)}v_{M-m}^{(j')} = \sum_{J,M} (jj'm\,M-m|JM)\,W_M^{(J)}, \tag{F.2}$$

so the row entries are *3-j* coefficients.

In Table F.1 for $D^{(\frac{1}{2})} \otimes D^{(\frac{1}{2})}$ note that u always has superscript $\frac{1}{2}$ and v always has superscript $\frac{1}{2}$ corresponding to the spin values. Similarly for the other tables.

The columns in the tables play the role of orthonormal vectors.

Table F.1 *3-j* coefficients for $D^{(\frac{1}{2})} \otimes D^{(\frac{1}{2})}$

	$W_1^{(1)}$	$W_0^{(1)}$	$W_0^{(0)}$	$W_{-1}^{(1)}$
$u_{1/2}^{(\frac{1}{2})}\,v_{1/2}^{(\frac{1}{2})}$	1			
$u_{1/2}^{(\frac{1}{2})}\,v_{-1/2}^{(\frac{1}{2})}$		$\sqrt{\frac{1}{2}}$	$\sqrt{\frac{1}{2}}$	
$u_{-1/2}^{(\frac{1}{2})}\,v_{1/2}^{(\frac{1}{2})}$		$\sqrt{\frac{1}{2}}$	$-\sqrt{\frac{1}{2}}$	
$u_{-1/2}^{(\frac{1}{2})}\,v_{-1/2}^{(\frac{1}{2})}$				1

306

Table F.2 *3-j* coefficients for $D^{(1)} \otimes D^{(\frac{1}{2})}$

	$W_{3/2}^{(\frac{3}{2})}$	$W_{1/2}^{(\frac{3}{2})}$	$W_{1/2}^{(\frac{1}{2})}$	$W_{-1/2}^{(\frac{3}{2})}$	$W_{-1/2}^{(\frac{1}{2})}$	$W_{-3/2}^{(\frac{3}{2})}$
$u_1^{(1)} v_{1/2}^{(\frac{1}{2})}$	1					
$u_1^{(1)} v_{-1/2}^{(\frac{1}{2})}$		$\sqrt{\frac{1}{3}}$	$\sqrt{\frac{2}{3}}$			
$u_0^{(1)} v_{1/2}^{(\frac{1}{2})}$		$\sqrt{\frac{2}{3}}$	$-\sqrt{\frac{1}{3}}$			
$u_0^{(1)} v_{-1/2}^{(\frac{1}{2})}$				$\sqrt{\frac{2}{3}}$	$\sqrt{\frac{1}{3}}$	
$u_{-1}^{(1)} v_{1/2}^{(\frac{1}{2})}$				$\sqrt{\frac{1}{3}}$	$-\sqrt{\frac{2}{3}}$	
$u_{-1}^{(1)} v_{-1/2}^{(\frac{1}{2})}$						1

$$\sum_m (j\,j'\,m\,M - m|J\,M)(j\,j'\,m\,M' - m|J'\,M') = \delta_{JJ'}\delta_{MM'} \qquad \text{(F.3)}$$

The rows also act as orthonormal vectors.

$$\sum_{J,M} (jj'm_1m_2|JM)(jj'm_1'm_2'|JM) = \delta_{m_1m_1'}\,\delta_{m_2m_2'} \qquad \text{(F.4)}$$

Table F.3 *3-j* coefficients for $D^{(1)} \otimes D^{(1)}$

	$W_2^{(2)}$	$W_1^{(2)}$	$W_1^{(1)}$	$W_0^{(2)}$	$W_0^{(1)}$	$W_0^{(0)}$	$W_{-1}^{(2)}$	$W_{-1}^{(1)}$	$W_{-2}^{(2)}$
$u_1^{(1)} v_1^{(1)}$	1								
$u_1^{(1)} v_0^{(1)}$		$\sqrt{\frac{1}{2}}$	$\sqrt{\frac{1}{2}}$						
$u_0^{(1)} v_1^{(1)}$		$\sqrt{\frac{1}{2}}$	$-\sqrt{\frac{1}{2}}$						
$u_1^{(1)} v_{-1}^{(1)}$				$\sqrt{\frac{1}{6}}$	$\sqrt{\frac{1}{2}}$	$\sqrt{\frac{1}{3}}$			
$u_0^{(1)} v_0^{(1)}$				$\sqrt{\frac{2}{3}}$	0	$-\sqrt{\frac{1}{3}}$			
$u_{-1}^{(1)} v_1^{(1)}$				$\sqrt{\frac{1}{6}}$	$-\sqrt{\frac{1}{2}}$	$\sqrt{\frac{1}{3}}$			
$u_0^{(1)} v_{-1}^{(1)}$							$\sqrt{\frac{1}{2}}$	$\sqrt{\frac{1}{2}}$	
$u_{-1}^{(1)} v_0^{(1)}$							$\sqrt{\frac{1}{2}}$	$-\sqrt{\frac{1}{2}}$	
$u_{-1}^{(1)} v_{-1}^{(1)}$									1

These tables are adapted from V. Heine, "Group Theory in Quantum Mechanics," Appendix I, Pergamon Press, New York, 1960, where additional tables are listed. Some references, including E. P. Wigner, "Group Theory and Its Application to the Quantum Mechanics of Atomic Spectra," p. 191, Academic Press, New York, 1959 and Heine p. 181 give general formulas for the *3-j* coefficients, but they are not easy to use. E. U. Condon and G. H. Shortley, "Theory of Atomic Spectra," pp. 76–77, Cambridge University Press, New York, 1935, give simplified formulas for the important cases $j' = \frac{1}{2}, 1, \frac{3}{2}, 2$ that hold for any possible values of the other variables.

Appendix G

Proof of the Wigner–Eckart Theorem

This proof uses Theorem 10 derived in Section 6.3.2.

Consider the constructed wave function $\mathcal{F}$:

$$\mathcal{F} = \sum_{m'q} T_q^{(k)} |\alpha', j,' m'\rangle (j' k m' q | j'' m'') \qquad \text{(G.1)}$$

$$I_z \mathcal{F} = \sum_{m',q} I_z T_q^{(k)} |\alpha', j,' m'\rangle (j' k m' q | j'' m'').$$

The commutator from Eq. (8.47) gives $[I_z, T_q^{(k)}] = q T_q^{(k)}$. Using the relation $I_z |\alpha', j', m'\rangle = m' |\alpha', j', m'\rangle$,

$$I_z T_q^{(k)} = T_q^{(k)} I_z + q T_q^{(k)}$$

$$I_z \mathcal{F} = \sum_{m'q} T_q^{(k)} (m' + q) |\alpha', j,' m'\rangle (j' k m' q | j'' m'').$$

$$\text{(G.2)}$$

The *3-j* coefficient in Eq. (G.2) = 0 unless $m' + q = m''$. Hence

$$I_z \mathcal{F} = m'' \sum_{m'q} T_q^{(k)} |\alpha', j,' m'\rangle (j' k m' q | j'' m'')$$

$$= m'' \mathcal{F}.$$

It follows that

$$\mathcal{F} = \langle \alpha', j'', m'' |$$

$$\langle \alpha', j'', m''| = \sum_{m' q} T_q^{(k)} |\alpha', j', m'\rangle \, (j' k m' q | j'' m'')$$

$$T_q^{(k)} |\alpha', j', m'\rangle = \sum_{j'' m''} |\alpha', j'', m''\rangle \, (j' k m' q | j'' m'')$$

$$\langle \alpha, j, m|T_q^{(k)}|\alpha', j', m'\rangle = \sum_{j'' m''} \langle \alpha, j, m|\alpha', j'', m''\rangle \, (j' k m' q | j'' m'').$$

In the sum over m'' only the term $m'' = m$ contributes, and in the sum over j'' only the term $j'' = j$ contributes. Hence

$$\langle \alpha, j, m|T_q^{(k)}|\alpha', j', m'\rangle = \langle \alpha, j, m|\alpha', j, m\rangle \, (j' k m' q | j m).$$

It remains to show that the scalar product,

$$\langle \alpha, j, m|\alpha', j', m\rangle \equiv \int \left(\psi_{\alpha,m}^{(j)}\right)^* \psi_{\alpha',m}^{(j')} \, d\tau,$$

is independent of m.

Write $m - 1$ in place of m and use Eq. (8.28).

$$\langle \alpha, j, m-1|\alpha', j', m-1\rangle \equiv \int \left(\psi_{\alpha,m-1}^{(j)}\right)^* \psi_{\alpha',m-1}^{(j')} \, d\tau$$

$$\psi_{\alpha,m-1}^{(j')} = \frac{I_- \psi_{\alpha,m}^{(j')}}{\sqrt{j(j+1) - m(m-1)}}$$

$$\int \left(\psi_{\alpha,m-1}^{(j)}\right)^* \psi_{\alpha',m-1}^{(j')} \, d\tau = \int \left(\psi_{\alpha,m-1}^{(j)}\right)^* \frac{I_- \psi_{\alpha'm}^{(j')}}{\sqrt{j(j+1) - m(m-1)}} \, d\tau$$

By Theorem 10 the equality is maintained if $I_-^\dagger$ operates on the other wave function.

$$\int \left(\psi_{\alpha,m-1}^{(j)}\right)^* \frac{I_- \psi_{\alpha'm}^{(j')}}{\sqrt{j(j+1) - m(m-1)}} \, d\tau = \int \left(I_-^\dagger \psi_{\alpha,m-1}^{(j)}\right)^* \frac{\psi_{\alpha'm}^{(j')}}{\sqrt{j(j+1) - m(m-1)}} \, d\tau$$

Recall that $I_-^{\dagger *} = (I_x - iI_y)^* = (I_x + iI_y) = I_+$. Using Eq. (8.28),

$$\left(I_-^\dagger \psi_{\alpha,m-1}^{(j)}\right)^* = I_+ \left(\psi_{\alpha,m-1}^{(j)}\right)^*$$

$$= \sqrt{j(j+1) - (m-1)m} \left(\psi_{\alpha,m}^{(j)}\right)^*$$

so that

$$\int \left(\psi_{\alpha,m-1}^{(j)}\right)^* \psi_{\alpha',m-1}^{(j')} \, d\tau = \int \left(\psi_{\alpha,m}^{(j)}\right)^* \psi_{\alpha',m}^{(j')} \, d\tau,$$

proving that $\langle \alpha,\ j,\ m | \alpha',\ j,\ m \rangle$ is independent of m; all the dependence on m, m', q is contained in the *3-j* coefficient.

The factor $\langle \alpha,\ j,\ m | \alpha',\ j\ m \rangle$ is usually written

$$\langle \alpha,\ j,\ m | \alpha',\ j\ m \rangle \equiv \langle \alpha,\ j || T_q^{(k)} || \alpha',\ j' \rangle.$$

The Wigner–Eckart theorem in standard form is thus

$$\langle \alpha,\ j,\ m | T_q^{(k)} | \alpha',\ j',\ m' \rangle = (j'\ k\ m'\ q | j\ m)\ \langle \alpha,\ j || T_q^{(k)} || \alpha',\ j' \rangle. \qquad (G.3)$$

Index

Printed in the United States
by Baker & Taylor Publisher Services

Printed in the United States
by Baker & Taylor Publisher Services